Problem Solving in CONCEPTUAL Physics

eleventh edition

Paul G. Hewitt

CITY COLLEGE OF SAN FRANCISCO

Phillip R. Wolf

MT. SAN ANTONIO COLLEGE

Addison-Wesley

Boston Columbus Indianapolis New York San Francisco Upper Saddle River
Amsterdam Cape Town Dubai London Madrid Milan Munich Paris Montréal Toronto
Delhi Mexico City São Paulo Sydney Hong Kong Seoul Singapore Taipei Tokyo

Publisher: Jim Smith
Director of Development: Michael Gillespie
Editorial Manager: Laura Kenney
Project Editor: Chandrika Madhavan
Director of Marketing: Christy Lawrence
Senior Marketing Manager: Kerry Chapman
Managing Editor: Corinne Benson
Production Supervisor: Mary O'Connell
Production Service: Progressive Publishing Alternatives
Cover Photo: NASA
Supplement Cover Designer: Seventeenth Street Studios
Text and cover printer: Edwards Brothers, Inc.
Manufacturing Buyer: Jeff Sargent

ISBN 10: 0-321-66258-X
ISBN 13: 978-0-321-66258-3

Copyright © 2010, 2006 Paul G. Hewitt and Phillip R. Wolf. All rights reserved. Manufactured in the United States of America. This publication is protected by Copyright and permission should be obtained from the publisher prior to any prohibited reproduction, storage in a retrieval system, or transmission in any form or by any means, electronic, mechanical, photocopying, recording, or likewise. To obtain permission(s) to use material from this work, please submit a written request to Pearson Education, Inc., Permissions Department, 1900 E. Lake Ave., Glenview, IL 60025. For information regarding permissions, call (847) 486-2635.

Many of the designations used by manufacturers and sellers to distinguish their products are claimed as trademarks. Where those designations appear in this book, and the publisher was aware of a trademark claim, the designations have been printed in initial caps or all caps.

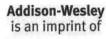

www.pearsonhighered.com   1 2 3 4 5 6 7 8 9 10 – EB – 15 14 13 12 10

Problem Solving in Conceptual Physics—*Eleventh Edition*
Table of Contents

PART ONE
Mechanics

3 Linear Motion	1	
Problems	9	
Show-That Problems	14	
4 and 5 Trigonometry	17	
Short Intro to Trigonometry	17	
Components Problems	23	
Show-That Problems	24	
4 Newton's Second Law of Motion	25	
Friction	31	
Inclined Planes	34	
Problems	36	
Show-That Problems	42	
Coefficients-of-Friction Problems	43	
Show-That Friction Problems	46	
Problems Involving Trigonometry	47	
Show-That Problems with Trig	50	
5 Newton's Third Law of Motion	51	
Problems	54	
Show-That Problems	58	
6 Momentum	59	
Problems	63	
Show-That Problems	70	

7 Energy	73	
Springs	77	
Problems	80	
Power Problems	87	
Efficiency & Machine Problems	88	
Springs & Elastic PE Problems	89	
Show-That Problems	93	
8 Rotational Motion	97	
Torque	101	
Circular Motion Problems	103	
Rotational Inertia Problems	104	
Torque Problems	105	
Centripetal Force Problems	107	
Angular Momentum Problems	110	
Problems with a Bit of Trig	111	
Show-That Problems	113	
9 Gravity	117	
Problems	120	
Show-That Problems	123	
10 Projectile and Satellite Motion	125	
Projectile Motion Problems	131	
Satellite Problems	135	
Show-That Projectile Problems	136	
Show-That Satellite Problems	137	

PART TWO
Properties of Matter

12 Solids	139	
Problems	142	
Show-That Problems	145	
13 Liquids	147	
Problems	151	
Show-That Problems	156	

14 Gases	159	
Problems	162	
Show-That Problems	166	

PART THREE
Heat

15 Temperature, Heat, & Expansion	169
Specific Heat Capacity Problems	172
Thermal Expansion Problems	176
Show-That Problems for Temperature, Heat, & Expansion	177

16	Heat Transfer	179	18	Thermodynamics	195
	Problems	182		Temperature and Gas Laws	196
	Show-That Problems	184		Heat Engines and Efficiency	198
				Problems	201
17	Change of Phase	185		Show-That Problems	205
	Problems	189			
	Show-That Problems	193			

PART FOUR
Sound

19	Waves and Vibrations	207	20	Sound	217
	Doppler Effect	208		Problems	218
	Shock Wave	210		Show-That Problems	222
	Problems	211			
	Show-That Problems	215			

PART FIVE
Electricity and Magnetism

22	Electrostatics	223	24 and 25	Magnetism	245
	Problems	225		Problems	249
	Show-That Problems	229		Show-That Problems	253
23	Electric Current	231			
	Problems	238			
	Show-That Problems	242			

PART SIX
Light

28	Reflection and Refraction	255		Ray Diagrams/Thin Lens Equation	259
	Reflection	255		Magnifying Glass	261
	Refraction	256		Diverging Lens	262
	Snell's Law	256		Problems	264
	Critical Angle	258		Show-That Problems	270

PART SEVEN
Atomic and Nuclear Physics

33	Atomic Nucleus & Radioactivity	273	34	Nuclear Fission and Fusion	281
	Half-Life	274		Problems	284
	Problems	277		Show-That Problems	286
	Show-That Problems	280			

PART EIGHT
Relativity

35	Special Theory of Relativity	287
	Problems	290
	Show-That Problems	294

To The Student

Concepts first; calculations later. This has always been the credo of *Conceptual Physics*. The intent of *this* book is to deepen your understanding of physics by applying its concepts to physics problems. Problems can be a way to appreciate the connections between concepts. Each chapter includes an introduction that highlights the important ideas and several sample problems to illustrate how these ideas can be applied. Number crunching is not the name of the game here. In fact, numbers are not given in the main part of the problems. Rather, problems are phrased in terms of mass m, speed v, force F, and so on, putting the focus on the concepts themselves. You'll derive a generalized solution expressed in symbols. In a second part of most problems, numbers are given so that you can transform a concept solution to a numerical one. Dealing with letters rather than numbers may take a bit of getting used to, but this is our way of helping you recognize that a wide variety of seemingly different problems are variations on the same root concept, which invite similar solutions. (This is thinking like a physicist!)

Each chapter also contains "Show-That" problems in which the answer is given as part of the problem and your goal is to use the given information to arrive at that answer. Use these problems to confirm for yourself that you are on the right track.

Generally, the problems in each chapter progress from simpler ones to more challenging ones, with a wide assortment to choose from.

Enjoy!

Paul G. Hewitt

Phillip R. Wolf

To The Instructor

The *Conceptual Physics* textbook is aimed at college students who are not majoring in a science. Its purpose is to provide a solid course in physics for future accountants, lawyers, medical technicians, business people, journalists, and teachers, to name a few. So that the course has a wide swath of physics, time not spent on the techniques of problem solving is time spent moving beyond mechanics to "rainbows." Although the problem sets in the textbook are few in number compared with the exercises, the mathematical structure of physics is an integral part of *Conceptual Physics*, with its many equations seen as "guides to thinking." Equations unambiguously show the connections between concepts. Their secondary role has been for solving algebraic problems. Because of its focus on concepts, *Conceptual Physics* is widely adopted—but mainly for classes of non-science students.

Many instructors have expressed interest in a problems supplement to *Conceptual Physics* so that it could be used as the text for an algebra-trig course—hence this problems supplement. The selection of problems herein emphasizes the physics rather than the algebra and trig. Phil and I have avoided puzzles that stress mathematical manipulation, which end up as more math than physics. Problems herein are meant to enlighten your average students, rather than merely to challenge your best and brightest. Some are challenging, and are so labeled with a bullet (•). Physics more than math is the thrust of this book.

It is often said that the number of problem-solving strategies is about equal to the number of instructors teaching them. In my algebra and calculus based courses I've tried several strategies over the years and reduced them to the simple method illustrated in the following sample solutions. Knowing how to get started is a primary difficulty for students. We address this by first focusing on what is being asked for. For example, if the problem asks for speed, the first step is writing "$v = ?$" I give my students 40% credit if they begin a solution this way. Then identify the physics concept that underlies the problem. If the problem involves collisions, then consider conservation of momentum. If the problem concerns forces and motion, then consider the work-energy theorem, etc. (This weans students away from the practice of categorizing problems as pulley problems, inclined plane problems, and so on.) Then write that relationship in equation form. Unless your equation directs otherwise, avoid searching for a perceived missing mass, acceleration, or whatever when it's not needed. When the equation has more than one term that represents unknown quantities, the procedure is repeated. Each step of a derivation is guided by the step preceding it.

Some conceptual physics instructors downplay equations and discourage the practice of grabbing for an equation when confronted with a problem. I take the opposite view. I think grabbing for equations should be encouraged—especially when equations are seen as abbreviated statements about the way variables connect and relate to one another. The terms in an equation are akin to notes on a musical scale. Just as notes guide fingers on a keyboard, equations can guide thinking. I teach my students to see central physics relationships in equation form, favoring those that state first principles: Newton's second law, conservation of momentum, conservation of energy, work-energy theorem, Newton's law of gravitation, Boyle's law, Ohm's law, Faraday's law, and so on. Equations can guide a student's thinking.

You'll note that sample solutions in this book follow this procedure, focusing first via $v = ?$, $t = ?$, $F = ?$, and so on. Then central principles are stated in equation form. From there derivation leads to a generalized solution, expressed in symbols. A follow-up step asks for a numerical solution, with appropriate units of measurement. So the procedure is DERIVE and SOLVE, with secondary emphasis on follow-up calculations.

Chapters 1 and 2 don't lend themselves to problems, so we begin with Chapter 3: Linear Motion. This is the only chapter with no physical laws. But it has definitions that lay the foundation for the laws of the following chapters. A chapter on elementary trigonometry follows, which we title 4&5 since it relates to Chapters 4 and 5 in the eleventh edition of *Conceptual Physics*. The chapter numbers in this book are those of respective chapters in the textbook.

Let's begin!

PGH

To The Instructor

Conceptual Physics is designed to help students understand and appreciate how the world works. There is a world (at least!) of physics embodied in equations such as $a = F_{net}/m$, and a ton of conceptual understanding that can occur without dealing with numbers and solving physics problems. Yet well-chosen physics problems *can* lead students to a deeper understanding of the physics, with the equations serving to guide their thinking.

    We have come up with good problems that will help students make that deeper conceptual connection in your *Conceptual Physics* course. Each chapter includes lots of straightforward problems as well as some more challenging ones. We have expanded upon material touched on in *Conceptual Physics* to include topics (for example, coefficients of friction, the energy of stretched and compressed springs, and resistors in parallel and series) that are traditionally covered in a problem-solving course. We gently introduce trigonometry and include appropriate trig-based problems.

    Our hope is that you and your students can have the best of *both* worlds—the richness of a concept-focused approach to physics *and* the clarification of understanding that comes from using concepts to solve problems.

PRW

Detailed solutions to all problems are on the website in the Instructor's Resource Area.

Acknowledgments

In this edition we acknowledge the help of Tom Helliwell, Peter Hopkinson, Evan Jones, Alex Lee, and John Perry for their invaluable feedback and suggestions. We are grateful to Martin Mason and Karen Schnurbusch for reviewing manuscript and for insightful input and many valuable suggestions. Special thanks to Lillian Lee Hewitt for editorial contributions and Yongxi Hu and Sung A. Kim for proofreading many of the chapters and problems solutions. We gratefully acknowledge the many students at Mount San Antonio College who provided valued feedback. We thank Phil's wife Mala Arthur, and son Zephram Wolf for loving support, encouragement, and patience.

For material transferred from the first edition we remain grateful to the intensive input of Ken Ford, Herb Gottlieb, David Housden, and Diane Riendeau. For valued suggestions and feedback we remain grateful to Tsing Bardin, Howie Brand, George Curtis, Marshall Ellenstein, Jim Hicks, Chelcie Liu, Fred Myers, Stan Schiocchio, and David Williamson. Thanks also to Jesse David Wall and daughter Ellender for adaptations of several original and insightful problems from their book *Introductory Physics—A Problem-Solving Approach* (Analog Press, 1997). Our appreciation to the late Ernie Brown, the designer of the chapter headers and the distinctive physics logo, to whom the eleventh edition of *Conceptual Physics* is dedicated.

3 Linear Motion

Kinematics is about *describing* motion. We define the quantities *distance, displacement, speed, velocity,* and *acceleration,* and the relationships among these quantities.

Physicists describe linear motion by imagining that it takes place along a one-dimensional coordinate system, really a number line. We assign a direction to be positive and the opposite direction to be negative. **Position** is your location on that number line.

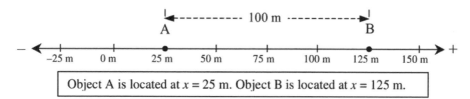

Object A is located at $x = 25$ m. Object B is located at $x = 125$ m.

Distance as used in physics is the same as distance in everyday usage—how far something moves. It's measured in meters (or some other length unit). Distance doesn't depend on direction. In the diagram above, we'd say that you traveled a distance of 100 meters whether you went from position A to B or from B to A.

Displacement is a little trickier. It's the straight-line *change* in position between the starting and ending points of motion. Displacement is a *vector* quantity since it involves both a distance and a direction. So if you go from position A to B in the diagram above, your displacement will be +100 m, but if you go from B to A, your displacement will be –100 m. If you toss a ball up into the air and then catch it again, the ball will have *zero* displacement, since it ends up where it started. (Its change in position is zero.) In this book we avoid tricky problems that confuse distance and displacement.

Average speed $= \dfrac{\text{total distance covered}}{\text{time interval}}$; $\bar{v} = \dfrac{d}{t}$.

We use the symbol v for speed. The bar over the v indicates "average" speed. For example, if you walk 100 meters in 40 seconds, your average speed is $\bar{v} = \dfrac{d}{t} = \dfrac{100 \text{ m}}{40 \text{ s}} = 2.5 \dfrac{\text{m}}{\text{s}}$.

If you know both the average speed and the time, distance can be calculated easily. So if a car averages 60 miles per hour for 2 hours, the distance traveled is $d = \bar{v}t = 60 \frac{\text{mi}}{\text{h}} \times 2 \text{ h} = 120$ mi. In each of these cases, you probably don't walk or drive at *exactly* the same speed the whole time—you speed up and slow down a little bit during the trip. So your speed at any instant may differ from your average speed.

Speed or instantaneous speed, v, is your speed at any given moment. This is what a car's speedometer indicates. As the car speeds up from rest, the speedometer may read 10 mi/h, then 20 mi/h, etc. When the speedometer reads 10 mi/h, *that* is the instantaneous speed of the car.

Average velocity and **instantaneous velocity** are also given the symbols $\bar{v}$ and v, respectively, and are expressed with the same equations, but involve direction. Velocity is a *vector* quantity. A rising elevator might have a velocity of 5 m/s upward = +5 m/s, while a descending elevator might have a velocity of 5 m/s downward = −5 m/s. In both cases the *speed* would be simply 5 m/s.

When the velocity (or speed) changes at a steady rate, average velocity (or average speed) is the sum of the initial and final speeds, or velocities, divided by 2. That is,

$$\bar{v} = \frac{v_0 + v_f}{2},$$ where v_0 is the *initial* velocity, and v_f is the *final* velocity.

Distance traveled is found by multiplying the average speed by the time interval, which in this case would be

$$d = \bar{v}t = \frac{v_0 + v_f}{2} t.$$

Acceleration is the rate at which velocity changes:

$$a = \frac{\text{change in velocity}}{\text{time interval}} = \frac{\Delta v}{\Delta t} = \frac{v_f - v_0}{\Delta t}.$$

In this chapter we consider only constant acceleration—where velocity changes at a steady rate. The standard unit of acceleration is meters per second per second, or meters per second squared (m/s²). For example, if an object accelerates from 3 m/s to 11 m/s in 4 seconds, the acceleration will be

$$a = \frac{\Delta v}{\Delta t} = \frac{v_f - v_0}{t} = \frac{11\frac{m}{s} - 3\frac{m}{s}}{4 \text{ s}} = \frac{8\frac{m}{s}}{4 \text{ s}} = \frac{2\frac{m}{s}}{1 \text{ s}} = 2\frac{m}{s^2}.$$

We can think of the motion as proceeding as follows:

Time	0 s	1 s	2 s	3 s	4 s
Speed	3 m/s	5 m/s	7 m/s	9 m/s	11 m/s

$\Delta v = 2$ m/s $\Delta v = 2$ m/s $\Delta v = 2$ m/s $\Delta v = 2$ m/s

In each second, the velocity changes by +2 m/s, so the acceleration is 2 m/s per second = 2 m/s².

Simple rearrangement of $a = \frac{v_f - v_0}{t}$ gives

$$v_f - v_0 = at,$$

which says the change in velocity = the acceleration × time,[*] or

$$v_f = v_0 + at,$$

which says that the instantaneous velocity of an object at a time t equals its initial velocity plus at, the additional velocity acquired due to acceleration during this time.

---

[*] For example, an acceleration of $2\frac{m}{s^2}$ for 3 seconds gives a change in velocity of $6\frac{m}{s}$. That is, $2\frac{m}{s^2} \times 3\text{s} = 6\frac{m}{s}$.

There are 3 other equations that are useful in solving kinematics problems:

$d = v_0 t + \frac{1}{2} a t^2$; * this eliminates final velocity from the equation.

$d = v_f t - \frac{1}{2} a t^2$; ♣ this eliminates initial velocity from the equation.

$2ad = v_f^2 - v_0^2$; ♠ this eliminates time from the equation.

The equations above are not laws of physics, but are simply definitions and relationships expressed in mathematical notation. It is sometimes useful to substitute the symbols x and y for horizontal and vertical distance, respectively, or Δx and Δy for horizontal and vertical displacement.

Free fall describes the case where an object is falling or rising and air resistance is negligible. Acceleration due only to gravity is given the symbol g, which has the constant value 9.8 m/s² at Earth's surface (although it is useful to use 10 m/s² for estimation). For a freely falling object dropped from rest, the speed acquired is given by the equation

$v_f = gt$

and the distance fallen is given by

$d = \frac{1}{2} g t^2$. ⊕

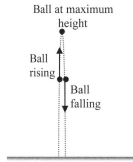

Here the assumption is that *downward* is the positive direction and the origin is at the point from which you are dropping the object.

If you throw an object upward, it is more convenient to call *upward* the positive direction and to place the origin at the ground. The downward acceleration due to gravity would be $-g$ and $d = v_0 t + \frac{1}{2} a t^2$ becomes $y = v_0 t - \frac{1}{2} g t^2$, where y is the height of the object above the ground.

Because there's so much good physics to cover in your course, your instructor may hurry you through this chapter. Don't fret, for the concepts of kinematics are employed in following chapters where you'll have opportunities to develop a deeper understanding of them. It would be a shame to get bogged down in this part of the course, which is devoid of physical laws. Grasp the ideas of speed, velocity, and acceleration, and then move on to where they're useful!

In the *Conceptual Physics* textbook, equations are seen as guides to thinking. In this problem-solving book, where derivations involve two or more successive mathematical steps, equations guide the thinking in choosing these steps. Most of the problems in this book begin by asking you to derive a solution in the form of an equation. A solution is achieved when all of the symbols in the equation are known quantities.

---

* $d = \overline{v} t = \frac{v_0 + v_f}{2} t = \frac{v_0 + (v_0 + at)}{2} t = \frac{2 v_0 + at}{2} t = v_0 t + \frac{1}{2} a t^2$.

♣ $d = \overline{v} t = \frac{v_0 + v_f}{2} t = \frac{(v_f - at) + v_f}{2} t = \frac{2 v_f - at}{2} t = v_f t - \frac{1}{2} a t^2$.

♠ From $d = \overline{v} t$ and $t = \frac{v_f - v_0}{a}$ $\Rightarrow d = \left(\frac{v_f + v_0}{2}\right)\left(\frac{v_f - v_0}{a}\right) = \frac{v_f^2 - v_0^2}{2a}$ $\Rightarrow 2ad = v_f^2 - v_0^2$.

⊕ If we use $g = 10$ m/s², this has the magnitude $d = 5t^2$. For $g = 9.8$ m/s², we get $d = 4.9\ t^2$.

© Paul G. Hewitt and Phillip R. Wolf

Here is a summary of useful linear motion equations (which apply when acceleration a is a constant):

Equation	Variables included
$d = \bar{v}t = \dfrac{v_0 + v_f}{2} t$	d, v_0, v_f, t
$a = \dfrac{v_f - v_0}{t}$	v_f, v_0, a, t
$d = v_0 t + \frac{1}{2} a t^2$	d, v_0, a, t
$d = v_f t - \frac{1}{2} a t^2$	d, v_f, a, t
$2ad = v_f^2 - v_0^2$	a, d, v_f, v_0

Which equation you use depends on what information you are given in the problem and what you are trying to find.

Most problems in this book begin by asking you to derive a generalized solution, expressed in symbols. This is usually followed by a step that asks for a numerical solution together with appropriate units of measurement. Let's consider some sample problems and their solutions.

Sample Problem 1

While driving along the highway at constant speed v, you sneeze, and your eyes close for a brief time t.
(a) Write an equation for the distance that you travel during your sneeze.

Focus: $d = ?$ (We focus on what is being asked for, which is distance traveled in this case.)

We start with $\bar{v} = \dfrac{d}{t}$ (beginning with a basic definition, in this case the equation that defines average or constant speed, which includes the distance we're looking for). Rearranging gives $d = \bar{v}t$, so the distance traveled with closed eyes is your average speed multiplied by the time of the sneeze.

Solution: The answer is $d = \bar{v}t$.

(b) Calculate the distance (in meters) that you travel while sneezing, given that your highway speed is 113 km/h and your eyes close for 0.70 s while sneezing.

Focus: Here we're asked to find the distance d in meters, while speed is given in km/h. Our task is to convert 113 km/h to m/s. We use a process called *units analysis*. We set up conversion factors so that the km and hours cancel, leaving us with m/s. The conversion factors we need are 1 h = 3600 s and 1000 m = 1 km.

$$113 \tfrac{km}{h} \times \left(\dfrac{1 h}{3600 s} \right) \times \left(\dfrac{1000 m}{1 km} \right) = 31.4 \tfrac{m}{s}$$

Solution: Now, $d = \bar{v}t = \left(31.4 \tfrac{m}{s} \right)(0.70 \text{ s}) = \mathbf{22\,m}$.

Again, we display the answer in **bold**.

Summary for converting units

- Start with the quantity to be converted—in this case, 113 km/h.
- Multiply by a *conversion factor* such as one of those listed on the inside back cover of *Conceptual Physics*. A conversion factor is a ratio of equivalent quantities that equals 1; the quantity in the numerator is equal to the quantity in the denominator but expressed in different units. In this case, the conversion factors are $\left(\frac{1000 \text{ m}}{1 \text{ km}}\right)$ and $\left(\frac{1 \text{ h}}{3600 \text{ s}}\right)$.
- The ratio is arranged so that multiplying and canceling eliminate the unwanted units and the wanted units remain.

The problem asks for distance in meters. But suppose it had instead asked for the distance in feet. We know that there are 3600 seconds in 1 hour, 1000 meters in 1 kilometer, 100 centimeters in 1 meter, 2.54 centimeters in 1 inch, and 12 inches in 1 foot. Then

$$113 \frac{\text{km}}{\text{h}} \times \left(\frac{1 \text{ h}}{3600 \text{ s}}\right) \times \left(\frac{1000 \text{ m}}{1 \text{ km}}\right) \times \left(\frac{100 \text{ cm}}{1 \text{ m}}\right) \times \left(\frac{1 \text{ in}}{2.54 \text{ cm}}\right) \times \left(\frac{1 \text{ ft}}{12 \text{ in}}\right) = 103 \frac{\text{ft}}{\text{s}}$$

$$\Rightarrow d = \overline{v} t = 103 \tfrac{\text{ft}}{\text{s}} (0.70 \text{ s}) = \textbf{72 ft}.$$

We see that the units themselves guide the mathematical operation. There are many ways to do the same conversion. We could have used 60 seconds = 1 minute and 60 minutes = 1 hour if we didn't use 3600 seconds = 1 hour. We could have used 3.28 ft = 1 m if we didn't use 2.54 cm = 1 in. Units guide the process.

Sample Problem 2

Mala enjoys exercise and jogs a distance x at a constant speed v.
(a) Derive an equation for the time it takes Mala to cover distance x.

Focus: $t = ?$

From $\overline{v} = \dfrac{d}{t} \Rightarrow t = \dfrac{d}{\overline{v}}.$

Solution: $t = \dfrac{d}{\overline{v}} = \dfrac{x}{v}.$

(b) Calculate the time in minutes for Mala to jog 1.7 km at a constant speed of 2.7 m/s.
First convert kilometers to meters:

$1.7 \text{ km} \times \left(\frac{1000 \text{ m}}{1 \text{ km}}\right) = 1700 \text{ m}.$ Then $t = \dfrac{x}{v} = \dfrac{1700 \text{ m}}{\left(2.7 \frac{\text{m}}{\text{s}}\right)} = 630 \text{ s}.$

Solution: By unit analysis, $630 \text{ s} \times \left(\frac{1 \text{ min}}{60 \text{ s}}\right) = \textbf{10.5 min}.$

Sample Problem 3

A bus starting from rest moves with constant acceleration along a level road.
(a) How far does the bus travel when accelerating from rest to speed v_f in a time interval t?

Focus: $d = ?$

From $\overline{v} = \dfrac{d}{t}$, $d = \overline{v} t = \left(\dfrac{v_0 + v_f}{2}\right) t$, and since v_0 is zero, $\boldsymbol{d = \dfrac{v_f t}{2}}.$

(We get the same answer if we use $d = v_0 t + \tfrac{1}{2} a t^2$, where $v_0 = 0$ and $a = \dfrac{\Delta v}{\Delta t} = \dfrac{v_f}{t}$. Then

$d = v_0 t + \tfrac{1}{2} a t^2 = 0 + \tfrac{1}{2}\left(\dfrac{v_f}{t}\right) t^2 = \dfrac{v_f t}{2}.)$

(b) Calculate the distance traveled by the bus when it starts from rest and accelerates at a constant rate to 12 m/s in a time of 5 s.

Solution as above: $d = \dfrac{v_f t}{2} = \dfrac{(12\frac{m}{s})(5s)}{2} = $ **30 m**.

Sample Problem 4

Two bicyclists ride toward each other on a long, unobstructed, straight road, each riding at a constant speed *v*. When the bikes are distance *x* apart, a bee begins flying from the front wheel of one bike to the front wheel of the other bike at a steady average speed of 3*v*. When the bee reaches each wheel it abruptly turns around and flies back to touch the other wheel, repeating the back-and-forth trip until the bikes meet, whereupon the bee is squashed.

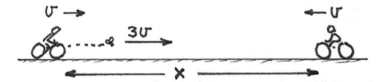

(a) What cumulative distance did the bee travel in its total back-and-forth trips?

Focus: $d = ?$

From $\bar{v} = \dfrac{d}{t}$, $d = \bar{v}t = 3vt$ for the bee.

We're told that the bee's average speed is 3*v*, but we're not told the time. The key to solving this problem is realizing that the time the bee spends flying is the *same* as the time it takes the bicycles to meet. Each bicycle goes a total distance $\dfrac{x}{2}$ in time *t*. From $\dfrac{x}{2} = vt$, time $t = \dfrac{x}{2v}$. So the bee travels a total distance $d = 3v\left(\dfrac{x}{2v}\right) = $ **1.5x**.

(b) Calculate the cumulative distance of flight for the bee to the squish point, given that the bikes each travel at 10 km/h and the bee begins its 30-km/h back-and-forth trips when the bikes are 20 km apart.

Solution: As above, $d = 1.5x = 1.5(20 \text{ km}) = $ **30 km**.

Check with all figures:

$$d = 3v\left(\dfrac{x}{2v}\right) = 3\left(10\tfrac{km}{h}\right)\left(\dfrac{20 \text{ km}}{2\left(10\tfrac{km}{h}\right)}\right) = \mathbf{30 \text{ km}}.$$

(Note that we can work directly with the units kilometers and hours without converting to meters and seconds.) The time of travel isn't given in the problem, but the equation for the distance traveled (average speed multiplied by time) alerts us to *time* being the important factor. Trying to solve this problem without considering time is a stumper. Let the terms in the equations guide the steps to the solution.

Sample Problem 5

Katie wants to know the height of a bridge, so Martin drops a rock over the edge of the bridge. The rock hits the stream below in time t.

(a) Derive an equation for the height of the bridge.

Focus: $d = ?$

The rock is freely falling. Since all of the motion in this problem is downward it is convenient to call downward the positive direction and to put the origin at the edge of the bridge. This means that $a = +g$. The initial velocity of the rock will be zero since Martin drops, rather than throws, the rock. We know v_0, a, and t, and we want d. The appropriate equation is $d = v_0 t + \frac{1}{2}at^2 = \frac{1}{2}gt^2$.

(b) Derive an equation for the speed of the rock as it hits the stream below.

Focus: $v_f = ?$

We know v_0, a, and t, and we want v_f. The appropriate equation is

$$a = \frac{v_f - v_0}{t} \Rightarrow v_f = v_0 + gt = \boldsymbol{gt} \text{ since } v_0 = 0.$$

(c) A rock dropped from a bridge takes 2.5 s to hit the stream below. Calculate the height of the bridge and the impact speed of the rock.

Solution: $d = \frac{1}{2}gt^2 = \frac{1}{2}\left(9.8\frac{m}{s^2}\right)(2.5 \text{ s})^2 = \boldsymbol{31 \text{ m}}$; $v_f = gt = 9.8\frac{m}{s^2}(2.5 \text{ s}) = \boldsymbol{25\frac{m}{s}}$.

Sample Problem 6

You toss a potato upward at speed v_0.

(a) What is the maximum height reached by the potato? Assume that air resistance is small enough to be ignored.

Focus: $y = ?$

We want to know the maximum height. Since we are tossing the potato upward, let's call upward the positive direction and put the origin at the ground. There are two things that we know, but which aren't given explicitly in the problem:

(1) We know the acceleration $a = -g$ since after the potato leaves your hand, the only acceleration is that due to gravity. The negative sign comes because the acceleration is downward, opposite the direction we have set as positive.

(2) We know the final instantaneous velocity $v_f = 0$.* At the potato's maximum height, its motion state is between rising and falling. For that instant, its instantaneous velocity is zero.

We have a, v_0, v_f, and we want a distance. The appropriate equation to use is

$$2ad = v_f^2 - v_0^2 \Rightarrow y = \frac{v_f^2 - v_0^2}{2a} = \frac{-v_0^2}{2(-g)} = \boldsymbol{\frac{v_0^2}{2g}}.$$

---

* This doesn't mean that the potato has stopped! *Stopped* would mean that the potato spends some finite amount of time in the same spot—that somehow you could look at its maximum height, turn away, and then look again later, and the potato would still be in exactly the same place. This only happens in cartoons! At the potato's maximum height, its state of motion is between rising and falling, never actually stopping, yet with an instantaneous velocity of zero. It spends no time at all *at* 0 m/s just as it spends no time at all at other values of instantaneous velocity. It keeps moving!

© Paul G. Hewitt and Phillip R. Wolf

(b) Derive an equation for the time it will take for the potato to reach its maximum height.

Focus: $t = ?$

We have a, v_0, v_f, and we want t. The appropriate equation to use is

$$a = \frac{v_f - v_0}{t} \Rightarrow t = \frac{v_f - v_0}{a} = \frac{0 - v_0}{-g} = \frac{v_0}{g}.$$

(c) Calculate the maximum height achieved and the time to get there for a potato that is initially tossed upward at 12.7 m/s.

Solution: $y = \dfrac{v_0^2}{2g} = \dfrac{\left(12.7\,\frac{m}{s}\right)^2}{2\left(9.8\,\frac{m}{s^2}\right)} = 8.2\,\text{m}; \quad t = \dfrac{v_0}{g} = \dfrac{12.7\,\frac{m}{s}}{9.8\,\frac{m}{s^2}} = 1.3\,\text{s}.$

(d) What is the height and velocity of the potato 2.0 seconds after being tossed?

Focus: $y = ?\ v_f = ?$

Now we have a, v_0, and t, and we want y and v_f. The appropriate equations to use are

$d = v_0 t + \tfrac{1}{2} a t^2 \Rightarrow y = v_0 t - \tfrac{1}{2} g t^2 = 12.7\,\tfrac{m}{s}(2\,\text{s}) - \tfrac{1}{2}\left(9.8\,\tfrac{m}{s^2}\right)(2\,\text{s})^2 = 5.8\,\text{m}$ above the ground and

$a = \dfrac{v_f - v_0}{t} \Rightarrow v_f = v_0 + at = 12.7\,\tfrac{m}{s} + \left(-9.8\,\tfrac{m}{s^2}\right)(2\,\text{s}) = -6.9\,\tfrac{m}{s}.$ The negative sign tells us that the potato's velocity is downward, which makes good physical sense since the potato has already passed its maximum height.

(e) When will the potato hit the ground? How fast will it be going when it does?

Focus: $t = ?\ v_f = ?$

When the potato hits the ground, $y = 0$. Now we have a, v_0, and y, and we want t and v_f. The appropriate equations to use are $d = v_0 t + \tfrac{1}{2} a t^2 \Rightarrow y = v_0 t - \tfrac{1}{2} g t^2 \Rightarrow 0 = t\left(v_0 - \tfrac{1}{2} g t\right)$, which has solutions when $t = 0$ (when we first let go of the potato), and $t = \dfrac{2 v_0}{g}$, exactly twice the time it took for the potato to reach its maximum height. To find the velocity with which the potato hits the ground, we can use $v_f = v_0 + at = v_0 + (-g)\left(\dfrac{2 v_0}{g}\right) = -v_0$. These results tell us that in the absence of air resistance, the trajectory of the potato is symmetric—it takes just as much time to fall as it does to rise, and it hits the ground with exactly the same speed at which it was thrown upward.

Solution: $t = \dfrac{2 v_0}{g} = \dfrac{2\left(12.7\,\frac{m}{s}\right)}{9.8\,\frac{m}{s^2}} = 2.6\,\text{s}; \quad v_f = -v_0 = -12.7\,\tfrac{m}{s}.$

Now have a go at the problems that follow!

Problems for Linear Motion

(*It may be useful in some of these problems to know that* 1 mi = 1.61 km *and* 1 ft = 0.3048 m.)

3-1. Paul hikes b km east to see a waterfall, and then he hikes c km west before stopping for a snack.
   (a) What distance does Paul walk?
   (b) What is Paul's displacement?
   (c) Calculate Paul's distance and displacement if he hikes 5 km east and then 2 km west.

3-2. The world's fastest train in commercial service is presently the magnetically levitated Transrapid's Shanghai Maglev in China.
   (a) Write an equation for the average speed of the train if it travels distance x in time t.
   (b) Calculate the train's average speed in m/s if it travels 30.0 km in 8.0 minutes. Then convert to mph (mi/h).

3-3. A tennis ball is served and travels the length of the court L in time t.
   (a) Write an equation for the ball's average horizontal speed.
   (b) Calculate the average speed of a ball traveling 24.0 m across the court in 0.60 s.

3-4. A baseball pitcher throws a fastball across home plate. The ball crosses the plate, which has a front-to-back length x, in time t.
   (a) Write an equation for the speed at which the baseball crosses the plate.
   (b) Calculate the speed of a baseball that takes 0.010 s to cross home plate, 0.30 m from front to back.

3-5. A race car races on a circular racetrack of radius r.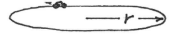
   (a) Write an equation for the car's average speed when it travels a complete lap in time t.
   (b) Calculate the average racing speed, given that the radius of the track is 400 m and the time to make a lap is 40 s.

3-6. A Taipei Tower in Taiwan, of height h, has the world's fastest elevators.
   (a) Write an equation for the time the elevator takes in rising from the ground floor to the top when the average elevator speed is $\overline{v}$.
   (b) Calculate the time of the upward ride to a height of 508 m at an average speed of 15 m/s.
   (c) The elevators are said to zip upward at peak speeds of about 16 m/s, greater than the average speed. Is it reasonable that peak speeds can be 16 m/s when average speed is only 15 m/s? Defend your answer.

3-7. Phil runs the length of an American football field, 100.0 yards long.
   (a) How much time is required for him to run the full length at speed v (in m/s)?
   (b) Calculate the time it takes for Phil, running at 6.0 m/s, to run the length of an American football field.

3-8. Light is incredibly fast and travels at speed c. Consider light traveling along a ruler of length L.
   (a) How long does it take light to travel the length of the ruler?
   (b) Calculate the time it takes for light to travel the length of a meterstick. (The speed of light is 3.00×10^8 m/s.) Give your answer in nanoseconds. (1 nanosecond = 10^{-9} s)

3-9. Lillian rides her bicycle along a straight road at an average speed v.
   (a) Write an equation for the distance she travels in time t.
   (b) Calculate the distance covered by Lillian if her average speed is 7.5 m/s for a time of 5.0 minutes.

3-10. A gecko, initially at rest, sprints to speed v in time t.
   (a) Write an equation for the average speed of the gecko, assuming steady acceleration.
   (b) Write an equation for the distance the gecko covers in its sprint.
   (c) Calculate the distance the gecko covers when sprinting from rest to 2.0 m/s in a time of 1.5 s.

3-11. A skier starts down a slope from rest and reaches speed v in time t.
   (a) Derive an equation for the distance the skier travels in this time, assuming a steady pickup of speed.
   (b) Calculate how far the skier travels down the slope when starting from rest and reaching a speed of 12 m/s in 8.0 s.

3-12. The cheetah is the fastest sprinter of all land animals. Suppose it starts from rest and accelerates uniformly to speed v in time t.
   (a) Derive a simple equation for the distance the cheetah covers.
   (b) Calculate the distance a cheetah covers if it starts from rest and accelerates uniformly to a speed of 100.0 km/h in a time of 8.0 seconds.

3-13. A moving van increases its speed from v_1 to v_2 in a time interval t.
   (a) Write an equation for the average acceleration of the van.
   (b) Calculate the average acceleration in m/s^2 that the van undergoes speeding up uniformly from 15 km/h to 40 km/h in 20 s.

3-14. A hybrid automobile traveling at speed v_1 steadily increases to speed v_2 in a time interval t.
   (a) Write an equation for the average acceleration of the automobile.
   (b) Calculate the average acceleration in m/s^2 when the automobile increases its speed from 5.0 km/h to 20.0 km/h in 10.0 s.
   (c) Calculate the distance the automobile travels during this period of acceleration.

3-15. Lonnie applies the brakes to his car moving at speed v. The car slows at a constant rate and is brought to rest in time t.
   (a) What is the acceleration?
   (b) Calculate the acceleration if the initial speed of the car is 26 m/s and the time to stop is 20 s.
   (c) Calculate the distance traveled while the car is decelerating.
   (d) Lonnie's *reaction time* is the time that passes between the instant he sees reason for braking and the moment he actually applies the brakes. Calculate the distance the car travels before Lonnie puts on the brakes, assuming a 26-m/s speed and a reaction time of 1.5 s.

3-16. A jet plane lands on a runway with a speed v, and after time t, it comes to a stop.
   (a) Assuming that its speed is reduced at a constant rate, write an equation for the acceleration of the plane.
   (b) Calculate the acceleration if the landing speed is 72 m/s and the stopping time is 12 s.
   (c) Calculate the distance the jet travels between the point of touchdown and the point of stopping.

3-17. A dart leaves the barrel of a blowgun at a speed v. The length of the blowgun barrel is L. Assume that the acceleration of the dart in the barrel is constant.
    (a) Derive an equation for the time the dart moves inside the barrel.
    (b) Calculate the time in the barrel if the dart's exit speed is 15.0 m/s and the length of the blowgun is 1.4 m.

3-18. A bullet leaves the barrel of a gun at speed v. The length of the gun barrel is L. Assume that the acceleration of the bullet in the barrel is uniform.
    (a) Write an equation for the bullet's average speed inside the barrel.
    (b) Calculate the average speed if the bullet leaves the gun at 350 m/s. The length of the gun barrel is 0.40 m.
    (c) Calculate the time the bullet is in the barrel.

3-19. To avoid hitting a stalled bus, Stan brakes his car and slows at a uniform rate from v_0 to v in time t.
    (a) Derive an equation for the distance the car travels while slowing to the lower speed.
    (b) Calculate the distance the car travels while braking from 25 m/s to 11 m/s in a time of 8.0 s.

3-20. Motion picture frames show that a rolling ball moves distance x between frames. The rate at which the frames are taken is 24 per second.
    (a) How fast is the ball moving?
    (b) Calculate the speed of a ball that moves 0.40 m in each frame.

3-21. An electron placed in an electric field accelerates uniformly from rest to speed v while traveling distance x.
    (a) Derive an equation for the acceleration of the electron.
    (b) Calculate the acceleration in m/s^2 for an electron that starts from rest and reaches a speed of 1.8×10^7 m/s over a distance of 0.10 m.
    (c) Calculate the time required for the electron to attain this speed.

3-22. A drag racer can cover distance d in time t, starting from rest.
    (a) Assuming constant acceleration, derive an equation for the dragster's final speed.
    (b) Derive an equation for the dragster's acceleration.
    (c) A drag racer can cover a quarter-mile (402 m) in 4.45 seconds. Calculate the final speed and the average acceleration of the dragster.

3-23. The speed of a toy rocket shooting straight upward increases from speed v to speed V at a uniform rate in time t.
    (a) Derive an equation for the distance the rocket travels during this time.
    (b) Calculate the distance (in meters) covered if the initial rocket speed is 110 m/s and it increases uniformly to 250 m/s in a period of 3.5 s.

3-24. Roger tosses a ball straight upward at speed v. Ignore air drag.
   (a) Derive an equation for the time the ball takes to reach its highest point.
   (b) Calculate the time in seconds that it takes for the ball to reach its maximum height when thrown straight upward at 32 m/s.
   (c) Calculate the maximum height of the ball.

3-25. A spud gun fires a potato straight upward. The potato hits the ground in time t.
   (a) Derive an equation for the initial speed of the potato, ignoring air drag.
   (b) Calculate the initial speed of a potato that is fired straight upward and takes 12 seconds to hit ground. What is this speed in mi/h?

3-26. George drops a stone from atop a cliff of height h that overlooks the ocean.
   (a) Derive an equation for the time it takes for the stone to hit the water.
   (b) Calculate the number of seconds to hit the water if the stone is dropped from a cliff 25 m high.
   (c) Calculate the speed of the stone when it hits the water.

3-27. Janet tosses a ball straight upward. The ball soon returns to her hand.
   (a) Derive an equation for the speed of her toss so that the ball returns to her hand a time interval t later.
   (b) Calculate the throwing speed straight upward for a 4.0-second time of flight.
   (c) Calculate the maximum height reached by the ball.

3-28. The ceiling of a school gymnasium is distance y above the floor.
   (a) Derive an equation for the maximum upward speed with which you can toss a ball from an elevation 2.0 meters above the floor to barely miss hitting the ceiling.
   (b) Calculate the maximum speed of the tossed ball if the floor-to-ceiling distance is 20.0 m.

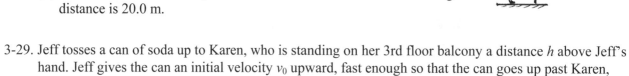

3-29. Jeff tosses a can of soda up to Karen, who is standing on her 3rd floor balcony a distance h above Jeff's hand. Jeff gives the can an initial velocity v_0 upward, fast enough so that the can goes up past Karen, who catches the can on its way down.
   (a) Derive an equation for the velocity of the can when it passes Karen on its way up.
   (b) What is the *velocity* of the can just as Karen catches it on its way down?
   (c) Derive an equation for the time between Jeff tossing the can and Karen catching it.
   (d) Calculate the velocity of the can and its time in the air for the instant before Karen grabs the can if Jeff tosses the can at 16 m/s and Karen is 8.5 m above Jeff when the can leaves his hand.

3-30. • Seth is standing on a diving platform when he tosses a water balloon upward at speed v_0. The water surface is distance h below Seth's hand when he releases the balloon. Assume air resistance can be neglected.
   (a) Derive an equation for the velocity of the water balloon when it hits the water below.
   (b) Derive an equation for the time it takes for the balloon to hit the water surface.
   (c) What will be the velocity of the water balloon when it hits the water if Seth instead throws the balloon *downward* with an initial speed v_0?
   (d) Calculate answers to the above for a water balloon thrown at 5.0 m/s and released 11.8 m above the water's surface.

3-31. •• A model rocket launches vertically from the ground with acceleration a for a time t_1, at which point the rocket runs out of fuel but continues rising to its maximum height, and then freely falls to the ground. Assume negligible air resistance.
(a) Find velocity v_1 of the rocket when it runs out of fuel.
(b) Find its height h_1 at this point.
(c) What *additional* height h_2 will the rocket gain after it runs out of fuel?
(d) How much time will this additional rise take?
(e) What is the rocket's maximum height?
(f) How long will it take for the rocket to hit the ground from its maximum height?
(g) What is the total time the rocket is in the air?
(h) Calculate answers to the above for a rocket whose initial acceleration is 120 m/s² for 1.70 s.

3-32. A motorist drives distance x from one city to another in time t, but makes the return trip in $0.75t$.
(a) What is the average speed for the total trip? (Recall that the definition of average speed is $\bar{v} = \frac{\text{total distance}}{\text{total time}}$.)
(b) Calculate the average speed for the round trip between cities 140 km apart if the initial outward trip takes 2.0 hours.

3-33. To reach her cabin, Tsing walks at average speed v for 30 minutes and then jogs at speed $2v$ for another 30 minutes.
(a) Write an equation for Tsing's average speed for her trip to the cabin.
(b) Calculate the average speed for the whole trip for a walking speed of 1.0 m/s.
(c) Calculate the distance between her starting point and the cabin.

3-34. Dennis drives for 1 hour at average speed v. Then he drives for another hour at average speed $4v$.
(a) Find the overall average speed. (Recall that the definition of average speed is $\bar{v} = \frac{\text{total distance}}{\text{total time}}$.)
(b) Calculate his overall average speed if his average speed for the first hour is 25 km/h, and his average speed during the second hour is 100 km/h.

3-35. Suppose that you drive a distance x at average speed v and then drive the same additional distance x at speed $1.5v$.
(a) Find your average speed.
(b) Calculate your average speed if you drive 1.0 kilometer at 28 km/h and then drive 1.0 kilometer back to your starting point at 42 km/h.

3-36. Judy and her dog, Atti, take their morning walk to the Vinoy Hotel, which is distance x away. Judy walks at a brisk speed v in a straight line while Atti runs back and forth at speed V between Judy and the hotel, until both reach the hotel.
(a) Find the total distance Atti runs.
(b) Calculate the total back-and-forth distance Atti runs if Atti's speed is 4.5 m/s, Judy's speed is 1.5 m/s, and the distance Judy walks is 150.0 m.

Show-That Problems for Linear Motion

The following problems supply numerical values and the answer to each problem. What's important is for you to show how you arrive at the stated answer. Sometimes more information is supplied than required. (Aren't problems in everyday life also accompanied by extraneous information? ☺) Please continue with the practice of finding the solution with symbols before you plug in numerical values.

3-37. A ball rolls down a 3-m inclined plane in 1.5 s.
Show that the average speed is 2 m/s.

3-38. A ball with a temperature of 22°C is thrown vertically upward at 14.7 m/s.
Show that its maximum height will be 11 m.

3-39. A car traveling along a level road uniformly increases its speed from rest to 27.5 m/s in 8.0 s.
Show that it travels a distance of 110 m in this time.

3-40. A 40-gram egg falls from its nest atop a 16-m tall tree.
Show that it takes 1.8 s for the egg to hit the ground below.

3-41. A ball rolling down an inclined plane starts from rest and reaches a speed of 12 m/s in 3 s.
Show that it has an acceleration of 4 m/s^2.

3-42. An F-14 Tomcat fighter jet goes from rest to a speed of 75 m/s in 2.5 s.
Show that the acceleration it undergoes is 30 m/s^2.

3-43. A motorboat accelerates from rest in a straight line at a constant 2.0 m/s^2 for a time of 8.0 s.
Show that it travels 64 m in this time.

3-44. Fred goes down a 5.0-m slide at a playground in 2.0 s. He starts from rest and accelerates uniformly.
Show that his acceleration while on the slide is 2.5 m/s^2.

3-45. Zephram is on roller blades on a hill. He starts from rest and skates for 5.5 s, accelerating uniformly at 3.5 m/s^2.
Show that the distance traveled is 53 m.

3-46. A red-colored ball tossed upward reaches a maximum height of 3.0 m.
Show that its initial speed was 7.7 m/s.

3-47. Will tosses a golf ball straight upward.
Show that if the ball is tossed upward at 18 m/s, it will be in the air for fewer than 4 s.

3-48. Carmelita tosses a baseball straight up into the air and it returns to her glove in 3.0 s.
Show that she has to give the ball a vertical speed of about 15 m/s if it is to return to her glove in that time.

3-49. Show that a dropped ball will cover 3 times as much distance during the 2nd second of its fall as it will during the 1st second of its fall.

3-50. When some volcanoes erupt, big rocks have been measured to shoot upward with speeds of about 1000 m/s.
Show that, neglecting air resistance, these rocks could reach heights exceeding 50 km.

3-51. An orange is tossed upward at 22 m/s. Show that 3.5 seconds later it will be 17 m above its initial launch height.

3-52. A grizzly bear has a top running speed of about 13 m/s (30 mi/h). Two campsites are 65 m apart.
Show that a grizzly bear can run from one campsite to the other in 5 s.

3-53. A spiffy sports car accelerates from rest to 28 m/s (100 km/h) at an average acceleration of 7.0 m/s^2.
Show that this acceleration takes 4.0 s.

3-54. A small plane starts from rest and reaches takeoff speed v while covering distance d.
(a) Show that the time for the plane to take off from rest is $2d/v$.
(b) Show that the acceleration of the plane is $v^2/2d$.
(c) If the plane has a takeoff speed of 28 m/s and a takeoff distance of 140 m, show that takeoff time is 10 s and that the acceleration of the plane is 2.8 m/s^2.

3-55. A car accelerates uniformly from rest to a speed of 25 m/s in a time of 5.0 s.
Show that the car covers a distance of 63 m.

3-56. The orbital speed of the space shuttle is approximately 28,000 km/h. Show that, at that speed, someone could make the trip from Los Angeles to New York, a distance of 2462 miles, in 8.5 minutes. (It is useful to note that 1 km = 0.621 mi.)

3-57. When the space shuttle is landing, it touches down at approximately 220 mi/h and comes to rest in approximately 800 m. Show that the acceleration of the shuttle is about –6 m/s^2. (It is useful to note that 1 km = 0.621 mi.)

3-58. A subway car accelerates from rest at 1.5 m/s^2 for 12 seconds, maintains a constant speed for 38 seconds, and finally has an acceleration of –1.5 m/s^2 for 12 seconds as it comes to rest in the next station.
Show that the total distance covered by the subway car is 900 m.

3-59. Phil and Mala run a 100-m race, and Mala gloriously wins in 12.8 s while Phil takes 13.6 s.
Show that Mala wins the race by a distance of 6 m.

3-60. While making a game-winning jump shot, Terrence rises about 0.6 m above the court floor.
Show that his hang time is 0.7 s.

3-61. At 60 mph it takes 60 s to travel a mile. At twice this speed, 120 mph, it takes 30 s.
Show that covering a mile in 45 s requires a speed of 80 mph.

3-62. Norma drives to a destination at an average speed of 40 km/h and returns at an average speed of 60 km/h.
Show that her average speed is 48 km/h (and not 50 km/h!) for the round trip.

Most of nature's rules can be expressed mathematically.

Elementary Trigonometry

A Short Introduction to Trigonometry

Trigonometry blends a bit of geometry with a lot of common sense. It lets you solve problems that would otherwise be undoable. The part of trigonometry that will be useful in this book deals with the relationships between the lengths of the sides of a right triangle.

Consider the following two similar right triangles[*]:

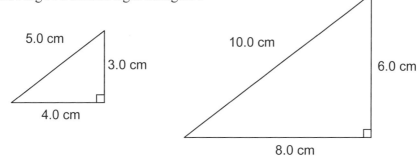

The symbol θ is the Greek letter "theta." We use θ to represent the size of the angle. Because the triangles are similar, θ is the same in both triangles.

It is an interesting fact (related to the similarity of the triangles) that the ratio of the length of the triangle's side opposite θ to the length of the hypotenuse gives the same result for both triangles:

$$\frac{\text{side opposite } \theta}{\text{hypotenuse}} = \frac{3.0 \text{ cm}}{5.0 \text{ cm}} = \frac{6.0 \text{ cm}}{10.0 \text{ cm}} = 0.60$$

EVERY right triangle that has an angle θ of this same size has the same value for this ratio. We give this ratio, $\frac{\text{side opposite } \theta}{\text{hypotenuse}}$, a special name. We call it the "sine of theta," usually written $\sin \theta$.

Two other ratios that turn out to be useful are:

$$\text{cosine of } \theta = \frac{\text{side adjacent to } \theta}{\text{hypotenuse}} = \frac{4.0 \text{ cm}}{5.0 \text{ cm}} = \frac{8.0 \text{ cm}}{10.0 \text{ cm}} = 0.80 \text{ in this example.}$$

and

$$\text{tangent of } \theta = \frac{\text{side opposite to } \theta}{\text{side adjacent to } \theta} = \frac{3.0 \text{ cm}}{4.0 \text{ cm}} = \frac{6.0 \text{ cm}}{8.0 \text{ cm}} = 0.75 \text{ in this example.}$$

So we can think of the sine, cosine, and tangent as properties of the angle, since they are the same for every right triangle that contains a particular angle.

---

[*] Two triangles are considered similar when the angle measurements in the first triangle are the same as the angle measurements in the second triangle.

If we choose a different right triangle, we get a different set of ratios:

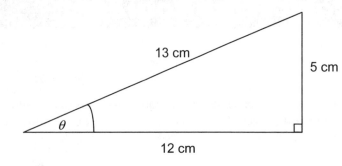

$$\sin\theta = \frac{5\text{ cm}}{13\text{ cm}}$$

$$\cos\theta = \frac{12\text{ cm}}{13\text{ cm}}$$

$$\tan\theta = \frac{5\text{ cm}}{12\text{ cm}}$$

It is useful to imagine that people have drawn every right triangle possible, measured all of these ratios for every possible angle, and stored all of that information in your calculator so that you can access it by pressing the **sin**, **cos**, and **tan** keys. (That's not how it's actually done, but it is useful to imagine it as such.)

Sample Problem 1

You lean your 7.0-foot ladder up against the wall such that it makes a 65° angle with the floor.

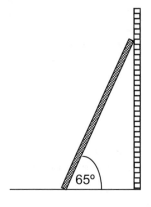
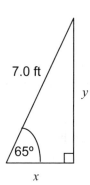

(a) How high on the wall is the top of the ladder?

Focus: $y = ?$ We can model the ladder against the wall as a right triangle.

Solution: $\sin 65° = \dfrac{y}{7.0\text{ ft}}$;

$y = (7.0\text{ ft})\sin 65° = 7.0\text{ ft} \times 0.906 =$ **6.3 ft**.

(b) How far from the wall is the base of the ladder?

Solution: $\cos 65° = \dfrac{x}{7.0\text{ ft}}$;

$x = (7.0\text{ ft})\cos 65° = 7.0\text{ ft} \times 0.423 =$ **3.0 ft**.

Sample Problem 2

You are lying on level ground and looking at a tree 45 meters away. You have to look up 32° above the ground level to see the top of the tree.

(a) How tall is the tree?

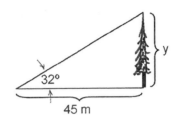

Focus: $y = ?$ (y is the height of the tree.) Model the situation as a right triangle. y is the side of the triangle opposite the angle. We know the angle and the side adjacent to the angle.

From tangent of $\theta = \dfrac{\text{side opposite to }\theta}{\text{side adjacent to }\theta} = \dfrac{y}{45 \text{ m}}$

$\Rightarrow y = \tan\theta \times 45 \text{ m}.$

Solution: $y = (\tan 32°)(45 \text{ m}) = (0.625)(45 \text{ m}) = 28.1 \text{ m} \approx \mathbf{28 \text{ m}}.$

(a) You put a post into the ground a little crooked so that the post makes an 83° angle with the ground. The protruding bit of the post is 180 cm long. When the sun is directly overhead at noon, how long is the shadow of the post?

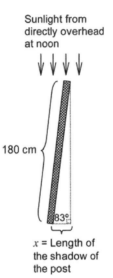

Focus: $x = ?$ (x is the length of the shadow of the post.)

Again, draw a right triangle, where x is the side of the triangle adjacent to the angle. We know the hypotenuse and the angle.

From cosine $\theta = \dfrac{\text{side adjacent to }\theta}{\text{hypotenuse}} = \dfrac{x}{180 \text{ cm}}$.

$\Rightarrow x = \cos\theta \times 180 \text{ cm}.$

Solution: $x = (\cos 83°)(180 \text{ cm}) = (0.122)(180 \text{ cm}) = \mathbf{22 \text{ cm}}.$

Inverse Trigonometric Functions

Consider again one of the triangles we've already looked at:

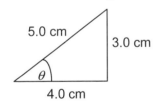

We know that for the above triangle, $\sin\theta = \dfrac{3.0 \text{ cm}}{5.0 \text{ cm}} = 0.60$. But how do we find θ?

What we are really asking here is

"What is the angle whose sine is 0.60?"

This question is answered by the inverse trigonometric functions. The inverse sine of 0.60, written $\sin^{-1}(0.60)$, gives as a result the angle whose sine is 0.60. Likewise, $\cos^{-1}(0.80)$ gives the angle whose cosine is 0.80, and $\tan^{-1}(0.75)$ gives the angle whose tangent is 0.75.

For most calculators you can get the inverse trig function by first pushing the "2nd" button followed by the desired trigonometric function. You should confirm on your own calculator that $\theta = \sin^{-1}(0.60) = \cos^{-1}(0.80) = \tan^{-1}(0.75) = 36.9° \approx \mathbf{37°}$.

Practice Problems:

Given a side and an angle in diagrams (a) and (b), find the other 2 sides.

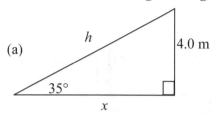
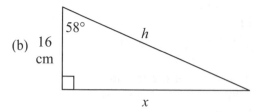

Given 2 sides in diagrams (c) and (d), find the missing angles and sides.

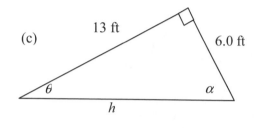

Solutions:

(a) $x = ?$ $h = ?$

$$\tan 35° = \frac{4.0\,\text{m}}{x} \Rightarrow x = \frac{4.0\,\text{m}}{\tan 35°} = \textbf{5.7 m}.$$

$$\sin 35° = \frac{4.0\,\text{m}}{h} \Rightarrow h = \frac{4.0\,\text{m}}{\sin 35°} = \textbf{7.0 m} \quad \text{or} \quad h^2 = \sqrt{(4.0\,\text{m})^2 + (5.7\,\text{m})^2} \Rightarrow h = \textbf{7.0 m}.$$

(b) $x = ?$ $h = ?$

$$\tan 58° = \frac{x}{16\,\text{cm}} \Rightarrow x = (16\,\text{cm})\tan 58° = 25.6\,\text{cm} \approx \textbf{26 cm}.$$

$$\cos 58° = \frac{16\,\text{cm}}{h} \Rightarrow h = \frac{16\,\text{cm}}{\cos 58°} = 30.2\,\text{cm} \approx \textbf{30 cm} \quad \text{or}$$

$$h^2 = \sqrt{(16\,\text{cm})^2 + (25.6\,\text{cm})^2} \Rightarrow h = \textbf{30 cm}.$$

(c) $\theta = ?$; $\alpha = ?$; $h = ?$

$$\tan\theta = \frac{6.0\,\text{ft}}{13\,\text{ft}} \Rightarrow \theta = \tan^{-1}\left(\frac{6.0}{13}\right) = 24.8° \approx \textbf{25°}.$$

$$\tan\alpha = \frac{13\,\text{ft}}{6.0\,\text{ft}} \Rightarrow \alpha = \tan^{-1}\left(\frac{13}{6.0}\right) = 65.2° \approx \textbf{65°} \quad \text{or} \quad \alpha = 90° - \theta = 90° - 25° = \textbf{65°}.$$

$$h^2 = \sqrt{(6.0\,\text{ft})^2 + (13\,\text{ft})^2} \Rightarrow h = 14.3\,\text{ft} \approx \textbf{14 ft}.$$

(d) $y = ?$; $x = ?$

$$\sin 24° = \frac{y}{23\,\text{km}} \Rightarrow y = (23\,\text{km})\sin 24° = 9.36\,\text{km} \approx \textbf{9.4 km}.$$

$$\cos 24° = \frac{x}{23\,\text{km}} \Rightarrow x = (23\,\text{km})\cos 24° = \textbf{21 km}.$$

Displacement

Sample Problem 3

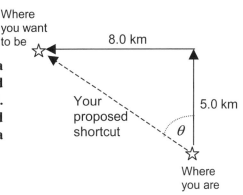

Suppose you are out in the desert in search of a town. The closest town is 5.0 km north of you, and the town that you want is 8.0 km west of that. Instead of going first to the town north of you and then to the second town, you decide to take a shortcut directly to the second town.
(a) In what direction should you proceed?

Solution: From the diagram $\tan\theta = \dfrac{8.0\,\text{km}}{5.0\,\text{km}} = 1.6$ so our θ is "the angle whose tangent is 1.6," which is the inverse tangent of 1.6, or $\tan^{-1}(1.6)$. So $\theta = \tan^{-1}(1.6)$. Your calculator should give you an answer of **58°**.

(b) How much distance do you "save" by taking the shortcut?

Solution: To find out how far you go, you could use the Pythagorean theorem, $a^2 + b^2 = c^2 \Rightarrow c = \sqrt{a^2 + b^2}$.

So you will go $\sqrt{(5.0\,\text{km})^2 + (8.0\,\text{km})^2} \approx 9.4$ km. This saves you $(5.0\text{ km} + 8.0\text{ km}) - 9.4$ km, or **3.6 km**, of hiking through the desert. Knowing some trig could save you an hour's walk in the hot sun!

Sample Problem 4

Suppose you come to a wide river that moves at a steady speed of 3.0 mi/h. A man will rent you a boat that can go 7.0 mi/h. The love of your life is waiting for you, directly across the river. In which direction do you steer the boat so that you can arrive there as directly as possible?

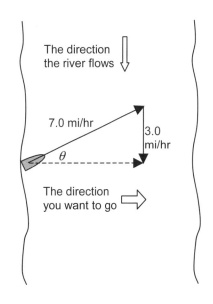

Solution: If you were to run the boat for 1 hour, you'd travel 7 miles relative to the river, but the river would carry you 3 miles downstream. You want to end up going straight across so you need to point your boat somewhat upriver as shown.

From the diagram, the angle you want has a sine equal to 3.0/7.0. That is,

$\sin\theta = \dfrac{3.0}{7.0} \Rightarrow \theta = \sin^{-1}\left(\dfrac{3.0}{7.0}\right) = 25.4° \approx \mathbf{25°}$.

© Paul G. Hewitt and Phillip R. Wolf

Sample Problem 5

Consider the following three situations. In each case we have a 1000-kg wagon moving to the right, pulled with a 100-N force in the directions shown.* **How does the applied force in each case affect the motion of the wagon?**

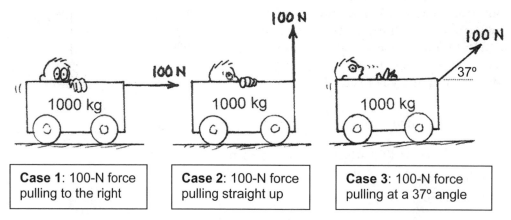

Case 1: 100-N force pulling to the right

Case 2: 100-N force pulling straight up

Case 3: 100-N force pulling at a 37° angle

Solutions:

Case 1:

All of the 100-N force pulls horizontally and contributes to accelerating the cart.

Case 2:

None of the 100-N force contributes to accelerating the cart. The force pulls up on the cart but neither raises it from the ground, nor speeds it up, nor slows it down.

Case 3:

The cart will accelerate less than in case 1, but more than in case 2, because only a fraction of the 100-N pull is forward and "useful" (as in case 1).

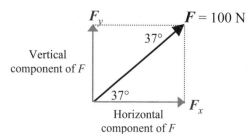

The 100-N force at 37° has two *components*—a horizontal x-component (which contributes to the acceleration of the cart) and a vertical y-component (which, in this case, doesn't accelerate the cart).

The magnitude of the horizontal component, F_x, can be found from $\cos 37° = \dfrac{F_x}{100 \text{ N}} \Rightarrow$
$F_x = (100 \text{ N})\cos 37° = \mathbf{80 \text{ N}}$. The horizontal component of the force on the cart is less than 100 N, so the acceleration of the cart here is less than in case 1.

Now have a go at the problems that follow!

---

* In this book we take a reasonable view of significant figures. A reasonable person would say that this cart has a mass of about 1000s kg, give or take 50 or 100 kg (so, 2 significant figures). We could write 1.0×10^3 kg, but this quantity is less easily grasped than "1000 kg." For a Conceptual Physics course we take the approach that the numbers should make understanding the concepts easier and that being overly picky about significant figures can get in the way of understanding the concepts in the problem.

Components Problems

For each of these, you should draw a vector diagram and sketch in the components before you solve the problem.

4&5-1. Joe pushes with a force F down along the handle of his lawnmower. His pushing force makes an angle θ with the surface of the ground.
   (a) What is the magnitude of the component of F that points parallel to the ground?
   (b) What is the magnitude of the component of F that points perpendicular to the ground?
   (c) Find the magnitude of the horizontal and vertical components of the force if Joe pushes down the lawnmower handle with a 55 N force directed at an angle of 53° below the horizontal.

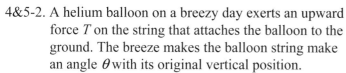

4&5-2. A helium balloon on a breezy day exerts an upward force T on the string that attaches the balloon to the ground. The breeze makes the balloon string make an angle θ with its original vertical position.
   (a) How large is the horizontal component of the balloon's pull on the ground?
   (b) How large is the vertical component of the balloon's pull on the ground?
   (c) Calculate the horizontal and vertical components of the balloon's 1.2-N pull when the string makes an angle of 20° with its original vertical position.

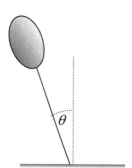

4&5-3. A block of ice of weight W slides down a slope of angle θ.
   (a) How large is the component of the weight that acts parallel to the slope?
   (b) How large is the component of the weight that acts perpendicular to the slope?
   (c) For a 200-N ice block on a 26° slope, find the components of the weight that are parallel and perpendicular to the slope.

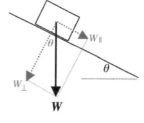

4&5-4. Juan throws a kiwi fruit at an initial speed v_0 at an angle θ above the ground.
   (a) What is the magnitude of the x (horizontal) component of the kiwi's initial velocity?
   (b) What is the magnitude of the y (vertical) component of the kiwi's initial velocity?
   (c) Calculate the x- and y-components of the initial velocity of a fuzzy fruit that is launched at 18.0 m/s at an angle 17° above the horizontal.

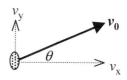

© Paul G. Hewitt and Phillip R. Wolf

Show-That Problems for Trigonometry

For each of the following, first draw a diagram and label the components before calculating your solution.

4&5-5. Milan, Caroline, and Andy all start at the southwest corner of a football field. Caroline runs 92 yards in a direction that makes a 57° angle with the southernmost goal line and then stops.
  (a) Show that Milan would have to go 50 yards along the southern goal line to end up directly south of Caroline.
  (b) Show that Andy would have to go 77 yards north along the field's western border to end up directly west of Caroline.

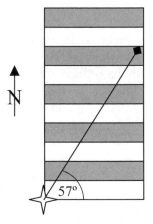

4&5-6. A submarine travels 15.0 km at an angle of 35.0° south of due west. Show that the submarine ends up 8.6 km south and 12.3 km west of its initial position.

4&5-7. Joanne drives 6.0 miles east, then 3.0 miles north, and then 2.0 miles west. Show that a crow could make the same voyage by flying 5.0 miles in a direction 37° north of east.

4&5-8. Three dogs are all pulling on a rubber chew toy but their pulls all balance out so that the chew toy doesn't move. Rex pulls east on the toy with a 60-N force while Rover pulls south on the toy with a 40-N force. Show that Fluffy must be pulling with a 72-N force at an angle of 34° north of west.

4&5-9. Giovanni stacks extra-extra-large pizzas into a tower 56 meters high. When he finishes he discovers to his horror that the stack leans at an angle of 4.0° with the vertical. Show that if a piece of pepperoni fell over the edge of the top pizza it would land about 3.9 meters from the edge of the bottom pizza.

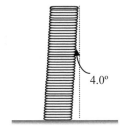

4&5-10. A soldier points his musket at an angle of 12° above the ground and fires a lead ball with an initial speed of 77 m/s. Show that the initial horizontal and vertical components of the velocity of the ball are 75 m/s and 16 m/s, respectively.

4&5-11. • A prairie dog runs 60 meters west, and then 80 meters northwest. Show that a second prairie dog using binoculars and located at the starting point could see the first prairie dog if he looked in a direction 26° north of west.

4 Newton's Second Law

In Chapter 2 we introduced the equilibrium rule—when the sum of all of the forces acting on an object is zero (in symbols, $\Sigma F = 0$, or $F_{net} = 0$), then no acceleration occurs. But in situations where the sum of the forces acting on the object is *not* zero (that is, $F_{net} \neq 0$), then the object will accelerate. This acceleration depends on the magnitude and direction of the net force and on the object's mass. This behavior is summarized in Newton's second law:

$$a = \frac{F_{net}}{m} \quad \text{or} \quad F_{net} = ma.$$

Both force and acceleration are vector quantities. The direction of acceleration is always in the direction of the net force.

The standard symbol for force is F. For some specific forces we use other symbols—f stands for friction, N for normal force, T for the force due to tension, and mg or W for weight, the force due to gravity.

Our unit for force is the *newton* (N), with 1 newton defined as the amount of force acting on a 1-kg mass that will produce an acceleration of 1 m/s², so $1 \text{ N} \equiv 1 \text{ kg·m/s}^2$.

In this book, units of measurement are displayed in standard typeface, while symbols are shown in italics. For example, we can say a normal force $N = 5$ N, or a block of mass m, travels a distance of 5 m. In the *Conceptual Physics* textbook and in the *Practicing Physics* book, we used the acceleration due to gravity $g = 10$ m/s² since multiples of 10 simplify mental math more than multiples of 9.8. In the problems in this book, we'll use the more precise $g = 9.8$ m/s².

The weight of an object, in its simplest form, is the force that gravity exerts on it. If an object is in free fall, then the only force acting on it is gravity so that $F_{net} = ma$ becomes $W = mg$. So the weight of an object is mg. (In Chapter 9 we extend the definition of weight to being the force that an object exerts on a supporting surface.)

For an object in free fall, $a = \frac{F_{net}}{m}$ becomes $a = \frac{W}{m} = \frac{mg}{m} = g$, so every object in free fall has acceleration g regardless of its mass. We can also see from $g = \frac{W}{m} = 9.8 \frac{\text{N}}{\text{kg}} = 9.8 \frac{\left(\text{kg} \cdot \frac{\text{m}}{\text{s}^2}\right)}{\text{kg}} = 9.8 \frac{\text{m}}{\text{s}^2}$ that the units $\frac{\text{m}}{\text{s}^2}$ and $\frac{\text{N}}{\text{kg}}$ are equivalent and we can use either for g.

Adding Forces in One Dimension

Consider two people fighting over a TV remote. One pulls to the right with a force of 50 newtons. (We'll write this as 50 N.) The other pulls to the left with a force of 20 N. We can represent this situation with a force diagram:

20 N ← [remote] → 50 N

We can see that the net force is 30 N to the right.

To sum the forces more formally as vectors, we'd say:
    Let's call "to the right" the positive direction. Then
    $\Sigma F = F_{net} = $ (50 N in the positive direction) + (20 N in the negative direction)
    $= (+50 \text{ N}) + (-20 \text{ N}) = 50 \text{ N} - 20 \text{ N} = \mathbf{30 \text{ N}}$.

    Since our answer is positive, the net force is in the positive direction, to the right.

Usually we simplify all of this by saying,

$$F_{net} = \begin{pmatrix} \text{sum of the magnitudes} \\ \text{of the forces in the} \\ \text{positive direction} \end{pmatrix} - \begin{pmatrix} \text{sum of the magnitudes} \\ \text{of the forces in the} \\ \text{negative direction} \end{pmatrix} = 50\text{ N} - 20\text{ N} = \mathbf{30\text{ N}}.$$

Sample Problem 1

A caveman drags a mammoth bone of mass *m* back to his cave. He exerts a horizontal force *F* on the bone in the forward direction. The ground exerts a horizontal friction force *f* on the bone in the opposite direction.

(a) Write an expression for the net force acting on the bone.

Focus: $F_{net} = ?$

First we draw a diagram illustrating the relevant forces acting on the bone:

Here the bone is moving horizontally along the ground, so we concern ourselves only with forces in the horizontal direction. Calling the direction of *F* positive, the direction of *f* is therefore negative. Summing the forces to get the *net* force on the bone, we get

$F_{net} = F - f.$

(b) What is the acceleration of the bone?

Focus: $a = ?$

From Newton's second law, $a = \dfrac{F_{net}}{m} = \dfrac{F - f}{m}.$

(c) Determine the acceleration of a 20-kg mammoth bone if the caveman's steady pull is 120 N and the friction force acting on the bone is 110 N.

Solution: From part (b) $a = \dfrac{F - f}{m} = \dfrac{120\text{ N} - 110\text{ N}}{20\text{ kg}} = 0.5\dfrac{\text{N}}{\text{kg}} = 0.5\dfrac{\text{kg}\cdot\frac{\text{m}}{\text{s}^2}}{\text{kg}} = \mathbf{0.5\dfrac{\text{m}}{\text{s}^2}}.$

Sample Problem 2

A crate of mass *m* sits on a factory floor. Find the acceleration of the crate when the net force acting on it is horizontal, with a magnitude equal to half the weight of the crate.

Focus: $a = ?$

Draw a diagram illustrating the situation:

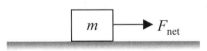

The physics is still Newton's second law.
In equation form:

$a = \dfrac{F_{net}}{m} = \dfrac{mg/2}{m} = \dfrac{g}{2} = 0.5g = 0.5\left(9.8\dfrac{\text{m}}{\text{s}^2}\right) = \mathbf{4.9\dfrac{\text{m}}{\text{s}^2}}.$

Note that we've simply substituted $mg/2$ for F_{net}. Then, after canceling the *m*, we have our solution (shown in boldface). Note that we did not consider friction separately in this problem. Friction is one of the forces that make up the net force acting on the crate.

Vertical-Acceleration Sample Problem 3

A box of mass *m* hangs by a string from the ceiling of an elevator.
(a) What is the tension *T* in the string when the elevator is at rest?

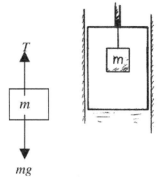

Focus: $T = ?$

Two forces act on the box—the tension force T and the gravitational force mg. We can illustrate this with a *force diagram*, which shows all of the relevant forces acting on the system.* In this case the system is simply our box of mass m.

The physics is still Newton's second law.

We'll call upward the positive direction. The box is at rest and the acceleration is zero:

$$a = \frac{F_{net}}{m} = \frac{T - mg}{m} = 0, \text{ so } \boldsymbol{T = mg}.$$

This answer makes sense from Newton's *first* law of motion. The box remains at rest because no net force acts upon it.

(b) What would be the tension in part (a) if the elevator were moving at constant velocity?

Answer: At constant velocity, $a = 0$ and the tension T would again be equal to mg.

(c) Derive an equation for the string tension *T* when the elevator accelerates upward with acceleration *a*.

Focus: $T = ?$ Two forces again act on the box—the tension force T and the gravitational force mg. This time we have

$$a = \frac{F_{net}}{m} = \frac{T - mg}{m} \Rightarrow T = mg + ma = \boldsymbol{m(g + a)}.$$

The quantity ma represents the additional upward force that the string must provide to accelerate the box.

(d) Calculate the tension in the string if the mass of the box is 5.0 kg and the upward acceleration of the elevator is 1.4 m/s².

Solution: Plugging in, $T = m(g + a) = 5.0 \text{ kg} \left(9.8 \frac{m}{s^2} + 1.4 \frac{m}{s^2}\right) = \boldsymbol{56 \text{ N}}.$

(e) Calculate the tension in the string if the mass of the box is 5.0 kg and the elevator accelerates *downward* at 1.4 m/s².

Focus: This time our acceleration is downward—in the *negative* direction (since we had decided earlier to call upward the positive direction), so the acceleration is -1.4 m/s².

Plugging in, $T = m(g + a) = 5.0 \text{ kg} \left(9.8 \frac{m}{s^2} + (-1.4 \frac{m}{s^2})\right) = \boldsymbol{42 \text{ N}}.$

---

* A force diagram is a useful representation that you should use for every force problem. It consists of the following:
  1. A sketch of the object of interest
  2. Each force drawn as an arrow, where the direction of the arrow represents the direction of the force and the length of the arrow represents the magnitude of the force
  3. An appropriate label for each of the forces

Vertical-Acceleration Sample Problem 4

When you stand on the floor, gravity is pulling you downward with a force equal to *mg*. But since you are not accelerating, the net force on you must be zero. There must be some upward force acting on you that is equal in magnitude to your weight.

In this case the upward force that is acting on you is supplied by the floor. As your feet press into the floor, the floor molecules are pushed a little closer together. Electrical forces between the molecules resist them being squeezed much closer together—the floor pushes back upward on your feet. This upward force, perpendicular to the floor's surface, is called the *normal force*. ("Normal" is an old word for "perpendicular.") We give it the symbol *N*.

(a) If the box in the previous problem were instead sitting on the floor of the same elevator with the elevator not moving, what would be the normal force on the box?

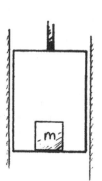

Focus: $N = ?$

> The box is at rest so the net force on the box is zero. The downward force is provided by gravity while the upward force is provided by the normal force. These forces must be equal in magnitude, so $N = mg$.
>
> More formally: Let's draw a force diagram for the box.*
>
> Now, applying Newton's second law and calling upward the positive direction:

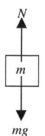

$$a = \frac{F_{net}}{m} = \frac{N - mg}{m} = 0, \text{ so } N = mg.$$

(b) Suppose the mass of the box is 5.0 kg and that the floor of the same elevator moves upward with an acceleration of 1.4 m/s². What is the normal force on the box?

Focus: $N = ?$

> There are still only two forces acting on the box—gravity (from Earth pulling down) and the normal force (from the floor pushing up). The force diagram is the same as above except that *N* is drawn a little bit longer than *mg* because the box is accelerating upward.

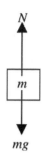

$$\text{From } a = \frac{F_{net}}{m} = \frac{N - mg}{m} \Rightarrow N = mg + ma = m(g + a) = 5.0 \text{ kg} \left(9.8 \frac{m}{s^2} + 1.4 \frac{m}{s^2}\right) = 56 \text{ N}.$$

---

* The vector representing *N* is drawn upward from the middle of the box rather than at the bottom of the box because it is conventional to draw all force vectors as though they act at the center of mass of the object. The net force on the box is the same whether you are pulling or pushing upward on the box.

(c) Calculate the magnitude of the normal force acting on the box if the same box rests on the elevator floor while the elevator accelerates *downward* at 1.4 m/s².

Focus: $N = ?$

The physics is the same as before, but this time the acceleration is negative because it is downward, opposite the direction we identified as positive, so $a = -1.4$ m/s².

From $a = \dfrac{F_{net}}{m} = \dfrac{N - mg}{m} \Rightarrow N = mg + ma = m(g + a) = 5.0 \text{ kg}\left(9.8\dfrac{m}{s^2} + (-1.4\dfrac{m}{s^2})\right) = 42 \text{ N}.$

Notice that the normal force is less when the elevator accelerates downward. This makes sense, for in the extreme case of the elevator accelerating downward at *g*, free fall, the normal force would be zero.

(d) What would be the magnitude of the normal force in part (c) if the elevator were moving at constant velocity?

Answer: At constant velocity, $a = 0$ and the supporting forces would simply be ***mg***.

Two-Block Vertical and Horizontal Acceleration Sample Problem 5

A cart of mass *m* on a horizontal friction-free air track is accelerated by a string attached to a weight, also of mass *m*, hanging vertically from a small pulley as shown.
(a) Find the acceleration of the two-mass system.

Focus: $a_{system} = ?$

If we consider the two masses as our system, the only unbalanced force acting on the system is the force of gravity on the hanging weight, *mg*, so this is the net force on the system. In this case we'll consider acceleration to be positive when the cart moves to the right and the weight drops.

$$a_{system} = \dfrac{F_{net\ on\ the\ system}}{m_{system}} = \dfrac{mg}{2m} = \dfrac{g}{2}.$$

(b) Show that the string tension is $\dfrac{mg}{2}$ (and NOT *mg*!).

Focus: Tension $T = ?$

Let's draw a force diagram for the hanging mass—*mg* pulls downward and *T* (the tension in the string) pulls upward. From $\Sigma F = ma$ for the falling weight we get,

$mg - T = m\left(\dfrac{g}{2}\right)$ (from $a = \dfrac{g}{2}$), so $T = \dfrac{mg}{2}$.

It's very important to realize that string tension is *less* than *mg* because otherwise the tension acting up would balance the gravitational force acting down and there would be no acceleration. If you held the sliding weight to prevent the masses from moving, then there would be no acceleration and *T* would be equal to *mg*.

(c) **Suppose the masses, instead of being equal, are 0.100 kg and 0.600 kg, respectively. Calculate the acceleration when the 0.100-kg mass dangles over the pulley.**

Step 1: *Focus*: $a_{system} = ?$

Step 2: Again, Newton's second law. Let's call the dangling mass *m* and the mass on the track *M*.

$$a_{system} = \frac{F_{on\ system}}{m_{system}} = \frac{mg}{(m+M)} = \left(\frac{m}{m+M}\right)g = \left(\frac{0.100\ kg}{0.100\ kg + 0.600\ kg}\right)9.8\frac{m}{s^2} = \frac{1}{7}\left(9.8\frac{m}{s^2}\right) = 1.4\frac{m}{s^2}.$$

(d) **Interchange the 0.100-kg and 0.600-kg masses, so the 0.600-kg mass dangles over the pulley. What acceleration occurs then?**

$$a_{system} = \left(\frac{m}{m+M}\right)g = \left(\frac{0.600\ kg}{0.600\ kg + 0.100\ kg}\right)9.8\frac{m}{s^2} = \frac{6}{7}\left(9.8\frac{m}{s^2}\right) = 8.4\frac{m}{s^2}.$$

Notice how different the accelerations are between the two situations. This leads to the next question.

(e) **For a variety of masses for this experiment, what range of accelerations is possible?** [*Hint*: Let one mass be nearly zero (like a feather) and the other mass huge; then switch.]

Solution: **The range of possible accelerations is from 0 to *g*.**

From $a_{system} = \left(\frac{m}{m+M}\right)g$, if *m* (the hanging mass) is very small and *M* (the pulled mass) is very large, the ratio $\left(\frac{m}{m+M}\right)$ is nearly zero and the acceleration is nearly zero. If the hanging mass *m* is very large and the pulled mass *M* is very small, the ratio $\left(\frac{m}{m+M}\right)$ is nearly one so the acceleration is nearly *g* (which makes sense, since a heavy weight pulling a feather across the track is essentially in free fall). These scenarios are featured on pages 14–16 of the *Practicing Physics* book.

Static Friction

Consider a wooden block at rest on a lab table. A string fastened to the side of the block passes over a pulley to a suspended cup, as shown in the diagram. You (patiently!) add water to the cup a little bit at a time, and finally one drop at a time. Friction between the surfaces of the block and table holds the block in place. This is *static friction*, which occurs because of some adhesion as well as irregularities in the surfaces of the block and table.

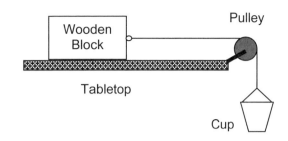

As more water is added to the cup, the tension in the string (and thus the horizontal pull on the block) increases. However, as long as the block is at rest, $\Sigma F = 0$, so increases in string tension are met with corresponding increases in the opposing static friction force. A force diagram shows all of the forces that act on the block when it is at rest.

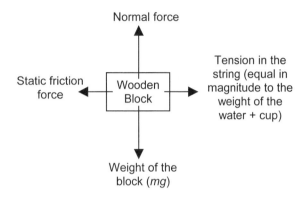

Since the block is at rest, we can write our second law equation for both the vertical and horizontal directions separately. Taking upward and to-the-right as our positive directions:

$$\Sigma F_y = 0 \Rightarrow N - mg = 0 \Rightarrow N = mg, \text{ and}$$

$$\Sigma F_x = 0 \Rightarrow T - f_{static} = 0 \Rightarrow T = f_{static}.$$

At some point the next drop of water increases the tension in the string enough to overcome some critical threshold and the block starts to slide. The pull on the block has *just* exceeded the maximum value that the static friction force can exert on this block. This threshold of motion is characterized by a *coefficient of static friction* μ_s, defined as

$$\mu_s = \frac{f_{static,\ maximum}}{N}.$$

The coefficient of static friction for a given pair of surfaces equals the *maximum* value of static friction between the surfaces divided by the normal force N that squeezes them together. Note that μ_s is a dimensionless ratio, one force divided by another force.

In general, the force of static friction can be expressed by

$$f_{static} \leq \mu_s N.$$

The "≤" symbol means that the static friction between the block and the table can range from zero (when nothing is pulling horizontally on the block) to some maximum value ($\mu_s N$), beyond which the block starts to slide.

Notice that in this experiment the maximum magnitude of the (horizontal) static friction force depends upon the magnitude of the (vertical) normal force. If we were to add a weight on top of the wooden block to double the value of the normal force N, the value of the tension required to make the block start moving would also double.

It turns out that for any pair of surfaces, the maximum value of the static friction force ($f_{static, max}$) increases in direct proportion with the normal force so that the value of μ_s ($= f_{static, max}/N$) is not affected. The coefficient of static friction, μ_s, is very nearly a constant for any two given materials in contact regardless of their individual size or weight. For example, the coefficient of static friction between two surfaces of wood is typically about 0.4. For rubber in contact with wood, it is about 1.0. A higher μ_s means that a larger force is required to get the surfaces to begin sliding against each other.

Friction, Sample Problem 6

A 1.8-kg block of wood rests on a table. A 5.0-N force pulls horizontally on the block but is not large enough to move the block.
(a) Draw a force diagram for the block.

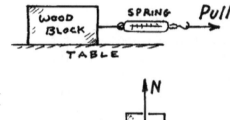

Comments: Since the block is at rest, $\Sigma F = 0$. Upward and downward forces cancel exactly, so we draw the N and the mg vectors with equal lengths. In the horizontal direction, pull P must have the same magnitude as f (in this case f_s).

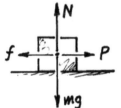

(b) What is the magnitude of the static friction force on the block?

Solution: The force of static friction must be equal in magnitude to the pulling force, **5.0 N**.

(c) What is the coefficient of static friction for the block and table surfaces?

Solution: We can't determine μ_s from the data given. μ_s can only tell us the *maximum* value that the static friction force can have. We know that 5.0 N won't get the box to move, but we don't know from the data given whether 5.0 N is that maximum value.

Suppose that we increase our pull on the block and find that a pull of 7.0 N is just enough to break the block free. Static friction at this point is 7.0 N*, and since the table is horizontal, the normal force equals the weight of the block, $mg = 17.6$ N. Then

$$\mu_s = \frac{f_{static}}{N} = \frac{7.0 \text{ N}}{17.6 \text{ N}} = 0.40.$$

We see that, in this case, the maximum friction force before breakaway is 0.40 times the weight of the block. That's because the weight of the block and the normal have the same magnitude.

---

* Perky Picky might argue that the *actual* maximum friction force must have been 6.999999 N since the block broke free with a pull of 7.0 N. If you calculate 6.999999 N/17.6 N, you still get, for all practical purposes, the same answer, 0.40. So we *say* that the maximum static friction force is 7.0 N.

Kinetic Friction

Let's return to our block on the table. We replace the cup and pulley arrangement with a spring scale. Now we give the block a little tap to get it moving and record how much horizontal force we must exert to keep the block moving across the table at a constant speed.

Experimentally, we find that the pull needed to keep the block moving at constant velocity is less than the force needed to start it moving. Further experimentation shows that this required pull is approximately the same no matter how slowly or quickly the block moves (unless the speed is very high). The required pull is also about the same no matter which side of the block is in contact with the table—friction doesn't depend on surface area (when on a narrow side, the smaller area is offset by a correspondingly greater pressure). Since the block is moving at a constant speed, we know that the net force on it is zero, so the friction force must be equal in magnitude to the pull.

We define the coefficient of kinetic (i.e., moving) friction by

$$\mu_k = \frac{f_{\text{kinetic}}}{N}.$$

This coefficient, like the coefficient of static friction, depends only on the surface materials, not on the weight of the object or its area of contact.

The force of kinetic friction can be written

$$f_{\text{kinetic}} = \mu_k N.$$

This looks a lot like the static friction equation except that now there's an equal sign. The kinetic friction force has a fixed value for a given N, regardless of the speed of the motion. For wood on wood, μ_k has a typical value 0.3, less than the static coefficient of friction.

Friction, Sample Problem 7

Sung must exert a horizontal 5.2-N force to pull a 2.5-kg box of chocolates across the dining room table at a constant speed.
(a) Make a force diagram of the box of chocolates.

(b) What is the coefficient of kinetic friction between the box of chocolates and the table?

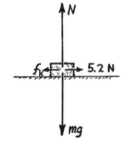

Focus: $\mu_k = ?$

We'll call upward the positive vertical direction and to-the-right the positive horizontal direction.
Since the box is moving in a straight line at constant speed,
$\Sigma F_y = 0$, $N - mg = 0$ $\Rightarrow N = mg$, and
$\Sigma F_x = 0$, $P - f_k = 0$ $\Rightarrow f_k = P$.
From the equation for kinetic friction, where $f_k = P$ and $N = mg$,

$$\mu_k = \frac{f_k}{N} = \frac{P}{mg} = \frac{5.2 \text{ N}}{(2.5 \text{ kg})(9.8 \text{ m/s}^2)} = \mathbf{0.21}.$$

(c) Derive the acceleration of the box if Sung exerts a more enthusiastic 7.3-N pull.

Focus: $a = ?$

The net force in the vertical direction is still zero, so we still have $N = mg$. The acceleration will be due to a net force in the horizontal direction.

Solution:

$$\Sigma F_x = ma_x \Rightarrow a_x = \frac{F_{net}}{m} = \frac{P - f_k}{m} = \frac{P - \mu_k N}{m} = \frac{P - \mu_k mg}{m} = \frac{P}{m} - \mu_k g.$$

(The first term in the solution (P/m) is the acceleration the box would have with no friction. The second term ($\mu_k g$) is the reduction in the acceleration caused by the kinetic friction force.)

With our numbers: $a_x = \dfrac{P}{m} - \mu_k g = \dfrac{7.3\left(\text{kg} \cdot \frac{\text{m}}{\text{s}^2}\right)}{2.5 \text{ kg}} - (0.21)\left(9.8 \dfrac{\text{m}}{\text{s}^2}\right) = 0.86 \dfrac{\text{m}}{\text{s}^2}.$

Inclined Planes

An inclined plane is simply a sloped surface. It could represent a hill, a ramp, a playground slide, or a sloped roof. Lots of interesting physics problems involve inclined planes!

Inclined Plane, Sample Problem 8

A box of mass m sits on a frictionless slope at an angle θ.
(a) Find the acceleration of the box down the slope.

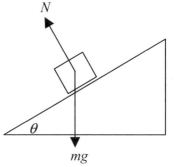

Focus: $a = ?$

$a = F_{net}/m$. So to find the acceleration we need to determine F_{net}. Two forces act on the box—gravity (mg) and the normal force from the slope (N). The vector sum of these two forces, the resultant, produces acceleration in a direction downward and parallel to the slope.

Usually we add forces in different directions by adding their components and then calculating the resultant. Here we'll do the same thing, but we'll take components along a *rotated* set of axes with the rotated x'-axis pointing parallel to the slope and the rotated y'-axis perpendicular to the slope.

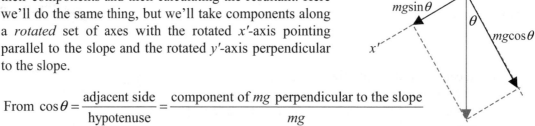

From $\cos\theta = \dfrac{\text{adjacent side}}{\text{hypotenuse}} = \dfrac{\text{component of } mg \text{ perpendicular to the slope}}{mg}$

$\Rightarrow$ component of mg perpendicular to the slope $= mg\cos\theta$, and

$\sin\theta = \dfrac{\text{opposite side}}{\text{hypotenuse}} = \dfrac{\text{component of } mg \text{ parallel to the slope}}{mg}$

$\Rightarrow$ component of mg parallel to the slope $= mg \sin\theta.$

Now we can sum our forces:
$\Sigma F_{y'} = 0 \Rightarrow N - mg\cos\theta = 0$ and
$\Sigma F_{x'} = F_{net} = mg\sin\theta$.

Now we have our answer! $a = \dfrac{F_{x'}}{m} = \dfrac{mg\sin\theta}{m} = g\sin\theta$.

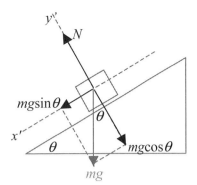

Let's check to see if this answer makes sense.

In the case of a vertical slope, $\theta = 90°$ and $a = g\sin 90° = g$ (as it should, for the box will be in free fall). On a level plane $\theta = 0°$ and $a = g\sin 0° = 0$ (which makes sense too). And just as in free fall, the acceleration is independent of the mass.

Look at the normal force, $N = mg\cos\theta$. When the slope is vertical, $N = mg\cos 90° = 0$ (the box is in free fall), and when the slope is horizontal, $N = mg\cos 0° = mg$ (which is just what we'd expect since the normal force equals the weight when the surface is horizontal).

(b) Suppose that the slope is *not* frictionless, that the coefficient of kinetic friction between the box and the slope is μ_k, and that the box still accelerates down the slope. Now derive the acceleration of the box.

Focus: $a = ?$

Now *three* forces act on the box—gravity (mg), the normal force from the slope (N), and the kinetic friction force (f_k) acting up and parallel to the slope (up because it tends to oppose the box's sliding motion).

The vector sum of these three forces will cause a smaller acceleration down the slope, while the net force will still be in a direction down and parallel to the slope.

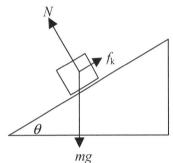

We can write our force equations:
$\Sigma F_{y'} = 0 \Rightarrow N - mg\cos\theta = 0$, and
$\Sigma F_{x'} = mg\sin\theta - f_k = mg\sin\theta - \mu_k N$.

Note from $\Sigma F_{y'} = 0$ that $N = mg\cos\theta$ (and *not* $N = mg$ as it would be on a surface with no slope). This makes sense because the box would not press as hard against the sloped surface as it would against a level plane.

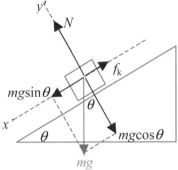

So, $a = \dfrac{\Sigma F_{x'}}{m} = \dfrac{mg\sin\theta - \mu_k N}{m} = \dfrac{mg\sin\theta - \mu_k mg\cos\theta}{m} = g\sin\theta - \mu_k g\cos\theta$

$= g(\sin\theta - \mu_k \cos\theta)$.

Notice the acceleration has two terms—the acceleration that the box would have if there were no friction ($g\sin\theta$) and a reduction in acceleration due to the kinetic friction ($\mu_k g\cos\theta$).

(c) Calculate the acceleration of a 120-kg box on a 22° slope where the coefficient of kinetic friction between the box and the slope is 0.15.

Solution: $a = g(\sin\theta - \mu_k \cos\theta) = (9.8 \text{ m/s}^2)(\sin 22° - 0.15\cos 22°) = 2.3 \text{ m/s}^2$.

Note that the mass of the box is irrelevant.

Problems for Newton's Second Law

4-1. Suzie exerts a horizontal force F on a box of mass m that sits on a frictionless horizontal surface.
 (a) Write an equation for the acceleration of the box.
 (b) Calculate the acceleration when her push is 15 N and the mass of the box is 10 kg.

4-2. The engines of a Boeing 747 jet of mass m produce an acceleration a along the runway during takeoff.
 (a) Write an equation for the net force that acts on the jet.
 (b) Calculate the force when the loaded jet of mass 3.2×10^5 kg has an acceleration of 3.5 m/s² down the runway.

4-3. When using a spring balance in the lab, you find that a force F accelerates a block across a nearly frictionless horizontal surface at acceleration a.
 (a) Write an equation for the mass of the block.
 (b) Calculate the mass of the block if a 5.0-N force accelerates it 2.5 m/s².

4-4. Helen exerts a horizontal force F on a puck of mass m on a horizontal surface where the force of friction is f.
 (a) Write an equation for the resulting acceleration.
 (b) Calculate the acceleration when her push is 5.0 N, the mass of the puck is 0.50 kg, and the friction is 1.0 N.

4-5. A motorcycle of mass m speeds up from speed v_0 to speed v_f in a time t.
 (a) Derive an equation for the average force F on the motorcycle. (You'll want to first come up with an equation for the acceleration of the motorcycle in terms of v_0, v_f, and t.)
 (b) Calculate the force that will accelerate a 210-kg motorcycle from 12.5 m/s to 20.8 m/s in 6.9 seconds.
 (c) How does this force compare with the weight of the motorcycle? Compute the ratio F/mg.

4-6. A certain net force applied to a vehicle of mass m causes it to accelerate at a.
 (a) In terms of a, how much acceleration will the same net force produce on a vehicle of mass $1.5\,m$?
 (b) Calculate the acceleration of the heavier vehicle if the lighter one has a mass of 4000 kg and the net force on each is 6000 N.

4-7. In serving, a tennis player accelerates a tennis ball of mass m horizontally from rest to a speed v. The acceleration is uniform when applied by the racquet over time t.
 (a) Derive an equation for the force exerted on the ball by the racquet.
 (b) Calculate the average force when the mass of the ball is 56 g and its speed from the racquet is 32 m/s for an impact time of 0.090 s.

Note that since our force unit is kg·m/s², the mass you plug in needs to be in kilograms.

4-8. Bronco Brown, of mass m, steps into the air from a high-flying helicopter. At a certain point in his fall, he experiences air resistance R.
   (a) Draw a force diagram for Bronco.
   (b) Derive an equation for Bronco's acceleration.
   (c) Calculate his acceleration if his mass is 100 kg and the air drag at that moment is 680 N.

4-9. When a horizontal force F is applied to a box having a mass m, the box slides on a level floor, opposed by a force of friction f.
   (a) Draw a force diagram illustrating this situation.
   (b) Write an equation for the acceleration of the box.
   (c) Calculate the acceleration if the horizontal force is 415 N, the friction force is 130 N, and the mass of the box is 75 kg.

4-10. Car A has mass m and experiences a net force F. Car B has mass $m/2$ and experiences a net force $F/2$.
   (a) Which car, if either, has the greater acceleration?
   (b) Calculate the accelerations when $m = 2000$ kg and $F = 4000$ N.

4-11. Jean produces a net force F on a cart of mass m that gives it an acceleration a.
   (a) In terms of a, how much acceleration will the same force produce on a cart of 0.60 m?
   (b) If the mass m of the original cart is 4.0 kg and its acceleration is 2.0 m/s^2, how much acceleration will the same force produce on the lighter cart?

4-12. Two boxes with masses m and M rest on a horizontal surface. Each box is mounted on tiny rollers and moves with negligible friction. A force applied to m gives it an acceleration a_m.
   (a) Find the acceleration M undergoes if pushed by the same amount of force.
   (b) The boxes have masses 2 kg and 5 kg. A force on the lighter one gives it an acceleration of 2 m/s^2. Calculate the acceleration of the 5-kg box when it is pushed by the same amount of force.

4-13. A stone of weight W rests on a horizontal friction-free surface.
   (a) Find the acceleration that occurs when a horizontal force F acts on the stone.
   (b) Calculate the acceleration of a 38-N stone on a friction-free surface when a 57-N horizontal force acts on the stone.

4-14. George pushes upward with a force F on a stone of weight W. F is greater than the stone's weight.
   (a) Draw a force diagram illustrating this situation.
   (b) What vertical acceleration occurs?
   (c) Calculate the acceleration when the weight of the stone is 38 N and the upward push is 57 N.

4-15. A bucket of water of mass m is pulled horizontally along a frictionless surface by a force equal in magnitude to 3 times the bucket's weight.
   (a) What is the bucket's acceleration?
   (b) What would be the acceleration if the pull were 4 times the bucket's weight?

4-16. A bucket of water of mass *m* is pulled straight upward by a force equal in magnitude to 3 times its weight.
  (a) Find its acceleration.
  (b) A friend says the answer is 3 g. What error did your friend likely make?

4-17. A paint bucket of mass *m*, initially at rest, is pulled upward with a steady force *P* that is greater than the weight of the bucket.
  (a) Find the net force acting on the bucket.
  (b) What is the acceleration of the bucket?
  (c) Calculate the acceleration if the mass of the paint bucket is 8.5 kg and it is pulled upward with a force of 95 N.

4-18. When a parachutist of mass *m* falls, the upward drag force on the chute is *R*.
  (a) Draw a force diagram for the parachutist.
  (b) Derive an equation for the acceleration of the parachutist.
  (c) *R* is greater than *mg* when the parachute first opens. What is the direction of the acceleration? What is the direction of the velocity?
  (d) Calculate the acceleration of the falling parachutist when the drag force is 1000 N and the mass of the parachutist is 80 kg. Neglect the mass of the parachute.
  (e) How large a drag force will produce terminal velocity?

4-19. George drops a stone of mass *m* from atop a high cliff of height *h* that overlooks the ocean. Just before the stone hits the water, air drag has built up to equal exactly one-fifth the weight of the stone.
  (a) Draw a force diagram for the stone just before it hits the water.
  (b) Derive an equation for its acceleration at this point, expressed as a fraction of *g*.
  (c) Calculate the acceleration of the stone in m/s^2 just before it hits the water.

4-20. A small puck of mass *m* rests on a horizontal surface that exerts no frictional force when the puck moves (an air table, for instance).
  (a) How much horizontal force must be applied to the puck to make it accelerate as much as it would if it were in free fall?
  (b) If the mass of the puck is 1.1 kg, calculate the horizontal force required to give the puck an acceleration of *g*.

4-21. A subcompact car of mass *m* and a brawny pickup truck of mass 3*m* are given equal accelerations.
  (a) How much greater is the force that acts on the more massive vehicle?
  (b) How would the accelerations compare if the same force acted on each vehicle? (That is, find the ratio a_{truck}/a_{car}.)

4-22. A truck of mass *m* slows from speed v_0 to speed *v* in a time *t*.
  (a) Write an equation for the applied braking force.
  (b) Calculate the braking force if the mass of the truck is 3.0×10^3 kg and it is slowed from 32 m/s to 12 m/s in 6.0 seconds.

4-23. A loaded pick-up truck of mass m moves with speed v. Its brakes are applied, and it comes to a stop in t seconds.
   (a) Relative to the direction of motion, in what direction does the braking force act?
   (b) Draw a force diagram showing the three forces acting on the slowing truck.
   (c) Derive an equation for the average braking force that slows the truck.
   (d) Derive an equation for the stopping distance x.
   (e) Calculate the braking force if the truck's mass is 2.0×10^3 kg, its initial speed is 18 m/s, and the stopping time is 8.0 s.
   (f) Calculate the stopping distance.

4-24. During a collision, a driver of mass m in a car moving at speed v is brought to rest by an inflated airbag in time t.
   (a) Derive an equation for the average force exerted by the airbag on the driver.
   (b) Calculate the average force if the mass of the driver is 55 kg, the initial speed of the car 28 m/s, and the contact time with the airbag is 0.20 s.

4-25. An average force is applied by a shoulder-strap seatbelt to a passenger of mass m in an emergency stop when the car's initial speed is v and it stops in t seconds.
   (a) Derive an equation for the magnitude of this average force.
   (b) Calculate the average force on a 55-kg passenger when the car slows from 86 km/h to a complete stop in 3.2 s.

> Recall that the unit of force, N, is kg·m/s². Check your units before you plug in numbers

4-26. A jet plane of mass m is initially at rest on the deck of an aircraft carrier. The jet catapults from rest to a horizontal speed v in time t.
   (a) Write an equation for the average net force acting on the jet.
   (b) Calculate the average force that propels a jet plane of mass 26,500 kg from rest to a horizontal speed of 72 m/s in a time 1.9 s.

4-27. A certain force accelerates an electron of mass m from rest to a speed v in time t.
   (a) Derive an equation for this force.
   (b) Calculate the force that accelerates an electron of mass 9.11×10^{-31} kg from rest to a speed of 3.0×10^6 m/s in a time of 1.5×10^{-7} s.

4-28. A golf ball of mass m leaves the tee at a speed v, which it gains in a brief time t.
   (a) Derive an equation for average force of impact of the club on the ball.
   (b) Calculate the average force exerted by the club on the ball of mass 0.045 kg to give it a speed of 72.0 m/s in an impact time of 0.0050 s.
   (c) Why is it permissible to ignore gravity and assume that this net force is provided entirely by the golf club?

4-29. A water skier of mass m is pulled at a constant velocity v by a boat of mass M. Tension in the rope held horizontally by the skier is T.
   (a) Find the total resistive force by the water and air on the skier.
   (b) How much upward force does the water exert on the skier?

4-30. Firefighter Fred of mass *m* has acceleration 0.25*g* when he slides
down a vertical pole.
   (a) How large is the friction force that acts on Fred?
   (b) What friction force would be needed for Fred to slide down at
       constant speed?

4-31. Suzy Skydiver, who has mass *m*, steps from the basket of a high-flying balloon of mass *M* and does a
sky dive.
   (a) What is the net force on Suzy at the moment she has stepped from the basket?
   (b) What is the net force on her when air drag builds up to equal half of her weight?
   (c) What is the net force on her when she reaches terminal speed *v*?
   (d) What is the net force on her after opening her parachute and reaching a new terminal speed 0.1 *v*?

4-32. A landing craft of mass *m* readies itself for a Moon landing. When the craft is a vertical distance *y*
above the Moon's surface, its speed downward is *v*. A retrorocket is fired to give the craft an upward
thrust to steadily slow its speed to zero as it meets the surface.
   (a) Draw a force diagram for the landing craft as it is slowing.
   (b) Derive an equation for this thrust, using g_{moon} to represent the acceleration due to gravity near the
       Moon's surface.
   (c) Calculate the needed thrust to decelerate the craft from a downward velocity of 15 m/s when 160 m
       above the lunar surface to rest at the lunar surface. The mass of the craft is 12,000 kg, and the
       acceleration due to gravity on the Moon is $g_{Earth}/6$.

4-33. An astronaut of mass *m* floating in space outside her spaceship receives a thrust *F* from a nitrogen spurt
gun. The duration of the spurt is *t*.
   (a) Derive an equation for her post-spurt speed relative to the spaceship.
   (b) Calculate her speed relative to the spaceship if her mass is 88 kg, the spurting force is 32 N, and the
       duration of the spurt is 2.1 s.

4-34. A cart of mass *m* is accelerated from rest by a net force *F* acting for a time *t*.
   (a) Find the speed of the cart at the end of that time.
   (b) Calculate the final speed of the cart of mass 5.0 kg when a 2.0 N force acts for 10 s.

4-35. Ramon, of mass *m*, stands on a scale in an elevator of mass *M* that is momentarily at rest.
   (a) What is the reading of the scale?
   (b) What is the reading when the elevator moves upward at constant velocity *v*?
   (c) What is the reading when moving upward with an acceleration *a*?
   (d) Calculate scale readings for Ramon, whose mass is 70.0 kg, the constant velocity in (b) is 1.5 m/s,
       and the constant acceleration in (c) is 2.0 m/s^2.

4-36. Consider the tension in the strands of cable that move an elevator upward in a skyscraper of height *y*.
The elevator with its load of people has mass *M* and moves upward at constant speed *v*.
   (a) What is the combined tension in the supporting cables?
   (b) Calculate the tension in the cables that moves the elevator of mass 1960 kg upward a distance of
       292 m at a constant speed of 10 m/s.

4-37. Maria Moonglow stands on a bathroom scale in the elevator of the physics building. When at rest, the scale shows a weight W. As the elevator starts to moves up, the scale reading is $1.2\,W$.
(a) Find the acceleration of the elevator.
(b) Maria normally weighs 550 N. What is her weight as recorded by the scale in the accelerating elevator?
(c) Calculate her "elevator weight" when the elevator is moving downward with the same magnitude of acceleration that you calculated in part (a) above.

4-38. An astronaut of mass m sits in a rocket that blasts off vertically upward from rest on Earth's surface, attaining a speed v in time t.
(a) Assuming constant acceleration, what force N does the spacecraft exert on the astronaut during takeoff?
(b) If the astronaut is sitting on a scale during the blast-off, what is the reading on the scale?
(c) Calculate the scale reading for a 75-kg astronaut during takeoff if the rocket gains a vertical speed of 162 m/s from rest in 14 s.
(d) How does this reading compare with the scale reading when the astronaut was sitting motionless before blast-off?

4-39. Tammy applies a constant horizontal force F to a block of ice on a smooth horizontal floor. Friction is negligible. The block starts from rest and reaches a speed v in time t.
(a) Derive an equation for the mass of the block of ice.
(b) If Tammy stops pushing at the end of 3.0 s when the speed of the block is 2.4 m/s, how far will the block slide in the next 3.0 s?

4-40. An iceboat of mass M with occupant of mass m begins at a rest position. A wind from directly behind exerts a force F on the sail.
(a) For negligible friction, find the acceleration of the iceboat.
(b) As the boat gains speed, the force of wind impact decreases. How does this affect acceleration?
(c) When the speed of the iceboat approaches wind speed, what will be the resulting acceleration of the iceboat?

4-41. A truck of mass M tows a car of mass m along a level road by a chain. The forward force on the truck's tires from the road is F.
(a) Write an equation for the acceleration of the car.
(b) How large is the tension in the connecting chain?
(c) Calculate answers to parts (a) and (b) if the mass of the truck is 2900 kg, the mass of the car is 1100 kg, and the forward force from the road on the truck tires is 4800 N.

4-42. Jackie is playing with vegetables in the lab. She screws a hook into a pumpkin, places the pumpkin (mass *M*) onto the lab table, and ties a string from the hook over a pulley to a smaller hanging mass *m*.

(a) Find the pumpkin's acceleration if the lab table is frictionless.
(b) Calculate the acceleration for a 4.00 kg pumpkin and a 2.20 kg hanging mass.
(c) Calculate the acceleration of the pumpkin if the positions are reversed, with the mass lying on the table and the pumpkin hanging from the string.

Show-That Problems for Newton's Second Law
(Please continue to find your solution with lettered concepts before you plug in numerical values. And most of all, show how you arrive at your answer.)

4-43. A 60.0-kg astronaut finds his weight to be 96.0 N on the Moon.
Show that the acceleration due to gravity on the Moon is 1.6 m/s^2.

4-44. A net force of 10.0 N is exerted by Irene on a 2.0-kg cart for 3.0 seconds.
Show that the cart will have an acceleration of 5.0 m/s^2.

4-45. Toby Toobad, who has a mass of 100 kg, is skateboarding at 9.0 m/s when he smacks into a brick wall and comes to a dead stop in 0.2 s.
Show that his deceleration is 45 m/s^2 (ouch!) and the net force on him is 4500 N.

4-46. A 5.0-kg cart that is moving with a velocity of 3.0 m/s is brought to a stop in 2.0 s.
Show that the magnitude of the stopping force is 7.5 N.

4-47. A 1.0-kg ball, traveling at a velocity of 45 m/s, collides with a wall and rebounds with a velocity of −33 m/s after an impact time of 0.20 s.
Show that the force of impact on the ball is 390 N.

4-48. A certain force exerted on an 8.0-kg block gives it an acceleration of 5.0 m/s^2.
Show that the same force exerted on a 2.0-kg block will give it an acceleration of 20 m/s^2.

4-49. A net force of 10.0 N on a block causes it to accelerate 2.0 m/s^2.
Show that the mass of the block is 5.0 kg.

4-50. Dan's truck has a mass of 2000 kg. When traveling at 22.0 m/s, it brakes to a stop in 4.0 s.
Show that the braking force acting on the truck is 11,000 N.

4-51. A 5.0-kg block accelerates at 0.80 m/s^2 along a horizontal surface when a forward horizontal 12.0-N force acts on it.
Show that the force of friction on the block is 8.0 N.

4-52. A falling 50-kg parachutist experiences an upward acceleration of 6.2 m/s^2 when she opens her parachute.
Show that the drag force is 800 N when this occurs.

Coefficients-of-Friction Problems

4-53. A minimum horizontal pull P is required to get a block of mass m to just start moving from rest on a horizontal surface.
   (a) How large is the normal force acting on the block?
   (b) How large is the friction force acting on the block just before it starts to move?
   (c) What is the coefficient of static friction between the block and the horizontal surface?
   (d) Calculate the value of μ_s if it takes a 13-N pull to get a 2.2 kg block to start moving from rest.

4-54. Malcolm can get his couch, mass M, to start sliding across the living room by pushing on it horizontally with force P.
   (a) What is the coefficient of static friction between the couch and the floor?
   (b) Megan, mass m, playfully decides to lie down on the couch. How large a force will Malcolm then have to exert to start the couch moving?
   (c) Calculate μ_s if it takes a 180-N force to get a 32-kg couch to start moving.
   (d) Calculate the size of Malcolm's new push if Megan's mass is 72 kg.

4-55. A block of mass M sits on a lab table. The block is tied to a string that passes over a pulley to a hanging mass. Chelsea increases the hanging mass 1 gram at a time until, at mass m, the block just starts to move.

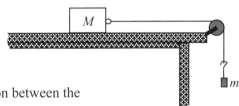

   (a) What is the approximate coefficient of static friction between the block and the table?
   (b) Calculate μ_s between a 256-gram block and the table if the block starts to move when the hanging mass becomes 119 grams.

4-56. • Maryam lays a brick on top of a horizontal pine board and then gives the *board* a sideways pull. The brick accelerates along with the board, without slipping.

   (a) Draw a force diagram showing the three forces acting on the accelerating *brick*. What force supplies the net force on the brick? (*Hint*: It isn't Maryam's pull.)
   (b) Given that the coefficient of static friction between the brick and the board is μ_s, what is the maximum acceleration the board and brick can have without the brick slipping?
   (c) Calculate this maximum acceleration if μ_s between the brick and the board is 0.60.

4-57. In order to hold a book of mass m stationary against a wall, Leslie finds she must push horizontally with a force of at least P.

   (a) First make a force diagram, and then find the coefficient of static friction.
   (b) Calculate the coefficient of static friction for the wall and book when at least a 35-N horizontal push is needed to hold a 1.2-kg physics book steadily at rest against a wall.

4-58. Carlos and Miranda are in the lab, determining the coefficients of friction between blocks of various materials and the floor. Each block has a hook in its side.

Miranda attaches the spring scale to each block of mass m and pulls each block at a constant speed across the floor while Carlos records the spring scale reading P.
(a) How will Carlos and Miranda calculate their values for μ_k?
(b) Complete the table for the results that they measured in the lab:

Material pulled along the floor	Mass of the block (kg)	Spring scale reading at constant speed (N)	Value of μ_k
Brick	2.20	13	
Pine	0.56	2.2	
Steel	1.20	5.5	

(c) How would their calculated values of μ_k have changed if they had done the experiment using blocks of material that had twice the mass?

4-59. A friction force f acts upon a horizontally moving box of weight W.
(a) What is the coefficient of kinetic friction between the box and the surface?
(b) Calculate the coefficient of kinetic friction between a moving box and the surface if the friction force is 22 N and the weight of the box is 56 N.

4-60. Bob finds that a constant horizontal pull P is needed to keep a crate of mass m moving across the floor at a constant velocity v.
(a) What is the coefficient of kinetic friction μ_k between the crate and the floor?
(b) Calculate μ_k if the pull is 87 N and the mass of the crate is 32 kg.

4-61. Alex finds that a constant horizontal pull P is needed to keep a box of mass m moving horizontally across the floor at a constant velocity v. The coefficient of kinetic friction between the box and the floor is μ_k.
(a) What is the magnitude of P?
(b) Calculate the horizontal pull to keep a 25-kg box moving horizontally across the floor at constant velocity when the coefficient of kinetic friction between the box and the floor is 0.31.

4-62. Stan the stuntman of mass m is dragged along a road at constant velocity v by a cable attached to a moving truck of mass M. The cable is parallel to the level road and the coefficient of kinetic friction between Stan and road is μ_k.
(a) Make a force diagram of Stan in this scenario.
(b) Find the tension in the cable.
(c) If the truck has a mass of 3200 kg and it is dragging 82-kg Stan at velocity 5.0 m/s, calculate the tension in the cable when μ_k between Stan and the road is 0.60.
(d) How would the tension differ if Stan were dragged at a constant velocity of 6.0 m/s?

4-63. Ludmila exerts a constant horizontal push P to keep a box of mass m moving with constant acceleration a across a level floor. The coefficient of friction between the box and the floor is μ_k.
   (a) What is the magnitude of P?
   (b) Calculate the horizontal push needed to make an 18-kg box slide across the floor with an acceleration of 0.50 m/s² if the coefficient of kinetic friction between the box and the floor is 0.34.

4-64. Megan finds that a constant horizontal pull P is needed to keep a box of mass m moving across the floor with a constant forward acceleration a.
   (a) What is the coefficient of kinetic friction between the box and the floor?
   (b) Calculate μ_k for the case of a horizontal pull of 143 N accelerating a 42-kg box across the floor at 1.3 m/s².

4-65. Ashley gives her calculator an initial horizontal velocity v_0 across her lab desk. The calculator slides a distance x along the table before coming to rest.
   (a) Draw a force diagram showing the three forces acting on the calculator while it is still sliding, but after Ashley has released it. What is the direction of the net force on the calculator?
   (b) What is the value of μ_k between the calculator and the desktop?
   (c) Calculate μ_k if she launches her calculator at 1.8 m/s and it slides 0.45 m before coming to rest.

4-66. A car of mass m is moving at speed v when its driver slams on the brakes, which lock. The car then skids to a stop in a given distance x.

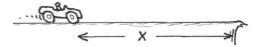

   (a) Draw a force diagram for the car. What force provides the net force on the car? What is the direction of the net force?
   (b) What is the coefficient of kinetic friction between the tires and the road?
   (c) Calculate μ_k for a 1500-kg car moving at 62 km/hr that skids a distance of 18 m to rest when its brakes are locked. (Remember that you will need the speed in m/s.)

4-67. Rabbit finds that his constant horizontal pull P on a large bag of carrots accelerates it across the ground at a constant rate a. The coefficient of kinetic friction between the bag and the ground is μ_k.
   (a) What is the mass of the bag of carrots?
   (b) Calculate the mass of the bag of carrots if μ_k between the bag and the ground is 0.60, and Rabbit's 14-N pull accelerates the bag at 0.40 m/s².

4-68. While Ari sits inside a wooden crate, Jamie gives the crate a steady horizontal push P, accelerating the crate from rest along the sidewalk.
   (a) Draw a force diagram showing the four forces acting on the moving system (Ari + crate). We'll call the mass of this system m.
   (b) Assume that we know the value of μ_k between the crate and the sidewalk. How large a push must Jamie exert so that the crate, starting from rest, accelerates steadily and ends with speed v after covering a distance d?
   (c) Calculate the size of Jamie's push if the Ari + crate system has a mass of 49 kg, $\mu_k = 0.52$, and the box is moving at 2.5 m/s after having covered a distance of 5.0 m.

4-69. Emily exerts a horizontal forward push P on a crate of oranges of mass m that is already sliding across a horizontal floor. Despite her push, the moving crate slows at a uniform rate a.
(a) Which is greater, the pushing force P or the frictional force f? Draw a force diagram for the crate.
(b) Find the coefficient of kinetic friction between the crate and the floor.
(c) Calculate μ_k when the mass of the crate is 22 kg, the pushing force is 75 N, and the crate slows by 0.80 m/s each second.

4-70. • Back in the days before antilock brakes, an accident investigator was summoned to a scene where a car had crashed into a brick wall. The length of the skid marks indicated that the car had slammed on its brakes a distance x before hitting the wall, and the damage to the car indicated that the car had a speed v_f when it collided with the wall. The investigator knew the value of μ_k for rubber sliding on the road, dry asphalt.
(a) Draw a force diagram for the car. What force provides the net force on the car while skidding? What is the direction of the net force?
(b) What was the initial speed v_0 of the car when its brakes were first applied?
(c) Calculate v_0 for a car that crashed into a wall at 15 mi/h (6.7 m/s) after leaving 26-m skid marks on the road, where μ_k for rubber on the road is 0.70.

4-71. • A crate of mass m rests on the flatbed of a truck of mass M. The coefficient of static friction between the crate and the surface of the flatbed is μ_s.
(a) Draw a force diagram showing the three forces acting on the accelerating crate. What is the maximum forward acceleration the truck can have on a level road before the crate starts to slide backward relative to the bed of the truck?
(b) Once the truck and crate move at the same speed v, what is the maximum deceleration the truck can have before the crate slides forward relative to the bed of the truck? (*Hint*: First draw a force diagram showing the three forces acting on the decelerating crate.)
(c) If the truck stops suddenly by hitting a wall, what is the maximum speed of the crate as it bashes against the cabin of the truck? (Why is Newton's first law sufficient to answer this, with no calculations needed?)

Show-That Problems for Coefficients of Friction

4-72. A 245-kg safe moves along a level floor. The coefficient of kinetic friction between the safe and the floor is 0.62.
Show that a horizontal force of about 1500 N will be required to pull it along at a constant speed.

4-73. While experimenting at lunchtime, Marjorie finds that a 1.4-N horizontal force is needed to start a 340-gram sandwich sliding across the cafeteria table.
Show that the coefficient of static friction between the sandwich and table is 0.42.

4-74. A dog sled with driver weighs 2000 N. Show that if the coefficient of kinetic friction between the sled and the snow is 0.050, the dogs must exert a 100-N force on the sled to keep the sled going at constant speed.

4-75. Matt places his 1.2-kg physics book on the lab table and ties a string around it. The string runs over a pulley to a hanging mass, which hangs over the edge of the table. Matt finds that a 500-gram hanging mass won't make the book move, but a 600-gram hanging mass will. Show that the coefficient of static friction between the book and table is between 0.42 and 0.50.

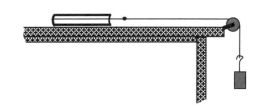

4-76. In an experimental mood, Dave finds that a root beer mug, slid across a bar top with an initial speed of 2.9 m/s, slows down but is still moving when it reaches its intended recipient, who is 4.0 m down the bar, 1.6 seconds later.
Show that the coefficient of friction between the (wet) mug and the bar top is 0.051.

4-77. Steve slides a hockey puck at 8.5 m/s across the ice. The puck comes to rest 46 meters from its starting point.
Show that the coefficient of kinetic friction between the puck and the ice is 0.080.

4-78. If a 130-gram calendar is pushed hard enough against a wall, the calendar won't slide. The coefficient of static friction between the calendar and the wall is 0.51.
Show that you'll have to exert at least a 2.5-N horizontal push on the calendar to prevent it from sliding down the wall.

4-79. Atti, a 2.5-kg poodle, is given a push so that she starts out sliding across a frozen lake at a speed of 6.3 m/s. The coefficient of kinetic friction between Atti and the ice is 0.088.
Show that Atti will slide for 7.3 seconds before coming to rest.

4-80. Consider the two-block acceleration problem (Sample Problem 5) on page 29. With no friction, the acceleration of both the horizontal and vertical blocks is 0.5 g. Suppose, however, that friction exists between the sliding cart and the track, with $\mu_k = 0.2$.
Show that the acceleration of the two-mass system will be 0.4 g.

Problems Involving Some Trigonometry

4-81. Two scales are used in a classroom demonstration to suspend a 10-N weight.
(a) Explain how each of the scales can register 10 N. That is, how can 10 N + 10 N = 10 N?
(b) Show your answer with a force diagram.

4-82. Gymnast Gracie of mass m is suspended by a pair of vertical ropes attached to the ceiling.
(a) What is the tension in each rope?
(b) What are the rope tensions if they comprise a V-shape, each at an angle θ with the ceiling?
(c) Suppose Gracie has a mass of 55 kg. Calculate answers for tension in the pair of vertical ropes, and for when each of the ropes is 53° to the ceiling.

4-83. The diagram shows the view from above of a boat being pulled along a canal. Each equal pull P is exerted at an angle θ to the left or right of the direction of the moving boat. The water exerts a drag force R against the motion of the boat.

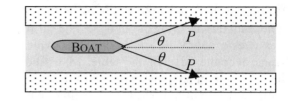

(a) What value of P is needed to keep the boat moving at a constant speed?
(b) Calculate the value of P for a drag force of 150 N and an angle of 22°.

4-84. Bronco Brown is mowing his lawn. He applies a force F along the handle of the lawnmower as shown. Bronco moves the mower at constant velocity.

(a) What is the magnitude of the resistance force acting horizontally on the mower?
(b) Why is "constant velocity" the key to solving this problem?
(c) Calculate the resistance force if Bronco applies 210 N along the handle, the mower moves at 1.2 m/s, and the angle that the handle makes with the ground is 55°.

4-85. The weight W of a box on an inclined plane can be resolved into two vector components, one parallel to the plane and the other perpendicular to it.

(a) At what angle of the plane relative to the horizontal would the components be equal?
(b) At what angle would the perpendicular component equal W?
(c) At what angle would the parallel component equal W?
(d) Assuming negligible friction, what is the acceleration of the box in each of the three cases?

4-86. Luigi wishes to suspend a lantern of mass m between his store and a telephone pole. The cable attached to his store makes an angle θ to the wall as shown. The cable to the telephone pole is horizontal.

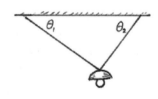

(a) What is the tension in the cable of angle θ?
(b) What is the tension in the horizontal cable?
(c) Calculate the two tensions for a 12-kg lantern with $\theta = 40°$.

4-87. • A street lamp is suspended by two cables, one at angle θ_1 and the other at angle θ_2 as shown.

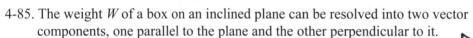

(a) Find the tension in each cable for a lamp of mass m.
(b) Calculate the two tensions when the mass of the lamp is 15 kg, θ_1 is 35°, and θ_2 is 55°.

4-88. Gracie slides down a frictionless slide of length L that makes an angle θ with the ground.
   (a) Gracie starts out at rest at the top of the slide. How fast will Gracie be going at the bottom of the slide?
   (b) Calculate Gracie's speed at the bottom of a 4.2-m slide angled at 39° with the horizontal.

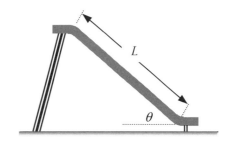

4-89. Bobby loses hold of his snow saucer, which starts to slide down a slope of length L and angle θ. The coefficient of kinetic friction between the saucer and the snow is μ_k.
   (a) What will be the acceleration of the saucer?
   (b) How long will it take for the saucer to get to the bottom of the hill?
   (c) Calculate the saucer's acceleration and time to get to the bottom of a 55-m long, 11° slope where the coefficient of friction between the saucer and snow is 0.081.

4-90. Snowboarder Sherry of mass m is pulled up a slope at constant velocity v by a tow bar. The force of the tow bar is parallel to the slope, which is inclined at θ with respect to the horizontal. The coefficient of kinetic friction between the snowboard and the snow is μ_k.
   (a) Find the force that the tow bar exerts on Sherry.
   (b) Calculate the force exerted by the tow bar on a 46-kg snowboarder being pulled at 3.0 m/s up a 20° slope if the coefficient of kinetic friction is 0.11.

4-91. Larry pulls a log of mass m up a ramp at constant speed using a rope that is parallel to the ramp's surface. The ramp is inclined at θ with respect to the horizontal, and the coefficient of kinetic friction between the log and the ramp is μ_k.
   (a) Find the tension in the rope.
   (b) Find the tension in the rope if the log undergoes a steady acceleration a.
   (c) Calculate the tensions for constant speed, and for acceleration 0.4 m/s² if the mass of the log is 48 kg and the ramp is inclined at 27° with respect to the horizontal. Assume $\mu_k = 0.30$.

4-92. • Tammy uses a rope to pull a box along the floor. The rope makes an angle θ with the floor. The coefficients of friction between the box and the floor are known, both μ_s and μ_k.
   (a) What tension in the rope is required to get the box to start moving?
   (b) A rope pulls at an angle of 20° above the horizontal on a 23-kg box. The coefficients of friction between the box and the surface are $\mu_s = 0.32$ and $\mu_k = 0.25$. Calculate the required tension in the rope to get the box to start moving.

Show-That Problems Involving Some Trigonometry

4-93. You take your 31-kg nephew to play on the swings at the park. You pull him 23° from the vertical and hold him stationary before letting him go.
Show that the combined tensions in the swing ropes equal 330 N. Then show that the horizontal force you exert in holding him stationary is 120 N.

4-94. Refer to the previous problem.
Show that the tension in each of the two ropes is $mg/(2\cos\theta) = 165$ N.

4-95. You exert a 91-N pull on a 25-kg wagon at an angle of 21°. The wagon moves at constant speed.
Show that the friction force is 85 N.

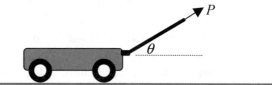

4-96. A 23-kg box slides down a frictionless inclined plane angled at 16°.
Show that the acceleration of the box is 2.7 m/s².

4-97. A 2.4-kg box is pushed up a 27°-inclined plane at constant speed, pushed by a force directed parallel to the slope. The coefficient of kinetic friction between the box and the slope is 0.33.
(a) Show that the normal force is 21 N.
(b) Show that the kinetic friction force is 6.9 N.
(c) Show that the required pushing force is a bit less than 18 N.

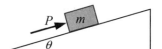

4-98. • As part of a lab experiment, a block is placed on a board on a lab table, and one end of the board is slowly lifted until the block just starts to slide.
Show that the block just starts to slide when $\theta = \tan^{-1}(\mu_s)$.

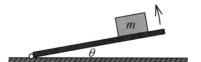

4-99. • Consider a force F directed at an angle θ below the horizontal on a box of mass m that sits on a level surface. Coefficient of static friction is μ_s.
(a) Show that the force required to get the box to move must be at least $\dfrac{\mu_s mg}{\cos\theta - \mu_s \sin\theta}$. (Hint: Note that the normal force is greater than mg.)
(b) Show that if $\theta \geq \tan^{-1}\left(\dfrac{1}{\mu_s}\right)$ then the box won't move no matter how hard you push.

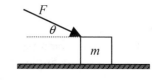

5 Newton's Third Law

This chapter progresses from Newton's second law to the third law, emphasizing the interactive nature of forces. Newton's third law tells us that when Body A exerts a force on Body B, Body B simultaneously exerts an equal and oppositely directed force on Body A. As a matter of fact, a body cannot exert a force on another body without this interaction!

Sample Problem 1

Esmeralda, an elephant of mass *M*, has a trunk full of water and stands on thick, frictionless ice. She extends her trunk horizontally and gives the water, mass *m*, an eastward acceleration a_{water}.

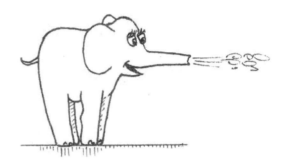

(a) What happens to Esmeralda?

> *Answer:* According to Newton's third law, when Esmeralda exerts a force on the water to accelerate it eastward, she experiences an equal and opposite westward force. Since the ice is frictionless, **Esmeralda will accelerate toward the west**.

(b) Derive an equation for Esmeralda's acceleration.

> *Focus:* $a_{Esmeralda} = ?$
>
> Esmeralda exerts a horizontal force on the water (magnitude ma_{water}). Newton's third law reminds us that the ejected water exerts an equal-magnitude force on Esmeralda in the opposite direction. That is,
>
> From $F_{\text{water on Esmeralda}} = F_{\text{Esmeralda on water}} \Rightarrow M_{Esmeralda} a_{Esmeralda} = m_{water} a_{water}$
>
> $$\Rightarrow a_{Esmeralda} = \frac{m_{water}}{M_{Esmeralda}} a_{water} = \frac{m}{M} a_{water}.$$
>
> Notice that the second law applies here. For the same magnitude of force, the larger mass undergoes a proportionately smaller acceleration.

(c) Calculate Esmeralda's acceleration if her mass is 3000 kg and she shoots 6 kg of water with an acceleration of 3 m/s².

> *Solution:* $a_{Esmeralda} = \frac{m}{M} a_{water} = \left(\frac{6 \text{ kg}}{3000 \text{ kg}}\right)(3 \text{ m/s}^2) = \mathbf{0.006 \text{ m/s}^2}.$
>
> This makes good sense—Esmeralda's mass is 500 times the water's mass, so the same force should cause 1/500th the acceleration.

Sample Problem 2

Rockets operate by burning fuel and accelerating the resulting exhaust gases. A rocket ejects m kg of fuel in a time t with a velocity change Δv.

(a) Write an equation for the force that the rocket exerts on the exhaust gases.

Focus: $F_{\text{rocket on exhaust}} = ?$

The exhaust speeds up by an amount Δv in time t, so its acceleration is $\Delta v/t$, and the force exerted on the exhaust is $F = ma_{\text{exhaust}} = m\dfrac{\Delta v}{t}$.

(b) How large is the resulting thrust force on the rocket?

Focus: Thrust is the force that the exhaust gases exert on the rocket. From the third law, in terms of magnitudes,

$\text{Thrust} = F_{\text{exhaust on rocket}} = F_{\text{rocket on exhaust}} = m\dfrac{\Delta v}{t}$. (Same amount of force in the opposite direction!)

(c) Calculate the thrust on a 280-gram model rocket that ejects 25 grams of exhaust material at a speed of 770 m/s over a period of 1.2 seconds.

Solution: Since we want force in newtons (1 N = 1 kg·m/s^2), the mass must be expressed in kilograms.

$$F = m\dfrac{\Delta v}{t} = \left(25 \text{ g} \times \dfrac{1 \text{ kg}}{1000 \text{ g}}\right)\dfrac{\left(770 \tfrac{\text{m}}{\text{s}}\right)}{1.6 \text{ s}} = 12 \text{ N}.$$

Sample Problem 3

A book of mass m sits on a table. You push with force P straight down on the book.

(a) Write an equation for the upward force that the table exerts on the book.

Focus: $F_{\text{table on book}}$ (we call this the *normal* force N) = ?

Let's first draw a force diagram for the book.
Since the book is not accelerating, $\Sigma F_y = 0$ so
$N - P - mg = 0 \Rightarrow F_{\text{table on book}} = N = mg + P$.

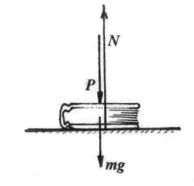

(b) Write an equation for the downward force that the book exerts on the table.

Focus: $F_{\text{book on table}} = ?$

From Newton's third law, the magnitude of $F_{\text{book on table}} = F_{\text{table on book}} = N = mg + P$.

(c) Calculate the upward force that the table exerts on a 1.2-kg book when you exert a 15-N downward push upon it.

Solution: $F_{\text{table on book}} = mg + P = (1.2 \text{ kg})\left(9.8\dfrac{\text{m}}{\text{s}^2}\right) + 15 \text{ N} = \mathbf{27 \text{ N}}$.

Sample Problem 4

Three blocks—A, B, and C—connected by strings of negligible mass, are pulled along a horizontal friction-free surface by a horizontal force F as shown.
Block A has a mass m; Block B has a mass $2m$; and Block C has a mass $3m$.

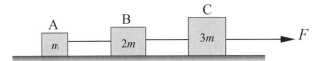

(a) Draw a force diagram showing the horizontal force acting on the system as a whole. Find the acceleration of the system.

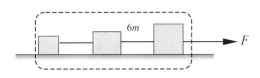

Solution: $a_{system} = \dfrac{\Sigma F}{\Sigma m} = \dfrac{F}{3m+2m+m} = \dfrac{F}{6m}$.

(b) Draw a separate force diagram for Block A. How much tension is in the string that pulls Block A?

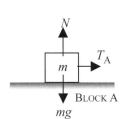

From $a_A = \dfrac{F_{net\,on\,A}}{m_A}$ with $T_A = F_{net\,on\,Block\,A} \Rightarrow T_A = m a_A$.

Because the blocks are connected and slide together, each has the same acceleration. So, $T_A = ma = m\left(\dfrac{F}{6m}\right) = \dfrac{F}{6}$.

(c) Draw a separate force diagram for Block B. How much tension is in the string that pulls backward on Block B, to the left?

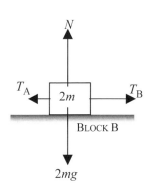

The tension in the string between Blocks A and B pulls equally hard on both blocks, but in opposite directions. So Block B is pulled with the same amount of tension, $\dfrac{F}{6}$, toward the left.

(d) How much tension is in the string that pulls Block B to the right?

$T_B = ?$ We know that $F_{net\,on\,B} = m_B a$. From the force diagram

$F_{net\,on\,B} = T_B - T_A = (2m)a \Rightarrow T_B = T_A + (2m)a = \dfrac{F}{6} + (2m)\left(\dfrac{F}{6m}\right) = \dfrac{F}{2}$.

(e) How much tension is in the string that pulls backward on Block C, to the left?

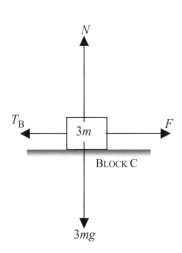

The tension in the string between Blocks B and C pulls equally hard on both blocks, but in opposite directions. So the force on Block C from the string is **0.5 F**, pulling to the left.

(f) Calculate the acceleration of the system if m is 2 kg and the pulling force is 24 N.

Answer: $a = \dfrac{F}{6m} = \dfrac{24\,N}{6(2\,kg)} = 2\,\dfrac{m}{s^2}$.

© Paul G. Hewitt and Phillip R. Wolf

Newton's Third Law Problems

5-1. Seth, mass m, and his sister Sarah, mass M, are wearing ice skates and are standing facing one another on a smooth, frozen lake. Seth pushes on Sarah with a force P.
  (a) What is Seth's acceleration?
  (b) Calculate Seth's acceleration for a 64-N push if Sarah's mass is 72 kg and Seth's mass is 80 kg.

5-2. A blob of clay of mass m, moving with speed v, impacts a wall and comes to rest in time t.
  (a) What is the average force that the clay exerts on the wall?
  (b) Calculate the force on a wall when a 1.9-kg chunk of clay slows from 7.0 m/s to rest in a time of 35 milliseconds. (1 millisecond = $\frac{1}{1000}$ s)

5-3. A car of mass m cruises along a level highway at a constant velocity v. The tires push backward on the road with a force f. The reaction to this force provides the forward force on the car. Wind resistance against the car is R.
  (a) Draw a force diagram showing the horizontal forces acting on the car.
  (b) What is the net force on the car?
  (c) What is the acceleration of the car?
  (d) A friend says that since the car is moving forward, there must be a net forward force, which means f must be greater than R, even at constant velocity. What can you say to enlighten your friend?

5-4. When Maria swims to the end of the pool, she wishes to increase her speed after turning around by pushing against the pool wall with force F. Her mass is m, and during her push the average water resistance is R.
  (a) On the diagram, show the direction of the force Maria exerts on the wall. (Does it act toward the left or toward the right?)
  (b) In what direction is the force that the wall exerts on Maria?
  (c) On a force diagram, show the forces acting on Maria during her push on the wall. What average acceleration is produced?
  (d) Is it possible for Maria to push against the wall *without* the wall simultaneously pushing back on her? Explain.

5-5. Nellie Newton pauses momentarily in her climb up a vertical rope attached to the ceiling. Consider the mass of the rope to be negligible, while Nellie's mass is m.
  (a) How much force does Nellie exert on the rope?
  (b) With how much force does the rope pull down on the ceiling?
  (c) With how much force does the ceiling pull up on the rope?
  (d) What is the net force on the rope? On Nellie?
  (e) A friend says the upward and downward forces exerted on the rope make an action-reaction pair. Do you agree or disagree? Defend your answer.

5-6. • Gymnast Gracie of mass *m* climbs a vertical rope attached to the ceiling. The mass of the rope is negligible.
   (a) Write an equation for the rope tension if she hangs motionless.
   (b) Write an equation for the rope tension if she climbs upward at a constant rate.
   (c) If she accelerates up the rope with acceleration *a*, what provides the upward force on Gracie?
   (d) Draw a force diagram for Gracie as she accelerates up the rope. Then write an equation for the tension in the rope.
   (e) Write an equation for the rope tension if she accelerates *down* the rope with the same magnitude of acceleration.

5-7. Block B of mass *m* rests on top of Block A of mass *M*, which rests on a horizontal table.
   (a) Make a force diagram for each block.
   (b) What is the net force on each block?
   (c) What is the force (magnitude and direction) exerted by Block B on Block A?
   (d) What is the force (magnitude and direction) exerted *on* Block B *by* Block A?
   (e) What is the force (magnitude and direction) exerted by the table on Block A?
   (f) Calculate answers to (c) through (e), for a mass of 5.0 kg for Block B and 7.0 kg for Block A.

5-8. • A third block, Block C of mass *m*/2, is placed on top of Block B of the previous problem.
   (a) Make a vector diagram showing forces on the blocks.
   (b) What is the net force on each block?
   (c) What is the force (magnitude and direction) exerted on Block B by Block C?
   (d) What is the force (magnitude and direction) exerted by Block B on Block C?
   (e) What is the force (magnitude and direction) exerted by Block B on Block A?
   (f) What is the force (magnitude and direction) exerted by the table on Block A?
   (g) Calculate answers to (c) through (f) for a mass of 2.5 kg for Block C, 5.0 kg for Block B, and 7.0 kg for Block A.

5-9. Pole-vaulter Phil of mass *m* falls from rest from a height *h* onto a rubber mat. The mat brings him to a halt in time *t*.
   (a) Make a diagram of the forces that act on Phil just as he hits the mat.
   (b) Derive an equation for the average force exerted on Phil by the mat as he is brought to a halt.
   (c) What is the average force on the mat as Phil is brought to a halt?
   (d) Calculate the average force on the mat if Phil has a mass of 50.0 kg, falls from a height of 4.9 m, and is brought to a halt in 0.30 s.

5-10. David's model rocket of mass *m* sits at rest on the launch pad. It then burns 0.050 kg of fuel in time *t*, ejecting it as a gas with velocity *v* relative to the rocket.
   (a) Derive an equation for the thrust on the rocket.
   (b) Calculate the thrust on the rocket if its mass is 1.8 kg, the time during which the fuel burns is 4.0 s, and the exhaust speed of the fuel is 1200 m/s relative to the rocket.
   (c) Why is David disappointed?

5-11. An astronaut of mass m, who is floating a distance d away from her spaceship, fires her jet pack so that she can move toward the ship. The jet pack provides a brief thrust F toward the ship for a quick time t_{thrust}.
   (a) Derive an equation for the astronaut's change in speed relative to the ship.
   (b) Derive an equation for the time it takes her to drift to her ship.
   (c) How would the drift time be affected if her jet pack provided twice the thrust in the same time t_{thrust}?

5-12. When two identical air pucks with repelling magnets are held together on an air table and released, they move in opposite directions at the same speed v. A third identical-mass nonmagnetic puck is secured to the top of one of the pucks (effectively doubling its mass) and the procedure is repeated.
   (a) How does the speed of the double-mass puck compare to the speed of the single puck?
   (b) Calculate the speed of the double-mass puck if the single puck moves away at 4 m/s.

5-13. Two astronauts, initially at rest in space, push against each other. The first has a mass m_1, and after the push, it moves backward at speed v_1 while the second one moves at speed v_2.
   (a) Derive an equation for the mass of the second astronaut.
   (b) Write an equation for the force they exert on each other when pushing for a time t.

5-14. • Mady with mass m and Sam with mass M stand x meters apart on frictionless ice. Sam pulls on a rope that connects him to Mady, giving her an acceleration a toward him.
   (a) Find Sam's acceleration.
   (b) Calculate Sam's acceleration given that his mass is 66 kg, Mady's mass is 55 kg, and her acceleration toward him is 0.90 m/s².
   (c) Calculate where Mady and Sam will meet if they are initially 11 m apart.

5-15. In a tug-of-war on a quite slippery floor, Jose of mass M pulls on the rope and imparts an acceleration a to Juanita of mass m.
   (a) Write an expression for the tension in the rope.
   (b) Calculate the tension if Jose's mass is 80 kg, Juanita's mass is 60 kg, and she accelerates toward Jose at 0.10 m/s².

5-16. A magnet of mass M is held firmly on frictionless ice while the magnetic force on a paper clip of mass m some distance away accelerates the paper clip with initial acceleration a.
   (a) If instead we were to hold onto the paper clip and release the magnet, what would be the initial acceleration of the magnet?
   (b) Calculate the initial acceleration of a 0.060-kg magnet if the 0.00040-kg paper clip was originally accelerated at 0.15 m/s².

5-17. A car of mass m and a truck of mass $2m$, both traveling at the same speed v, have a head-on collision. During the collision the truck undergoes an average deceleration a.
   (a) What deceleration does the car undergo?
   (b) Calculate the car's deceleration during the collision if its mass is 970 kg and the truck's average deceleration was –36 m/s².
   (c) Calculate the average force on the car during collision.
   (d) If you simulate this collision with smaller carts in the lab, each with a force sensor at its "front bumper," how will the force readings for both compare during the collision?

5-18. • One newer method of spaceship propulsion is called an *ion drive*. Charged atoms (ions) of some inert gas (usually krypton or xenon) are propelled out the back of the ship, driving the ship forward. Suppose that a mass m of gas is accelerated each second by an amount Δv.
   (a) Write an equation for the force exerted on the ions to accelerate them.
   (b) How much force is being exerted on the ship?
   (c) If the ship started out with a mass m_{fuel} of fuel, for how long could it continuously fire its engines?
   (d) If we assumed that the ship has a constant average total mass M, approximately how fast would the ship be moving after all of its fuel had been consumed?
   (e) Calculate the force exerted if 1.6×10^{-6} kg of fuel are accelerated by 30,000 m/s every second.
   (f) If the ship started out at rest in deep space with 40 kg of fuel and an average total mass of 600 kg, calculate the speed the ship would have when it runs out of fuel.

5-19. The tires on a four-wheel-drive car of mass m have a coefficient of static friction μ_s between them and the road.
   (a) Describe in words the particular force that is pushing the car forward.
   (b) What is the maximum forward acceleration for the car?
   (c) Calculate this maximum acceleration for a coefficient of static friction of 0.70 between the tires and the road surface.

Tires push backward on the road surface

5-20. A giant frog of mass m is placed on a skateboard of mass M at rest. When the frog jumps in a forward direction from the skateboard, the skateboard rolls freely in the opposite direction at speed V.
   (a) Find the frog's horizontal jumping speed.
   (b) Calculate the frog's horizontal jumping speed if its mass is 2.0 kg, the skateboard's mass is 3.0 kg, and the skateboard recoils at 1.4 m/s immediately after the jump.
   (c) A force of friction is needed for the frog to propel horizontally from the skateboard. (A slippery surface won't do.) Can the skateboard surface provide a friction force on the frog without an equal and opposite force of friction on the skateboard? Defend your answer.

5-21. Three blocks of equal mass m, each connected by a string of negligible mass, are pulled along a frictionless surface by a horizontal force F as shown in the sketch.

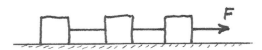

   (a) Write an equation for the acceleration of the system.
   (b) What is the tension in each string? (*Hint*: Draw a force diagram for the system, and for each block separately.)
   (c) Calculate the acceleration and the string tensions if the mass of each block is 10 kg and the force F is 60 N.

5-22. Dean exerts a force F on two crates, one in front of the other, as shown. Crate A has a mass m, while Crate B has a smaller mass, $m/2$. The crates are mounted on tiny rollers and move with negligible friction.

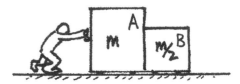

(a) Make a force diagram showing the horizontal force(s) on the system consisting of Crate A + Crate B.
(b) What is the acceleration of the two-crate system?
(c) Make a force diagram for Crate B. How great a force acts on Crate B while Dean continues pushing?
(d) Make a force diagram for Crate A. How many horizontal forces act on it, and what net force acts on it while Dean continues pushing?
(e) What would be different in this problem if Crates A and B were interchanged?
(f) A friend says that Crate A accelerates because Dean's hands push it, but Crate B has no force on it so just rides along with Crate A. What physics is your friend missing?

5-23. An athlete of mass M stands on a large scale. He lifts a barbell of mass m, giving it an upward acceleration a.
(a) Draw a force diagram for the barbell and write an expression for the upward force the athlete exerts on the barbell.
(b) Draw a force diagram for the athlete during the lift and write an expression for the upward force that the scale exerts on the athlete.
(c) Calculate what the scale reads when an 87-kg athlete lifts a 32-kg barbell with an acceleration of 1.0 m/s^2.

Show-That Problems for Newton's Third Law

5-24. Amanda looks at a 1.00-kg bag of jellybeans at rest on a table.
Show that the amount of force that the table exerts on the bag of jellybeans is 9.8 N. How does this compare with the force the bag of jellybeans exerts on the table?

5-25. Suppose the table in the previous problem was somehow accelerated upward at one-half g. Show that the force of the table on the bag of jellybeans would be 14.7 N.

5-26. Suppose the table in the previous problem was accelerated downward at one-half g.
Show that the force of the table on the bag of jellybeans would be 4.9 N.

5-27. When 60-kg Victor on rollerblades pushes against a wall with a force of 30 N, he recoils.
Show that the recoil acceleration is 0.5 m/s^2.

5-28. A 7.00-kg bowling ball moving at 8.00 m/s strikes a 1.00-kg bowling pin and slows to 7.00 m/s in 0.035 s.
Show that the force of impact is 200 N.

5-29. Two people, one with three times the mass of the other, attempt a tug-of-war on frictionless ice.
Show that the heavier person will gain a speed one-third that of the lighter person.

5-30. Carts A and B are connected by a compressed spring on an air table. Cart A has a mass of 0.25 kg and Cart B has a mass of 0.75 kg. The spring is then released.
Show that Cart A moves at three times the speed of Cart B.

6 Momentum

Momentum, inertia in motion, is a vector quantity equal to mass × velocity. The direction of the momentum vector is always the same as the direction of the velocity vector. In the absence of a net external force, the momentum of a body (or system of bodies) is unchanged—that is, the total momentum is *conserved*.

Force acting for an interval of time is called an *impulse*. Just as force changes the velocity of a body, an impulse—force × time—changes the momentum of a body. A net impulse acting on a body (or system of bodies) causes an equivalent change in the momentum of that body (or system). That is:

$$F_{net} \times \Delta t = \Delta(mv)$$

Usually a change in momentum involves an object of fixed mass changing velocity, in which case $\Delta(mv) = m\Delta v = m(v_f - v_0)$. The v's in this equation stand for velocities—the direction matters as well as the speed.

Sample Problem 1

Big Joe applies a forward force F for a time t on a frictionless cart of mass m, initially moving at speed v_0.

(a) Derive an equation for the final velocity of the cart.

Focus: $v_f = ?$

Big Joe applies an impulse $F \times t$, which changes the momentum of the cart by an amount $m\Delta v = m(v_f - v_0)$.

From $Ft = m\Delta v = m(v_f - v_0) = mv_f - mv_0 \Rightarrow mv_f = mv_0 + Ft$

$\Rightarrow v_f = v_0 + \dfrac{Ft}{m}$.

↑ Initial velocity   ↖ Change in velocity due to the applied impulse

(b) Calculate the final velocity if Big Joe applies a 90-N forward force for 4.0 seconds to a 120-kg cart that was initially moving at 3.5 m/s.

Solution: Call the initial direction of the cart the positive direction. So $v_0 = 3.5$ m/s, $F = 90$ N,

and $v_f = v_0 + \dfrac{Ft}{m} = 3.5\dfrac{m}{s} + \dfrac{(90\text{ N})(4.0\text{ s})}{120\text{ kg}} = 6.5\dfrac{m}{s}$.

Had the force been applied in a direction opposite the motion of the cart ($F = -90$ N), the applied impulse would have *reduced* the velocity of the cart.

(c) Little John can only supply a force of $\frac{1}{3}F$ on the cart. For how long a time will he have to push to give the cart the same final speed as Big Joe gave it?

Focus: $t_{new} = ?$ For the cart to have the same final speed, it will have to experience the same change in momentum due to the same impulse acting on the cart. Since the force is one-third as much, the time will have to be three times as much, **3t**. That is, from

$m\Delta v = Ft = \left(\dfrac{1}{3}F\right)t_{new} \Rightarrow t_{new} = 3t = 3 \times 4.0\text{ s} = \mathbf{12\text{ s}}$.

© Paul G. Hewitt and Phillip R. Wolf

Sample Problem 2

A horizontal stream of water is directed against the locked blade of a Pelton wheel, as shown. Suppose that the water strikes the blade at velocity v and bounces at $-v$. In time t a mass m of water rebounds.

(a) Derive an equation for the force on the blade (and on the water).

Focus: $F = ?$ Consider the force on the water. From

$$Ft = \Delta(mv) \Rightarrow F_{\text{on water}} = \frac{\Delta(mv)_{\text{water}}}{t} = \frac{m_{\text{water}}\Delta v_{\text{water}}}{t}.$$

Since the water strikes the turbine blade with initial velocity v and rebounds with velocity $-v$, its velocity changes by an amount $2v$.* So the magnitude of the force on the water is

$$F_{\text{on water}} = \frac{m_{\text{water}}(2v_{\text{water}})}{t} = \frac{2m_{\text{water}}v_{\text{water}}}{t}.$$ From Newton's third law, **the force on the blade has the same magnitude (but is opposite in direction).**

(b) Calculate the force on the blade if 22 kg of water impinges on the blade each second at a speed of 25 m/s.

Solution: $F = \dfrac{2m_{\text{water}}v_{\text{water}}}{t} = \dfrac{2(22\,\text{kg})\left(25\,\dfrac{\text{m}}{\text{s}}\right)}{1\,\text{s}} = 1100\,\text{N}.$

(c) Will this force be more, less, or the same, on a *moving* blade?

Answer: The relative speed of impact of the water on the blade will be less when the blade moves in the same direction as the water, so the force of impact will be less. To exaggerate: If the blade moved as fast as the water, there would be no force of impact.

Sample Problem 3

A giant frog of mass m is placed on a skateboard of mass M, initially at rest. When the frog jumps horizontally in a forward direction from the skateboard, the skateboard rolls backward at speed V.

(a) From momentum conservation, derive an equation for the frog's forward speed.

Focus: $v = ?$ No horizontal *external* forces act on the (frog + skateboard) system, so the momentum of the system is conserved.* Everything was initially at rest, so the initial momentum of the system was zero. From conservation of momentum, the final momentum of the frog + skateboard system must also be zero. From

Initial momentum of the system = Final momentum of the system

$$\Rightarrow 0 = m_{\text{frog}}\vec{v}_{\text{f, frog}} + M_{\text{board}}\vec{v}_{\text{f, board}} \quad (\vec{v}\text{'s are velocities: one is +, the other is }-)$$

Using speeds (that is, the magnitude of the velocities, so both v's are positive):

$$\Rightarrow m_{\text{frog}}v_{\text{frog}} = M_{\text{board}}V_{\text{board}} \Rightarrow v_{\text{frog}} = \frac{M_{\text{board}}V_{\text{board}}}{m_{\text{frog}}} = \frac{MV}{m}.$$

---

* The initial velocity is in the positive direction, so $v_0 = +v$. The final velocity is in the negative direction, so $v_f = -v$. $\Delta v = v_f - v_0 = (-v) - (+v) = -2v$. The magnitude of $\Delta v = 2v$.

* If we consider only the skateboard (that is, if the system of interest is only the skateboard), we will say that the force exerted by the frog on the skateboard is an external force and thus the momentum of the skateboard changes. If we consider just the frog, we will say that the skateboard exerts an external force on the frog, changing the frog's momentum. But since we are considering the frog + skateboard system, any interaction between the frog and skateboard is an *internal* force, so it doesn't change the overall momentum of the system.

(b) Calculate the frog's horizontal jumping speed if its mass is 2.0 kg, and the 3.0-kg skateboard recoils at 1.4 m/s immediately after the jump.

$$v_{frog} = \frac{MV}{m} = \frac{(3.0 \text{ kg})(1.4 \text{ m/s})}{2.0 \text{ kg}} = 2.1 \text{ m/s}.$$

Note that this is a problem from Chapter 4! Conservation of momentum provides a quicker route to the solution.

(c) Why don't we have to worry about gravity in this problem?

Since we are only looking at the horizontal motion of the skateboard + frog, we only have to consider the effect of forces and impulses that act in the horizontal direction. If we were interested in the vertical motion of the system, then we would look at forces and impulses acting in the vertical direction. This is another reminder that momentum is a vector quantity.

Sample Problem 4

A bullet of mass *m* is fired into a block of wood of mass *M*, conveniently located on a frozen, frictionless lake. The block + bullet system moves together at speed *V* after the collision.

(a) Derive an equation for the initial speed of the bullet.

SYSTEM BEFORE THE COLLISION	SYSTEM AFTER THE COLLISION
Mass *m*, velocity *v* Mass *M*, at rest	Mass (*m* + *M*), velocity *V*

Focus: $v = ?$ Our system is the block + bullet. No net external forces act on the system, so its momentum is conserved, or

$$\begin{pmatrix}\text{Initial momentum of the moving} \\ \text{bullet and stationary block}\end{pmatrix} = \begin{pmatrix}\text{Final momentum of bullet and block} \\ \text{moving together}\end{pmatrix}$$

From $mv + M(0) = (m+M)V \Rightarrow v = \left(\frac{m+M}{m}\right)V = \left(1+\frac{M}{m}\right)V.$

(b) By how much does the momentum of the bullet change? By how much does the momentum of the block change?

$$\Delta(mv)_{bullet} = m(v_f - v_0) = m(V - v) = m\left(V - \left(1+\frac{M}{m}\right)V\right) = m\left(V - V - \frac{M}{m}V\right) = -MV.$$

$$\Delta(MV)_{block} = M(V_f - V_0) = M(V - 0) = +MV.$$

Whatever momentum was lost by the bullet was gained by the block. Again, the total momentum of the system was unchanged.

(c) Calculate the initial speed of the 2.6-gram bullet if the 2.80-kg block ends up moving at 0.40 m/s.

Answer: $v = \left(1+\frac{M}{m}\right)V = \left(1+\frac{2.80 \text{ kg}}{0.0026 \text{ kg}}\right)\left(0.40\frac{\text{m}}{\text{s}}\right) = 430\frac{\text{m}}{\text{s}}.$

(d) How does the impulse acting on the block compare to the impulse on the bullet?

Answer: They are equal and opposite. There are two ways to see this. The first is to recognize that the third law applies here (as always!): $F_{\text{on bullet}} = -F_{\text{on block}}$. These forces act for exactly the same time, so the impulses $F_{\text{on bullet}}t$ and $-F_{\text{on block}}t$ are equal and opposite.

Another way to see this is to notice that the change in momentum for the block and the bullet [found in part (b)] are equal and opposite, so the impulses acting on each are also equal and opposite. Conservation of momentum is intimately tied to Newton's third law.

•Sample Problem 5

A white hockey puck of mass m moves on frictionless ice with speed v_0 at an angle θ with the $+x$ axis as shown. The puck collides with a black puck of mass M, initially at rest. After the collision, the original puck moves along the $+x$ direction while the second puck moves along the $+y$ direction.

(a) Find the final velocities of each puck.

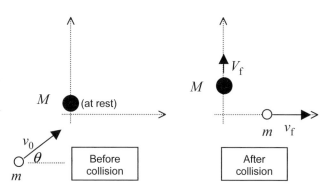

Focus: $v_f = ?$ $V_f = ?$

We identify our system as being the two hockey pucks and apply the law of conservation of momentum separately to the x and y directions since no net external forces act on the system in either direction. Our process is to

1. Write an expression for x- and y-momentum of each puck before and after the collision.

2. Write an expression for the total momentum of the system in the x-direction before and after the collision and in the y-direction before and after the collision.

3. Write and solve the conservation of momentum equations separately for the x- and y-directions.

	MOMENTUM BEFORE COLLISION		MOMENTUM AFTER COLLISION	
	x-direction	y-direction	x-direction	y-direction
m	$mv_0 \cos\theta$	$mv_0 \sin\theta$	mv_f	0
M	0	0	0	MV_f
TOTAL MOMENTUM	$mv_0 \cos\theta$	$mv_0 \sin\theta$	mv_f	MV_f

Solving:

For the x-direction:

From *total momentum before collision = total momentum after collision*
$\Rightarrow mv_0 \cos\theta_f = mv \Rightarrow v_f = v_0 \cos\theta$.

For the y-direction:

$\Rightarrow mv_0 \sin\theta = MV_f \Rightarrow V_f = \dfrac{m}{M} v_0 \sin\theta$.

(b) Calculate the final puck velocities if the initial velocity of the 0.80-kg white puck is 5.0 m/s at an angle of 37° above the x-axis, and the mass of the initially stationary black puck is 1.2 kg.

$v_f = v_0 \cos\theta = \left(5.0\,\dfrac{\text{m}}{\text{s}}\right) \cos 37° = 4.0\,\dfrac{\text{m}}{\text{s}}$ in the $+x$ direction, and

$V_f = \dfrac{m}{M} v_0 \sin\theta = \dfrac{0.80\text{ kg}}{1.20\text{ kg}} \left(5.0\,\dfrac{\text{m}}{\text{s}}\right) \sin 37° = 2.0\,\dfrac{\text{m}}{\text{s}}$ in the $+y$ direction.

Momentum Problems

6-1. A dog of mass m runs at a speed v.
   (a) What is the momentum of the dog?
   (b) Calculate the momentum of a 12-kg dog moving at 4.0 m/s.

6-2. A cat of mass m initially strolls to the right at a speed v_0 and accelerates to a speed v_f.
   (a) By how much does the momentum of the cat change?
   (b) Calculate the change in momentum for a 2.8-kg cat that initially moves to the right at 1.1 m/s and accelerates to 4.2 m/s.

6-3. A ball of mass m initially moves in the $+x$-direction at a speed v_0 but ends up moving in the $-x$-direction at a speed v_f.
   (a) By how much does the momentum of the ball change?
   (b) Calculate the change in momentum for a 0.50-kg ball that initially moves to the right at 1.0 m/s and ends up moving to the left at 2.6 m/s.

6-4. A superball of mass m approaches a wall at a horizontal speed v and bounces back at the same speed.
   (a) Find the magnitude of the change in momentum of the superball.
   (b) What is the magnitude of the impulse that acts on the ball?
   (c) Calculate the magnitude of the change in momentum for 50-gram superball that approaches and bounces from a wall at an initial speed of 7.0 m/s.

6-5. An average net force F applied for time t is required to accelerate an elevator of mass m from rest up to a steady speed v.
   (a) Write an expression for the impulse applied.
   (b) Write an expression for the momentum of the elevator when it reaches a steady speed.

6-6. Lonnie's truck of mass m moves at speed v into a headwind of speed V. The temperature of the road surface is T.
   (a) Write an expression for the momentum of the truck.
   (b) Calculate the momentum of Lonnie's 3500-kg truck moving at 18 m/s when it drives into a 5 m/s headwind, and when the temperature of the road is 24°C.
   (c) How does this problem show the importance of letting equations guide your thinking?

6-7. A block of ice of mass m slides from rest down an inclined plane. At the bottom of the incline, it slides onto the floor at speed v.
   (a) Write an expression for the block's gain in momentum in sliding down the incline.
   (b) Find the momentum gained by a 25-kg block of ice that starts from rest and slides off the end of an inclined plane at a speed of 6.0 m/s.

6-8. A bullet of mass m, initially at rest, exits a gun barrel at a speed v.
   (a) Write an expression for the impulse that acts on the bullet.
   (b) Calculate the impulse that acts on a 9.6-g bullet that leaves the gun barrel at a speed of 280 m/s.
   (c) Impulse = force × time, neither of which is given information in this problem. Why does this absence of data not affect your calculation?

6-9. A soccer player kicks a ball of mass m, initially at rest, applying an average force F. The player's foot remains in contact with the ball for a brief time t.
  (a) Write an expression for the impulse that acts on the ball.
  (b) Write an expression for the change in momentum of the ball.
  (c) Derive an equation for the final speed of the ball.
  (d) The ball has a mass of 0.42 kg, the average kicking force is 1350 N, and the player's foot contacts the ball for 0.0080 s. Calculate the speed of the ball after the kick.

6-10. In the hammer throw contest at a track and field event, Dean picks up a hammer of mass m and then whirls the hammer in a circle, releasing it at speed v.
  (a) Write an expression for the impulse that acts on the hammer.
  (b) Calculate the impulse if the mass of the hammer is 7.3 kg and it is released at a speed of 28 m/s.

6-11. Katie tosses a coconut of mass m straight up into the air by exerting a force F upward on the coconut for a time t.
  (a) What forces act on the coconut while being tossed? Draw a force diagram.
  (b) What is the net impulse on the coconut during the toss?
  (c) Derive an equation for the coconut's speed when it leaves Katie's hands.
  (d) Calculate the tossing speed of a 2.2-kg coconut if Katie exerts a 32-N force for 0.25 s.

6-12. A diver of mass m atop a high cliff dives straight down into the water below. Just before striking the water, her speed is v_1. After a brief time t in the water, her speed is reduced to v_2.
  (a) Derive an equation for the average net force exerted on the diver as she slows down.
  (b) Calculate the net force that slows the 48.0-kg diver when her speed just before hitting the water is 9.9 m/s and reduces to 0.5 m/s after 0.60 s.

6-13. When an abrupt average force F is exerted on a system, the system's momentum increases by mv.
  (a) Derive an equation for the time during which the force acts.
  (b) If a different force acts on the system for twice as much time and produces the same change in momentum, how will the "new" average force compare with the original force?

6-14. A baseball is tossed straight upward with an initial momentum mv.
  (a) If air drag is negligible, what will be the ball's momentum when caught at the same elevation at which it was thrown?
  (b) If air drag is not negligible, how does the magnitude of the baseball's momentum just before it is caught compare to the magnitude of its initial momentum?

6-15. An unfortunate bird with momentum mv flies into a glass window, which brings the bird to an abrupt stop.
  (a) What is the magnitude of the impulse that stopped the bird?
  (b) The bird recovers. If it had instead bounced from the window, would its recovery be more likely or less likely? Defend your answer.

6-16. Diane spikes a volleyball of mass m so that its incoming speed v is changed to an outgoing speed V in the opposite direction.
  (a) Write an expression for the impulse that acts on the volleyball.
  (b) Calculate the impulse when the mass of the volleyball is 0.25 kg, its incoming speed is 4 m/s, and its outgoing speed in the opposite direction is 20 m/s.

6-17. A golf ball of mass *m*, initially at rest, is given a speed *v* when the time of contact between the club and ball is *t*.
   (a) Write an equation for the average force on the ball.
   (b) Calculate the average force for the 0.045-kg golf ball that gains a speed of 24.0 m/s when the time of contact between club and ball is 2.0 milliseconds. (1 ms = $\frac{1}{1000}$ s)

6-18. Pedro throws a baseball of mass *m* at a wall at a speed *V*. It rebounds from the wall with a speed *v*.
   (a) If the contact time of the ball with the wall is *t*, what is the magnitude and direction of the average force that the ball exerts on the wall?
   (b) Calculate the average force on the wall from a ball of mass 0.15 kg, thrown at a speed of 35 m/s, that leaves the wall at 23 m/s after a 0.0030-s impact.
   (c) How does the magnitude of the force that the ball exerts on the wall compare with the magnitude of the force that the wall exerts on the ball?

6-19. In time *t*, a mass *m* of exhaust gas is ejected at speed *v* from a model rocket of mass *M*, initially sitting on its launch pad.
   (a) What is the thrust on the rocket?
   (b) Calculate the thrust on the rocket if its mass is 110 grams and the rocket exhausts 8.3 grams of the fuel in 1.1 seconds at a speed of 570 m/s relative to the rocket.

6-20. When hit with a bat, a baseball of mass *m* changes velocity from v_1 toward the bat to v_2 away from the bat in time *t*.
   (a) How much impulse acts on the ball?
   (b) Calculate the impulse acting on a 0.15-kg baseball that changes speed from 45.0 m/s to 55.0 m/s in the opposite direction.
   (c) Calculate the average force on the baseball for a ball-bat contact time of 0.0020 s.
   (d) How does the average force on the ball compare with the average force on the bat?

6.21. A jet engine gets its thrust by taking in air, heating and compressing it, and then ejecting it at a high speed. A particular engine takes in a mass of air *m* per time *t* at speed *v*, and ejects it at speed 10*v*.
   (a) Find the thrust of the engine.
   (b) Calculate the thrust of the engine if it takes in 20 kg of air per second at 100 m/s and ejects it at 1000 m/s.

6-22. A chunk of ice of mass *m* breaks loose from a suspension bridge and falls for time *t*.
   (a) Neglecting air resistance, what will be its momentum when it hits the water below?
   (b) What would be its momentum falling from a higher bridge where the falling time is 2*t*?

6-23. Manuel drops a water balloon of mass *m* from the roof of a building of height *h*.
   (a) Derive an expression for its momentum when it hits the street below.
   (b) The water balloon has a mass of 1.5 kg and the building is 12 m high. Calculate the momentum of the balloon the instant before it hits the street.
   (c) Why can't the force of impact be found using only the information given above?

6-24. • A steel ball of mass m is dropped from a height h above the flat, horizontal surface of a massive block of iron. The ball rebounds to a height $0.80h$ above the iron.
  (a) How large is the impulse that acts on the ball during impact?
  (b) If the ball-iron contact time is t, find the average impact force on the ball.
  (c) Calculate the average force if the mass of the ball is 0.41 kg, the impact time is 0.0018 s, and the height h is 1.0 m.

6-25. A lump of clay of mass m_1 and velocity v_1 catches up with and bumps into a slower lump of clay of mass m_2 and velocity v_2 heading in the same direction. They share a common velocity after they stick together.
  (a) Derive an equation for this common velocity.
  (b) Calculate the final velocity of the stuck-together lumps when a lump of mass 2.2 kg moving at 3.2 m/s catches up with and sticks to a 2.7-kg lump moving at 1.2 m/s.

6-26. Lynda of mass m_1 is on roller skates and moves at speed v_1 when she crashes into and hugs Duncan of mass m_2, initially at rest and also on skates.
  (a) Derive an equation for their resulting velocity as they roll off into the sunset.
  (b) Calculate their sunset velocity if Duncan's mass is 64.0 kg, Lynda's mass is 45.0 kg, and she initially moves toward Duncan at speed 4.6 m/s.

6-27. A blob of putty of mass m has a head-on collision with another blob of putty of mass $2m$ moving toward it. After the collision, the combined putty blobs don't move.
  (a) What can you conclude about the relative speeds of the blobs before collision?
  (b) If the smaller blob was moving at a speed of 2 m/s before the collision, what was the speed of the other blob before the collision?

6-28. A dart of mass m moves horizontally at speed v and strikes and sticks to a wooden block of mass M, initially at rest on a horizontal friction-free surface.
  (a) Derive an equation for the resulting speed of the dart-block system.
  (b) If the dart's mass is 1.0 kg, its initial speed is 8.0 m/s, and the mass of the wooden block is 12 kg, calculate the "after-collision" speed for the dart-block system.

6-29. Duncan's car of mass m, moving at speed v on an icy road, collides with a stationary truck of mass M.
  (a) What is the speed of the interlocked vehicles as they slide along the icy road?
  (b) Calculate the speed of interlocked vehicles if the car's mass is 1500 kg, the truck's mass is 3200 kg, and the initial speed of the car is 22 m/s.

6-30. Two identical air pucks with repelling magnets are held together on a frictionless air table. When they are released, the pucks move away from each other at equal speeds. Now a third identical nonmagnetic puck is secured to the top of one of the pucks (resulting in twice the mass) and the procedure is repeated.
  (a) How will the speed of the double puck compare to the speed of the single puck?
  (b) Calculate the speed of the double-mass puck if the single puck moves away at 4 m/s.

6-31. When a bullet of mass m strikes and is embedded in a wooden block of mass M initially at rest, the bullet-block system moves at speed V in the direction of the moving bullet.
  (a) With this information, derive an equation for the initial speed of the bullet.
  (b) Calculate the bullet's initial speed if its mass is 0.050 kg, the mass of the block is 5.0 kg, and the post-impact speed of the block-bullet system is 4.0 m/s.

6-32. Suppose a car of mass M, moving at speed V, has a head-on collision with a smaller car of mass m moving at speed v toward it.
   (a) Find the speed and direction of the coupled cars immediately after collision.
   (b) Calculate the velocity for the coupled cars if the mass of the larger car is 1500 kg with an initial speed of 18 m/s, and the mass of the smaller car is 1100 kg with an initial speed of 29 m/s in the opposite direction.

6-33. Sumo wrestler Akebono of mass m collides head on with Konishiki of mass M. Konishiki moves with a speed of V toward Akebono and the collision brings their combined speed to zero.
   (a) Derive an equation for Akebono's speed before the collision.
   (b) For the speed to be zero after the wrestlers collide, how fast must the 227-kg Akebono move if 267-kg Konishiki moves at 2.0 m/s toward Akebono?

6-34. Two astronauts, initially at rest in space, push on each other. After the push, the first one of mass m_1 moves backward at a speed v_1 while the second one moves backward at speed v_2.
   (a) Derive an equation for the mass of the second astronaut.
   (b) If they push on each other for a total time t, what average force does each astronaut exert on the other?

6-35. Carmelita loosely holds and fires a rifle of mass M. The bullet has momentum mv as it exits the gun.
   (a) Derive an equation for the recoil velocity of the gun.
   (b) How would recoil velocity be affected if Carmelita holds the rifle tightly? Defend your answer.

6-36. A launcher of mass M, resting on a horizontal, frictionless surface, fires a projectile of mass m horizontally at speed v relative to the surface.
   (a) Derive an equation for the recoil speed of the launcher.
   (b) Calculate the recoil speed for a 2400-kg launcher that fires a 38-kg projectile horizontally at 520 m/s.

6-37. A quarterback of mass m is moving north at speed v when he is tackled by an opponent of mass M running south at speed V.
   (a) Find their combined speed when they tangle. (Ignore friction with the ground.)
   (b) A 89-kg quarterback moving north with a speed of 3.5 m/s is tackled by a 110-kg opponent running south at 5.0 m/s. Calculate their combined speed after they become tangled, ignoring their grunts and friction with the ground.

6-38. A hockey puck moving at a speed V_1 on a frictionless surface collides head on with a second identical puck moving toward it at speed V_2. After the collision, the first puck slows down to speed v_1 without changing direction.
   (a) Derive an equation for the velocity v_2 of the second puck after the collision.
   (b) Calculate the velocity v_2 of the second puck when the first puck had an initial speed of 18 m/s that was changed to 2.0 m/s by the collision, and the initial speed of the second puck was 12.0 m/s. Both pucks have a mass of 0.16 kg.
   (c) Do your answers change if the masses of both pucks are doubled?

6-39. • An accident reconstruction expert is called out to a crash scene. A car of mass m and speed v had crashed into the rear of an identical parked car whose parking brake was set. The skid marks indicate that the joined cars traveled a distance x before coming to rest. The coefficient of friction between the tires and the road is μ_k.
(a) Derive an equation for the initial speed of the incoming car.
(b) Calculate the initial speed of the incoming car if $m = 1400$ kg, $\mu_k = 0.65$, and $x = 12$ m.

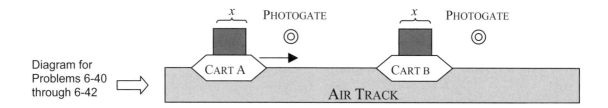

Diagram for Problems 6-40 through 6-42

6-40. A frictionless air track has two carts on it, A and B, with masses m_A and m_B. On top of each cart is a vertical metal plate of width x. The speed of the moving cart can be determined by measuring the time interval for the plate to pass across a photogate.

Cart A is given a brief push to the right. The plate on top of Cart A passes across the photogate in a time t_1. Cart A then collides with and sticks to Cart B. The joined carts continue moving to the right. The plate on top of Cart B eventually passes across the second photogate in a time t_2.
(a) Write an equation for the speed of Cart A before the collision.
(b) Derive an equation for the speed of Cart B after the collision.
(c) What is the time t_2 it will take for the plate on top of Cart B to pass across the second photogate?
(d) Calculate the time t_2 for $m_A = 0.240$ kg, $m_B = 0.280$ kg, $\Delta x = 10.0$ cm, and $t_1 = 0.360$ s.

6-41. Continuing with the air track from Problem 6-40, Cart B is replaced with Cart C, whose mass you don't know. You again give Cart A a push to the right and it passes across the photogate in a time t_1. After Cart A hits and sticks to Cart C, the plate on top of Cart C takes time t_2 to pass across the photogate.
(a) Derive an equation for the mass of Cart C.
(b) Calculate the mass for $m_A = 0.240$ kg, $\Delta x = 10.0$ cm, $t_1 = 0.320$ s, and $t_2 = 0.830$ s.

6-42.• Continuing with the air track from Problem 6-41, Cart C is replaced with Cart D with mass m_D. This time when you give Cart A a push to the right, it collides with and bounces back from Cart D. Cart A passes through the photogate a second time (moving to the left) in a time t_2.
(a) What is the time t_3 it will take for the plate on top of Cart D to pass across the second photogate?
(b) Calculate the time for $m_A = 0.240$ kg, $m_D = 0.480$ kg, $\Delta x = 10.0$ cm, $t_1 = 0.240$ s, and $t_2 = 0.720$ s.

6-43. A baseball of mass m leaves the bat at a speed v and makes an angle θ with the horizontal.
(a) Find the horizontal and vertical components of the ball's momentum.
(b) Calculate the horizontal and vertical components of momentum if the ball's mass is 0.15 kg, its speed is 45 m/s, and the angle θ is 37°.

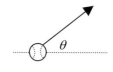

6-44. A golf ball of mass m is given a speed v at an angle θ with the horizontal.
  (a) Find the horizontal and vertical components of the ball's momentum.
  (b) Calculate the horizontal and vertical components of momentum if the ball's mass is 0.050 kg, speed is 35 m/s, and angle θ is 53°.

6-45. • A blue convertible of momentum mv traveling north collides with a red jeep having the same mass and speed, but traveling west. The cars collide, stick, and move off together with final speed V.
  (a) Draw a diagram illustrating the cars' motion before and after the collision.
  (b) Write expressions for the x- and y-components of momentum for each vehicle before the collision, and for the combined vehicles after the collision.
  (c) Momentum is conserved in the x-direction. Write an expression equating the total momentum in the x-direction before and after the collision.
  (d) Momentum is conserved in the y-direction. Write an expression equating the total momentum in the y-direction before and after the collision.
  (e) What is the magnitude and direction of the resulting momentum of the vehicles if they stick? (It may be useful to remember that $\frac{\sin\theta}{\cos\theta} = \tan\theta$.)
  (f) Calculate the resulting momentum if the mass of each vehicle is 1800 kg and their initial speeds were 11 m/s.
  (g) Calculate the resulting speed of the joined cars.

6-46. • A truck of mass M, traveling east at a speed v, collides with a car of mass m moving north at equal speed v. After collision both vehicles remain tangled together.
  (a) Draw a diagram illustrating the vehicles' motion before and after the collision.
  (b) Write expressions for the x- and y-components of momentum for each vehicle before the collision, and for the combined vehicles after the collision.
  (c) With what speed and in what direction does the wreckage move?
  (d) If the coefficient of kinetic friction between the wreck and the road is μ_k, how far does the wreck slide along the road?

6-47. • An exploding hockey puck, initially at rest, breaks into three pieces of mass m, mass $2m$, and mass $3m$, respectively. The piece of mass $2m$ heads across the ice in the x-direction at a speed $2v$, while the piece of mass $3m$ heads off along y-axis direction at a speed v.
  (a) What will be the momentum (magnitude and direction) of the piece of mass m?
  (b) How fast will the m piece be moving?

6-48. • A black hockey puck of mass m, sliding across frictionless ice at speed v_0, collides with an otherwise identical white hockey puck initially at rest. After the collision, the black puck moves at an angle θ_b to the right of its original direction, while the white puck moves at θ_w to the left of the incoming puck's original direction.
  (a) Find the final speeds of the two pucks.
  (b) Calculate the final speeds if the black puck has mass 0.16 kg and initial speed 32 m/s, $\theta_b = 65°$, and $\theta_w = 25°$.

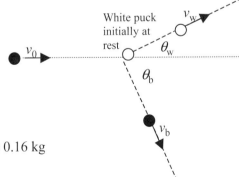

Show-That Problems for Momentum

6-49. Show that the units for impulse (N·s) and momentum (kg·m/s) are equivalent.

6-50. A force of 10.0 N acts for 0.010 s on an object.
Show that the change in momentum of the object will be 0.10 kg·m/s.

6-51. A 20.0-kg mass moving at a speed of 3.0 m/s is stopped by a constant force of 15.0 N.
Show that the stopping time required is 4.0 s.

6-52. A braking force is needed to bring a 1200-kg car moving at 25 m/s to rest in 20.0 s.
Show that the braking force is 1500 N.

6-53. A 50-gram egg is thrown at 4 m/s at a bed sheet and is brought to rest in 0.2 s.
Show that the average force of egg impact is 1 N.

6-54. An 84.0-kg passenger in a car moving at 24.0 m/s is brought to rest in 1.20 s by an airbag.
Show that the approximate force exerted by the airbag on the passenger is 1680 N.

6-55. A golf ball of mass 0.045 kg, traveling horizontally at 28 m/s, hits a brick wall and bounces back at the same speed after a contact time of 0.040 s.
Show that the force of impact is 63 N.

6-56. A 115-kg astronaut is floating at rest relative to her spacecraft in deep space. She throws an 18-kg tool kit at 4.6 m/s away from the spacecraft.
Show that the astronaut will recoil at a speed of 0.72 m/s.

6-57. Sam (90.0 kg) and Maddie (60.0 kg) stand 10.0 m apart on ice with a rope stretched between them. Big Josh sits on the side and watches. As Sam and Maddie pull on the rope hand over hand, Big Josh sees Sam move 2.0 m along the ice.
Show that Big Josh sees Maddie move 3.0 m along the ice toward Sam.

6-58. A 2.0-kg object traveling 10.0 m/s north has a perfectly elastic collision with a 5.0-kg object traveling 4.0 m/s south.
Show that the combined momentum after the collision is zero.

6-59. Comic-strip heroes Superman and Superdog meet an asteroid in outer space. Superdog floats nearby as Superman exerts a force on the asteroid, sending it away from Superdog at 100 m/s. The asteroid is 1000 times more massive than Superman is. Usually the comic strip would show Superman as being at rest after the throw. Taking physics into account, show that Superman would recoil away from Superdog at 100,000 m/s.

6-60. A 60-gram dart is moving horizontally at 5.0 m/s when it strikes and sticks into a 540-gram roller skate.
Show that the speed of the darted roller skate after the collision is 0.50 m/s.

6-61. A 300,000-kg spacecraft is floating in space when it senses a 40-kg enemy missile coming toward it at 20 m/s. The spacecraft fires a 60-kg lump of clay at 30 m/s directly at the missile. The clay sticks to the missile.
(a) Show that the spacecraft recoils at about 6 mm/s.
(b) Show that the final velocity of the clay-laden missile is 10 m/s in a direction away from the spacecraft.

6-62. An 80-gram puck slides eastward at 24 m/s across frictionless ice when it collides directly with a 160-gram puck at rest on the ice. After the collision, the 160-gram puck is moving eastward at 16 m/s.
Show that the final velocity of the 80-gram puck is 8 m/s westward.

6-63. Comic-strip bad guys fire a 5.0-gram bullet moving 400 m/s at 100-kg Superman's chest while Superman is standing at rest on frictionless ice. The bullet bounces harmlessly off Superman's chest at essentially the same speed and opposite direction.
Show that Superman moves back at about 4 cm/s.

6-64. • A car of mass 1200 kg travels north at 18 m/s and has an inelastic collision with a SUV of mass 2400 kg traveling east at 12 m/s.
Show that the entangled wreck slides at 53° east from a northerly direction.

6-65. • A 100-gram firecracker, sliding across the ice at 5.0 m/s, breaks up into two unequal pieces. A 60-gram piece moves off at 7.8 m/s at an angle of 50.2° relative to the firecracker's initial direction, while the 40-gram piece moves at 10.3 m/s at an angle of 60.9° relative to the firecracker's initial direction.
Show that momentum has been conserved in this explosion.

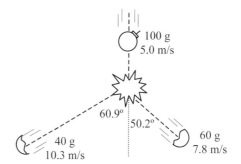

7 Energy

We move from the vector quantity momentum to the scalar quantity energy. Mechanical energy in a gravitational field involves both kinetic energy (KE $=1/2mv^2$)[*] and gravitational potential energy (PE $= mgh$[♦]). Work (W) done to raise or lower an object changes its gravitational PE. Work done to change an object's speed changes its KE. The rate at which work is performed is called Power (P). Stretched and compressed springs involve a different kind of PE and are treated later in the problem set.

Unless outside work is done on a system, the total energy of the system is *conserved*. Energy transforms from one form to another, but the total amount of energy is unchanged.

Ask a professional physicist to solve a typical mechanics problem and he or she will most likely use the energy concept. Many of the problems we've previously treated, whether in kinematics or dynamics, are more crisply solved by use of the work-energy theorem or conservation of energy.

Sample Problem 1

A rollercoaster car with mass m starts from rest at height h_1 on a frictionless rollercoaster.

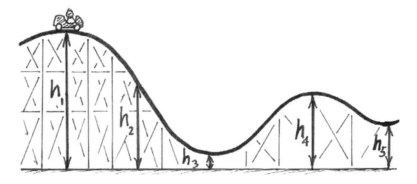

(a) Derive an equation for the speed of the car when its height is h_2.

Focus: $v_2 = ?$

Energy conservation applies to this problem. The PE of the car decreases as the car descends from height h_1 to h_2, and its KE increases by exactly the same amount.

From decrease in PE = increase in KE $\Rightarrow mgh_1 - mgh_2 = \frac{1}{2}mv^2$

$\Rightarrow \frac{1}{2}mv^2 = mg(h_1 - h_2) \Rightarrow v = \sqrt{2g(h_1 - h_2)}$.

---

[*] The amount of work done to accelerate an object of mass m from speed v_0 to speed v is $W = Fd = (ma)d$

$= m\left(\dfrac{v-v_0}{t}\right)\left(\dfrac{v+v_0}{2}t\right) = \dfrac{1}{2}mv^2 - \dfrac{1}{2}mv_0^2$. This work changes the kinetic energy of the object.

[♦] The amount of work done to lift an object of mass m up by a distance h from some reference position is $W = Fd = mgh$. This work is stored in the object as gravitational potential energy. Gravitational PE is always calculated relative to some convenient reference (say, the ground or a table top). What is important in solving problems is the *change* in gravitational PE.

© Paul G. Hewitt and Phillip R. Wolf

(b) Why can't we use Newton's second law and kinematics to solve this problem?

Answer: Since the slope of the hill is changing, the acceleration is NOT constant. All of the equations that we derived in Chapter 3 apply to constant acceleration and so don't apply in this case.

(c) At what point on the track is the car moving fastest?

Answer: At the low point, height h_3, the potential energy is lowest, so at that point the kinetic energy must be greatest and thus the car must be moving fastest.

Sample Problem 2

Phil pushes Mala horizontally across the frictionless snow on her sled, starting from rest. He applies a force F for a distance d.

(a) How much work does Phil's applied force do?

Focus: $W = ?$

Phil applies a force F on the sled in the direction of motion. So the work he does is simply $W = $ **Fd**.

(b) Derive an equation for the final speed of the sled.

Focus: $v = ?$

The work done by the applied force goes into changing the KE of the sled.

From $W = \Delta KE \Rightarrow Fd = \frac{1}{2}mv^2 \Rightarrow v = \sqrt{\frac{2Fd}{m}}$.

(c) Calculate the speed of the 62-kg sled, initially at rest, after being pushed a distance of 6.5 m by a 35-N horizontal push. Assume negligible friction.

Solution: $v = \sqrt{\dfrac{2Fd}{m}} = \sqrt{\dfrac{2(35\ \text{N})(6.5\ \text{m})}{62\ \text{kg}}} = 2.7\sqrt{\dfrac{\text{kg} \cdot \frac{\text{m}}{\text{s}^2} \cdot \text{m}}{\text{kg}}} = 2.7\ \dfrac{\text{m}}{\text{s}}$.

Sample Problem 3

This time, when Phil pushes Mala across the snow on her sled, he pushes at a downward angle θ as shown.

(a) How much work does Phil's applied force F do on the sled while pushing the sled a horizontal distance d?

Focus: $W = ?$

Note that the solution is NOT simply Fd, because F is not in the direction of the sled's motion. When an applied force is at an angle to the direction of motion, work is calculated by multiplying the *component* of the force parallel to the direction of motion, by the distance moved.

In this case,

$W = F_x d = (F\cos\theta)d$, or **$Fd \cos\theta$**.

(b) Calculate the speed of the 62-kg sled, initially at rest, after being pushed a distance of 6.5 m by a 35-N push directed at an angle of 25° below the horizontal. Assume negligible friction.

Focus: v = ?

If this problem were given in Chapters 4 or 5, before we learned about the connection between work and energy, our solution would likely go thusly:

From $v_f^2 - v_0^2 = 2ad$, with $v_0 = 0$ $\Rightarrow v = \sqrt{2ad}$. From $a_x = \dfrac{F_x}{m}$ $\Rightarrow a = \dfrac{F\cos\theta}{m}$.

$$v = \sqrt{\dfrac{2F\cos\theta\, d}{m}} = \sqrt{\dfrac{2(35\text{ N})\cos 25°(6.5\text{ m})}{62\text{ kg}}} = 2.6\,\dfrac{\text{m}}{\text{s}}.$$

Note the shorter route to a solution using the work-energy theorem:

Focus: v = ?

Work done on the sled changes its KE. Only the horizontal component of the force contributes to the work.

From $W = \Delta KE$ $\Rightarrow (F\cos\theta)d = \dfrac{1}{2}mv^2$ $\Rightarrow v = \sqrt{\dfrac{2F\cos\theta\, d}{m}}$

Again, $v = \sqrt{\dfrac{2(35\text{ N})\cos 25°(6.5\text{ m})}{62\text{ kg}}} = 2.6\,\dfrac{\text{m}}{\text{s}}$.

Can you see why physicists often prefer to use energy in solving problems that involve force and distance?

Sample Problem 4

Janet cruises along a level road at speed v and slams on the car's old-fashioned, non-antilock brakes. The car slides to a stop. The coefficient of kinetic friction between its tires and the road is μ_k.

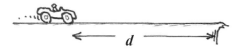

(a) Derive an equation for the distance that Janet's car slides.

Focus: d = ?

We again consider the work-energy theorem. The net force here is the force of friction, which acts in a direction opposite to the car's direction of motion to *decrease* the KE of the car, slowing it from speed v to rest.

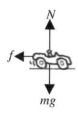

The amount of work done by the friction force is fd, while the KE changes by an amount $\dfrac{1}{2}mv^2$.

So from $W = \Delta KE$ $\Rightarrow fd = \dfrac{1}{2}mv^2$ $\Rightarrow d = \dfrac{\tfrac{1}{2}mv^2}{f}$.

Now $f = \mu_k N$, where $N = mg$ since the road is horizontal. So the solution is

$$d = \dfrac{\tfrac{1}{2}mv^2}{f} = \dfrac{\tfrac{1}{2}mv^2}{\mu_k N} = \dfrac{\tfrac{1}{2}mv^2}{\mu_k mg} = \dfrac{v^2}{2\mu_k g}.$$

Note how the terms in the equation dictate subsequent steps. Again, equations are more than recipes for plugging in values; they can also be guides to thinking.

(b) Does the solution make sense? And what does it tell us?

Answer: The solution tells us that the stopping distance is proportional to speed squared, which is consistent with KE—a car going twice as fast will have four times as much KE, and so will require four times the distance for the same friction force to bring it to a stop. It also tells us that if μ_k or g is greater, the stopping distance will be less because the force of friction will be greater. That makes sense. And we see the mass cancels in the equations, which tells us that the mass of the car doesn't matter. All cars skidding with the same initial speed and with equal coefficients of kinetic friction will skid the same distance. And as for units, note that $\frac{v^2}{2\mu_k g}$ has units $\frac{m^2}{s^2} / \frac{m}{s^2} = m$, which is a distance, as it should be. Good physics!

Sample Problem 5

A toy car of mass m has a motor that can supply it with power P. Assume that the car begins at rest on a level floor and the motor is activated.

(a) Derive an equation for the speed of the car after a time t.

Focus: $v = ?$ From the work-energy theorem, we know that the work W done by the motor will equal the gain in the car's kinetic energy.

That is, from $W = \frac{1}{2}mv^2 \Rightarrow v = \sqrt{\frac{2W}{m}}$. We know m. What is W?

Power is work per unit time. From

$P = \frac{W}{t} \Rightarrow W = Pt$. Substituting this for W gives $v = \sqrt{\frac{2Pt}{m}}$.

(b) Show that the units of $\sqrt{\frac{2Pt}{m}}$ are m/s.

Solution: The unit of power is in Watts (J/s), time is in seconds, and mass is in kg. So

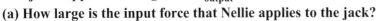

$\sqrt{\frac{2Pt}{m}} \Rightarrow \sqrt{\frac{\frac{J}{s} \cdot s}{kg}} = \sqrt{\frac{J}{kg}} = \sqrt{\frac{N \cdot m}{kg}} = \sqrt{\frac{\left(kg \cdot \frac{m}{s^2}\right) \cdot m}{kg}} = \sqrt{\frac{m^2}{s^2}} = \frac{m}{s}$, just the units we expect for speed.

Sample Problem 6

Nellie Newton lifts a car with a car jack as shown. She pushes downward a distance d and the car rises by a lesser distance h. The jack supplies a lifting force F_{output} to raise the car.

(a) How large is the input force that Nellie applies to the jack?

Focus: $F_{input} = ?$

Machines multiply force rather than work. Ideally, with no friction, a machine's work input equals a machine's work output.

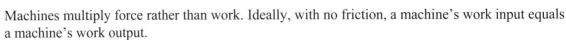

From $F_{input} \times d_{input} = F_{output} \times d_{output} \Rightarrow F_{input} = \frac{d_{output}}{d_{input}} F_{output} = \frac{h}{d} F_{output}$.

So Nellie need only push downward with a fraction (h/d) of the force needed to lift the car directly.

(b) The lifting force F_{output} required to raise the car is 5000 N. Calculate F_{input} when the lever is pushed downward 27 cm to raise the car 0.36 cm.

Solution: $F_{input} = \dfrac{h}{d} F_{output} = \dfrac{0.36 \text{ cm}}{27 \text{ cm}} (5000 \text{ N}) = 67 \text{ N}.$

(c) Nellie finds that she actually needs to supply an input force of 125 N to lift the car. What is the efficiency of the car jack?

Solution: Efficiency = ?

Because of friction in the jack, some of the work input is transformed into thermal energy. The efficiency of a machine tells us what fraction of the work input turns to useful work output. That is,

$$\text{Efficiency} = \dfrac{\text{actual } W_{output}}{W_{input}} = \dfrac{F_{output} \times d_{output}}{F_{input} \times d_{input}} = \dfrac{(125 \text{ N})(27 \text{ cm})}{(5000 \text{ N})(0.36 \text{ cm})} = 0.53 \text{ or } 53\%.$$

(d) Why would anybody want to use a machine that wastes 47% of the energy put into it?

Answer: Otherwise you wouldn't be able to lift the car! When you have a flat tire, your focus is on replacing it, not on how much energy you lose to the environment. The car jack multiplies your input force and makes possible an otherwise impossible task. The same reasoning applies to most people taking out loans to buy a home—they can't supply the money it takes to buy the house all at once, so they agree to pay an amount that they can supply over a longer period of time—even though the overall cost is greater.

Springs

The treatment of springs in the textbook is minimal, with Hooke's law first appearing in Chapter 12. Since stretching or compressing a spring requires work, we introduce a brief treatment of springs here.

Suppose you're in lab, measuring the stretch and force of a vertical spring suspended from a support. You place a ruler behind the spring and then hang a 50-gram mass at the end of the spring. The spring stretches 2.0 cm. Then you hang an additional 50 grams on the spring and find that it stretches another 2.0 cm.

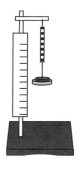

If you were to enter your lab data in a table, it might (ideally) look like this:

Measured Quantities		Calculated Quantities		
Total mass hanging on the spring (grams)	Total stretch of the spring (cm)	Total mass hanging on the spring (kg) $\left[g \times \left(\dfrac{1 \text{ kg}}{1000 \text{ g}} \right) \right]$	Force pulling on the spring (N) $m_{hanging} g$	Total Stretch of the spring (m) $\left[\text{cm} \times \left(\dfrac{1 \text{ m}}{100 \text{ cm}} \right) \right]$
0	0	0	0	0
50	2.0	0.050	0.49	0.020
100	4.0	0.100	0.98	0.040
150	6.0	0.150	1.47	0.060
200	8.0	0.200	1.96	0.080
250	10.0	0.250	2.45	0.100

If you graph the last two columns of the data, you can see that you'll get a straight line.* The straight line tells you that the two quantities, applied force and stretch, are directly proportional. Double one and the other doubles, and so on. The equation of a straight line through the origin is

$$y = slope \cdot x.$$

The proportionality of force to stretch (or compression) was discovered by English physicist Robert Hooke, a contemporary of Isaac Newton. This is Hooke's law:

Applied force $F = kx$

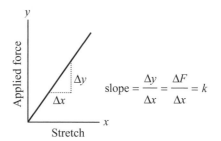

The constant k is the *spring constant* (typically in N/m), which tells us how much force corresponds to how much stretch. Many springs and other "springy" objects exhibit a stretch that is proportional to the applied force, at least for small deformations.

Springs in physics problems have three properties:

1. They are *linear*—the amount of stretch or compression is proportional to the amount of applied force.
2. Their behavior is *symmetric*—a given amount of force will compress or stretch a spring the same amount.
3. They are light compared with the objects we suspend or place on them, so we can *approximate* them as having zero mass.

Springs Sample Problem 1

A certain spring stretches a distance x when it carries a load of mass m.
(a) What is the spring constant of the spring?

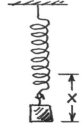

Focus: $k = ?$

The spring constant tells us the amount of force per unit stretch. The applied force here is the weight mg of the hanging mass, and the stretch is x. So $k = \dfrac{F}{x} = \dfrac{mg}{x}$.

(b) How much will the spring stretch when supporting a load $3m$ (assuming that the spring doesn't reach its elastic limit*)?

Focus: $x_{3m} = ?$

Three times the mass will be pulled downward by gravity with three times as much force, so we'd expect the stretch to be three times as much.

From $F = kx \;\Rightarrow\; x_m = \dfrac{F_m}{k} = \dfrac{mg}{k}$. Likewise, $x_{3m} = \dfrac{F_{3m}}{k} = \dfrac{3mg}{k} = 3\left(\dfrac{mg}{k}\right) = 3x_m$.

---

* It is common practice to place the experimental variable we control—the *independent variable* (in this case the hanging weights)—on the x-axis, and the experimental variable that we measure—the *dependent variable* (the stretch of the spring)—on the y-axis. Here we reverse the practice so that the slope will be the spring constant and have units of N/m.
* For small deformations, elastic items will snap back to their original length when the stress causing the deformation is removed. If an object is stretched or deformed too much, beyond its *elastic limit*, the item will *not* snap back. If you have ever seen a Slinky that has been stretched a little too much you have seen something stressed beyond its elastic limit!

(c) Calculate the spring constant of a spring that stretches by 4.0 cm when you hang a 100-g mass on it. How much will the spring stretch if you hang 300 g on it?

$$k = \frac{mg}{x} = \frac{(0.100 \text{ g})\left(9.8 \frac{m}{s^2}\right)}{0.040 \text{ m}} = 24.5 \frac{N}{m}.$$

Three times the hanging mass would stretch the spring by 3(4.0 cm) = **12.0 cm**.

Springs Sample Problem 2

When you stretch (or compress) a spring, it has potential energy—the ability to do work.
(a) How much work is required to stretch a particular spring a distance x by a force F?

Focus: $W = ?$

At first thought the solution might be $W = Fd = Fx$.
This is not so, for the full value of the force F doesn't occur until the very end of the stretch. Ordinarily, when you stretch a spring by a distance x, the initial force is zero and builds linearly to kx. So the average force exerted is $kx/2$. That means the work done in stretching the spring is

$$W = F_{average} x = \left(\frac{1}{2}kx\right)x = \frac{1}{2}kx^2.$$

The work done in stretching or compressing a spring is converted to *elastic potential energy*. This stored energy can be transformed to some other form.

(b) Calculate the work done in stretching a spring when a force of 50 N stretches it by 20 cm (0.20 m).

Solution: The solution is *not* $W = Fd = (50 \text{ N})(0.20 \text{ m}) = 10$ J. Rather, the work done is the *average* force × the distance (where the average force is one-half 50 N);

$$W = \frac{1}{2}(50 \text{ N})(0.20 \text{ m}) = 5 \text{ J, or using } k = \frac{50 \text{ N}}{0.20 \text{ m}} = 250 \frac{N}{m},$$

$$W = \frac{1}{2}kx^2 = \frac{1}{2}\left(250\frac{N}{m}\right)(0.20 \text{ m})^2 = 5 \text{ J}.$$

Notice we get the solution two ways. It's nice that there are more ways than one to solve many physics problems!

Springs Sample Problem 3

A mass m on a frictionless surface is compressed against a spring with a spring constant k. Derive an equation for the speed v of the mass when it is released by the spring.

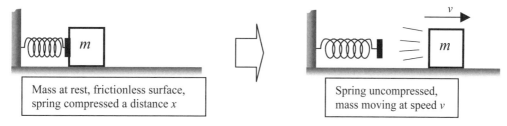

Mass at rest, frictionless surface, spring compressed a distance x

Spring uncompressed, mass moving at speed v

Solution: The work required to compress the spring is $\frac{1}{2}kx^2$. This work is stored as elastic potential energy in the spring and is transformed into KE of the moving mass.

$$(\text{Elastic PE in the Spring})_{initial} = (\text{KE of the mass})_{final}$$

$$\frac{1}{2}kx^2 = \frac{1}{2}mv^2 \Rightarrow v^2 = \frac{k}{m}x^2 \Rightarrow v = \sqrt{\frac{k}{m}}x.$$

Work and Energy Problems

7-1. A cart of mass m moves at speed v in the $+x$-direction.
    (a) Write an equation for the kinetic energy of the cart.
    (b) What would be the KE of the cart if it were instead moving in the $-x$-direction?
    (c) Calculate the KE of a 0.80-kg cart moving at 3.2 m/s.

7-2. John applies a constant horizontal force F to move a desk a distance d across the floor.
    (a) How much work is done on the desk by the applied force?
    (b) Calculate the work done by a 120-N force applied to move the desk a distance of 2.5 m.
    (c) Why doesn't the mass of the desk come into the calculation?

7-3. Paul's box of wrenches of mass m is located on a shelf of height y above the floor.
    (a) What is the gravitational potential energy of the box relative to the floor?
    (b) Calculate the potential energy of a 2.8-kg box of wrenches if you know that the wrenches are made of high-quality steel and the box is 1.2 m above the floor.

7-4. You lift a watermelon of mass m from the floor to a tabletop of height h.
    (a) Write an equation for the amount of work you do on the watermelon.
    (b) Calculate the work to lift a 5.5-kg watermelon from the floor to a 93-cm high table top.

7-5. Tsing does work W in pushing a box of mass m over a distance x on horizontal surface.
    (a) How much force does Tsing exert in pushing the box?
    (b) Calculate the steady force that Tsing exerts when doing 660 J of work to move a 28-kg box over a horizontal distance of 11 m.
    (c) Calculate the net force on the moving box if friction between the box and the floor is a steady 60 N.

7-6. A truck of mass m moves at speed v when moving against a headwind of speed V, while the temperature of the road surface is T.
    (a) Write an equation for the kinetic energy of the truck.
    (b) Calculate the kinetic energy of a 3300-kg truck moving at 18 m/s when a wind of speed 5 m/s hits it head on, and when the temperature of the road is 24°C. (*Hint*: Do you need all the information provided here?)

7-7. A battleship of mass M travels a distance x at constant speed in time t.
    (a) What is the KE of the battleship?
    (b) Calculate the KE of a 9.0×10^7-kg battleship that travels 250 km in 4.1 hours at a constant speed.

7-8. Adam bats a baseball that weighs w and leaves the bat with a speed v.
    (a) What is the baseball's KE?
    (b) Calculate the KE of a 1.5-N baseball when it leaves the bat with a speed of 38 m/s.

7-9. Milo pushes a lawn mower of mass m, exerting a force F on the handle that makes an angle θ with the horizontal.

  (a) How much work does Milo do on the mower when pushing it a horizontal distance x?
  (b) Calculate the work Milo does on the mower if he pushes the 25-kg mower a distance of 15 m with a force of 52 N while the angle that the mower handle makes with the ground is 51°.

7-10. Martin puts his beagle in a wagon and takes him to the store, a horizontal distance x away, to purchase a Japanese frying pan. Martin's pull P on the handle of the wagon makes an angle θ with the ground.
  (a) How much work does Martin's pull do in taking his dog for a wok?
  (b) Calculate the work done by Martin's pull does if he pulls with a 56-N force at an angle of 31° and the store is located 480 m away.

7-11. A baseball of mass m is accelerated from speed v_0 to speed v_f.
  (a) How much work was done on the baseball?
  (b) Calculate the work done to accelerate a 0.145-kg baseball from 2 m/s to 25 m/s.

7-12. Karen does work to accelerate Jeff of mass m in a frictionless shopping cart, from rest to a speed v over a distance x.
  (a) Derive an equation for the constant horizontal force F that Karen applies.
  (b) Calculate the force that Karen applies to accelerate the 73-kg Jeff from rest to 5.0 m/s over a 24-m distance.

7-13. When an average net force F is exerted over a certain distance on a system of mass m, initially at rest, the system's kinetic energy increases by $\frac{1}{2}mv^2$.
  (a) Derive an equation for the distance over which the net force acts.
  (b) If twice the force is exerted over twice the distance, how does the resulting increase in kinetic energy compare with the original increase in kinetic energy?

7-14. A constant horizontal net force F acts through a horizontal distance x on a cart of mass m, initially at rest.
  (a) What is the final kinetic energy of the cart?
  (b) Use work-energy considerations to derive an equation for the final speed of the cart.
  (c) Calculate answers to the above questions if the constant net force is 75 N, the mass of the cart is 8.4 kg, and the force is applied through a distance of 0.80 m.

7-15. A particle of mass m is accelerated from rest to a speed v by a steady force F. Assume that friction is insignificant in this case.
  (a) Derive an equation for the distance over which the force acts.
  (b) Calculate the distance over which a 14-N force must act to change the speed of a 0.60-kg particle from rest to 12 m/s.

7-16. A car of mass m undergoes an increase in speed from v_0 to v_f in a time interval t.
  (a) Write an equation for the increase in the car's kinetic energy.
  (b) Calculate the change of kinetic energy if the mass of the car is 1250 kg, its initial speed is 13 m/s, and 5 seconds later, its speed is 19 m/s.

7-17. A friction force f acts on a textbook sliding across the lab table. The book slows from speed v_0 to speed v_f.
 (a) Over what distance d does the friction force act?
 (b) Calculate the distance for a 7.1-N friction force that slows a 1.4-kg book from 2.2 m/s to 1.0 m/s.

7-18. Michael tosses laughing baby Gracie, mass m, into the air. Gracie rises a distance h.
 (a) With what speed did Gracie leave Michael's hands?
 (b) Calculate 5.5-kg Gracie's initial speed if she rises 0.60 m.

7-19. A vertical average force F applied over a height h is required to accelerate an elevator of mass m from rest up to a steady speed v.
 (a) How much work is done on the elevator by the applied force?
 (b) What is the kinetic energy of the elevator at steady speed v?
 (c) What is the potential energy of the elevator at height h?
 (d) Use the concepts of work and changes in total energy to write an expression for F.
 (e) Calculate the average force that acts on an 850-kg elevator to accelerate it from rest to a speed of 2.0 m/s over a distance of 2.5 m.

7-20. A bullet of mass m, traveling at speed v, hits a tree and slows down to a stop while penetrating a distance x into the tree trunk.
 (a) Derive an equation for the average force exerted on the bullet in bringing it to rest.
 (b) Calculate the average stopping force for a 5.2-gram bullet that hits the tree at a speed of 340 m/s and penetrates 12 cm into the trunk.

7-21. Ellyn throws a dart at a speed v into a thick dartboard. It penetrates a distance x.
 (a) If she wants the dart to penetrate twice as far, how fast does it have to be thrown?
 (b) How far would the dart penetrate if thrown at twice its initial speed? Why are your answers different?
 (c) A dart thrown at 3.2 m/s penetrates 0.90 cm into the dartboard. Calculate the speed of a dart that would penetrate 1.8 cm into the dartboard.
 (d) Calculate how far the same dart would penetrate if it were thrown at 6.4 m/s.

7-22. An oil tanker of mass m, moving at speed v_0, turns off its engines. After a distance x, friction with the water has slowed the tanker to a speed v_f.
 (a) Derive an equation for the average friction force acting on the oil tanker.
 (b) Calculate the friction force for a 200,000-metric ton tanker (1 metric ton = 1000 kg) initially moving at 20 km/hr that slows to 10 km/hr over a 5.0-km distance.
 (c) If you were to calculate a "coefficient of friction" between the tanker and the water ($= f/mg$) what would it be equal to?

7-23. An automobile traveling on a level road at speed v brakes to a stop in a distance x.
 (a) Assuming the same braking force, what would be the stopping distance for an initial speed of $4v$?
 (b) Calculate the new stopping distance if the braking force brings a car with an initial speed of 20 km/h to rest in a distance of 12 m.
 (c) From the above information, can the mass of the automobile be found?

7-24. A high diver of mass *m* dives from height *h* to the water below.
   (a) What is her kinetic energy when she has freely fallen halfway to the water?
   (b) What is her kinetic energy just before hitting the water?
   (c) Calculate these values for a diver of mass 55.0 kg and an initial height of 20.0 m above the water.

7-25. A window washer suspended from a scaffold is at the top of the 35th floor of a skyscraper when she accidentally drops her scrub brush of mass *m*. The vertical distance between floors is *y* (so she is at height 35*y* when she drops the brush).
   (a) What are the gravitational PE (relative to the ground below) and KE of the brush when it reaches the top of the 32nd floor? (Neglect air resistance.)
   (b) If each story is 3.2 m, calculate the KE and PE of the 0.51-kg brush relative to the ground when it reaches the top of the 32nd floor.

7-26. To throw a baseball of mass *m*, Umberto brings it from rest to speed *v* over a distance *d*.
   (a) Find the average force that Umberto exerts on the ball.
   (b) Calculate the average force Umberto exerts to bring a 0.145 kg ball from rest to 29 m/s, throwing through a distance of 1.3 meters.

7-27. A bucket of cement of mass *m* is lifted straight up by a crane to the third floor of a building under construction. The bucket is raised at a slow constant velocity *v* to an elevation *h* in the absence of air resistance and friction.
   (a) How much work is done in raising the bucket? (Neglect its initial short acceleration.)
   (b) The cable breaks just as the bucket reaches elevation *h*. What are the PE and the KE of the bucket at that point?
   (c) Assuming no friction or air resistance as it falls, what is the total energy of the bucket when it has fallen halfway to the ground?
   (d) What is the total energy of the bucket just before it hits the ground?
   (e) What becomes of its energy upon hitting the ground?
   (f) Calculate answers to parts (b), (c), and (d) for a 650-kg bucket that is moving at 0.5 m/s when it reaches a height of 15 meters.

7-28. Paul Doherty is rock climbing in the Grand Teton mountains and has just climbed to a height *h* when he drops his hammer of mass *m*. (Neglect air resistance.)
   (a) Assuming Paul and the hammer are at rest at the time of the drop, and that we set the zero of PE at the starting elevation of Paul's climb, what are the hammer's KE and gravitational PE at the moment when the hammer is dropped?
   (b) What are the PE and the KE of the hammer after it has dropped 3/4 the way to the ground?
   (c) What is the velocity of the hammer after it has dropped 3/4 the way to the ground?
   (d) What are the gravitational PE and the KE of the hammer just before it hits the ground?
   (e) What is the velocity of the hammer just before it hits the ground?
   (f) What happens to the energy of the hammer after it hits the ground?
   (g) Calculate answers to parts (b) through (e) for a 0.86 kg hammer dropped from a height of 44 m.

7-29. Marilyn the diver of mass m steps off a tower of height y above the water. After hitting the water, she comes to a stop at distance d beneath the surface.
  (a) What average net force does she experience while stopping beneath the water?
  (b) Calculate the force if her mass is 50 kg, the tower is 10 m above the water, and the distance to stop beneath the water is 3.1 m.

7-30. A block of ice of mass m is initially at rest at the top of an inclined plane of vertical height h. The block slides down the incline and has speed v when it reaches the floor.
  (a) Find this speed, assuming friction can be ignored.
  (b) Find the speed that a 27-kg block of ice will have when it reaches the bottom of an inclined plane of height 1.5 m. Again, ignore friction.
  (c) Why doesn't the slope of the incline affect your answer?

7-31. A chunk of ice of mass m breaks loose from a suspension bridge of height h.
  (a) What will be the chunk's kinetic energy when it hits the water below?
  (b) What would be its kinetic energy if it fell from a bridge twice as high?
  (c) A 0.80-kg chunk of ice falls from a suspension bridge 45 m above the water. Calculate the KE of the ice right before it hits the water.

7-32. Diane tosses a baseball straight upward with a certain speed, and therefore, a certain kinetic energy.
  (a) If air drag is negligible, what will be the ball's kinetic energy when it is caught at the same elevation from which it was thrown?
  (b) If air drag is not negligible, how will the kinetic energy differ when being caught?
  (c) Could you calculate the mass of the ball in part (a) if you're given its initial speed and the height to which it goes? Defend your answer.

7-33. A ball of mass m is thrown vertically upward with an initial speed v from a starting point y_0 above the ground. (Ignore air resistance.)
  (a) Derive an equation for the maximum height reached by the ball.
  (b) What is the potential energy of the ball at its maximum height relative to the ground?
  (c) What is its potential energy relative to its launch point?
  (d) Calculate answers to the above questions for a 0.15-kg ball thrown upward with an initial speed of 22.0 m/s from a starting point 1.2 m above the ground.

7-34. A pendulum bob of mass m is initially at a vertical distance y_1 above a table. After it is released, it swings to a lowest point, at distance y_2 above the table, and then almost to its initial height y_1 on the other side.
  (a) By how much does the bob's PE at its highest point exceed its PE at its lowest point?
  (b) What is the kinetic energy of the bob as it passes through its lowest point?
  (c) Calculate answers to the above questions if the mass of the bob is 1.0 kg, the highest point is 72 cm above the table, and the lowest point is 31 cm above the table.

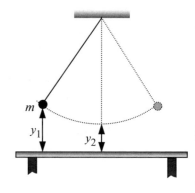

7-35. Alex drops a steel ball of mass m to the floor from a height h_1. The ball rebounds to a lesser height h_2.
 (a) How much kinetic energy does the ball have when it first bounces up from the floor?
 (b) How much potential energy did the ball have initially?
 (c) In your answers to parts (a) and (b), how do you account for the differences in energy?
 (d) A 0.50-kg ball falls from a height of 1.2 m, hits the floor, and bounces to a height of 0.90 m. Calculate the kinetic energy of the ball immediately after it bounces from the floor.

7-36. Gracie stands on a bridge of height h above a river and drops a stone of mass m.
 (a) What is the stone's potential energy at the time of release relative to the river below?
 (b) Neglecting air resistance, with what speed does the stone hit the water?
 (c) What are the stone's kinetic and potential energies after it has fallen $h/2$?
 (d) Calculate your answers to the above questions for a stone of mass 1.1 kg and a bridge height of 120 m.

7-37. A railroad car of mass m and speed v collides with a stationary car of equal mass. The cars couple and move together.
 (a) Beginning with the conservation of momentum, derive the speed of the coupled cars after the collision.
 (b) What is the kinetic energy of the coupled cars?
 (c) Calculate the kinetic energy of the coupled cars if each of them has a mass 10^4 kg and the initial speed of the moving car before collision was 5 m/s.
 (d) How does the kinetic energy of the coupled cars compare with the initial kinetic energy of the first car?
 (e) What explanation can you offer if there is an energy difference?

7-38. An acrobat of mass m stands on the left end of a seesaw. A second acrobat of mass M jumps from a height h onto the right end of the seesaw, thus propelling the first acrobat into the air.
 (a) Neglecting inefficiencies, how will the PE of the smaller acrobat at the top of his trajectory compare with the PE of the person just as he jumps?
 (b) Neglecting inefficiencies, how high does the first acrobat go?
 (c) Calculate the maximum height that the first acrobat reaches if his mass is 40 kg, the mass of the second acrobat is 70 kg, and the initial height jumped from was 4 m.

7-39. A nice demonstration utilizes a test tube of mass M suspended by a pair of parallel wires. The tube is lightly sealed by a stopper of mass m. Inside the tube is a bit of water that is heated with a flame. When the water turns to steam, the cork pops out, and the test tube swings back. The test tube rises a vertical distance h.
 (a) What must have been the speed of the test tube immediately after the cork popped out?
 (b) With what speed did the cork fly out? (*Hint*: Use conservation of momentum.)

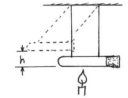

7-40. • A pendulum bob of mass m at the end of a string of length L is pulled back by an angle θ and then released.
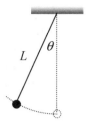
(a) Derive an equation for the speed of the bob when it reaches the lowest part of its swing.
(b) Calculate the speed of a 55-gram lead bob at the end of a 1.1-m string at the lowest point of its swing if it is released from a 25° angle.

7-41. A roller coaster at rest atop a hill of height h begins its downward descent.
(a) Derive an equation for the speed of the coaster at the bottom of the hill.
(b) For a bigger thrill, you might wish the car to be going twice as fast as in part (a) at the bottom of the run. How high does the hill need to be?
(c) How fast would the coaster be going at the bottom if the height of the hill were $2h$?
(d) Calculate the speed of a roller coaster at the bottom of a hill if the coaster starts from rest from the top of a 43-m hill.

7-42. A roller coaster of mass m starts from rest at an elevation h_1 and rolls down and then up another incline of lower height h_2.
(a) What is the kinetic energy of the coaster when it reaches the top of the second incline?
(b) Calculate the kinetic energy of a 2500-kg coaster atop the second incline of height 21 m if it starts from rest at the top of the first incline, 32 m high.

7-43. A roller coaster car of mass m is pulled from rest up to the top of a hill of height h where it is released with an initial speed very close to zero.
(a) What is the speed of the car when it rolls down to a distance $h/2$ above the ground?
(b) What force F will be necessary to stop the car in a distance d once it reaches the bottom of the hill?
(c) What would be the speed $h/2$ above the ground if the car had an initial speed v_0 at the top? (*Hint*: You can't just add v_0 to your earlier answer!)

7-44. • Little Andre of mass m climbs a vertical ladder of height h to the top of a 35°-inclined plane. The coefficient of friction between Andre and the incline as he slides down the plane is μ_k.
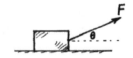
(a) Write an equation for the friction force acting on Andre. (Note that the normal force $\neq mg$.)
(b) Write an expression for L, the length of the slope.
(c) Write an equation for Andre's speed when he reaches the bottom of the inclined plane.
(d) Calculate Andre's speed at the bottom of the incline if the height h is 2.0 m and the coefficient of sliding friction μ_k is 0.30.
(e) If Andre's mass were 20% more than it is, how would this affect the speed?

7-45. • A force F pulls at an angle θ above the horizontal on a block of mass m that is initially at rest. The value of μ_k is known, and F is large enough so that the block starts to move.
(a) Write an expression for the normal force N. (*Hint*: It isn't $N = mg$.)
(b) How fast will the box be moving after having been pulled for a distance x?
(c) Calculate the speed of a 12-kg box pulled 4.9 m by a 46-N force at an angle of 35° above the horizontal when the box travels from rest along a surface with $\mu_k = 0.30$.

Power Problems

7-46. A diesel engine lifts a pile-driver hammer of mass m to its maximum height h in a time t.
   (a) What is the average power output of the engine?
   (b) Calculate the power delivered by the engine if the mass of the pile-driver hammer is 190 kg, and the height raised is 25 m in a time of 5.9 s.

7-47. A loaded elevator is lifted a distance of h in a time t by a motor delivering power P.
   (a) What average force does the motor exert to lift the elevator?
   (b) Calculate the force exerted if the distance risen is 20 m, the time of lift is 30 s, and the power of the motor is 6.0 kW.

7-48. Elvia of mass m, initially at rest, climbs a flight of stairs of overall height h in a time t. She reaches the top of the stairs with speed v.
   (a) How much work does Elvia do?
   (b) What is Elvia's average power output?
   (c) Calculate Elvia's average power output if her mass is 71 kg and she climbs a vertical distance of 6.1 m in 7.9 s, ending up moving at 2.2 m/s.

7-49. An electric motor lifts an elevator of mass m from ground level to the top of a building of height h in time t.
   (a) What average power does the motor deliver?
   (b) Calculate the power the motor delivers when it lifts a 10,000-kg elevator from ground level to the top of a 30-m building in 44 s.

7-50. A small hydroelectric generator draws water through a large pipe from a lake at a height h above the generating turbine. There are V liters of water passing through the turbine each second. (Note that 1 liter of water has a mass of 1 kg.)
   (a) Show that the maximum power developed by the turbine is Vgh in watts.
   (b) Calculate the maximum power that could be obtained from a hydroelectric generator that has 25 liters of water per second coming from a lake 49 m above the turbine.

7-51. Equipment having mass m is raised a vertical distance h at a constant speed v to the top of the world's tallest lighthouse.
   (a) How much power is required to raise the equipment?
   (b) Calculate the power developed if the mass of the equipment is 25 kg, the vertical speed 2.0 m/s, and the height of the lighthouse is 106 m.

7-52. A motor of power P raises an elevator of mass m.
   (a) What is the maximum speed at which the motor can raise the elevator?
   (b) The elevator has a mass 900 kg and is powered by a 21-kW motor. Calculate the maximum speed at which the elevator can be raised.

Efficiency and Machines Problems

7-53. Benjamin applies a force F over a distance x to a rope connected to a set of pulleys. This raises a block of weight w to height h above its initial level.
   (a) What is Benjamin's work input?
   (b) What work is done on the block?
   (c) What is the efficiency of the pulley system?
   (d) Calculate answers to the above questions given that the force applied is 75 N over a 12-m distance, the weight of the block is 300 N, and the block is raised a vertical distance of 2.0 m.

7-54. The block in the previous problem is raised a vertical distance h by means of a device having an efficiency E.
   (a) What is the work output?
   (b) What is the work input?
   (c) Calculate answers to the above questions when the height raised is 5.0 m and the efficiency of the device is 65%.

7-55. Howie pushes a crate of mass m up a plank of length L into a truck whose platform is a vertical distance h above the road. His applied force F acts parallel to the plank.
   (a) How much work does Howie do in pushing the crate up the plank?
   (b) Calculate Howie's work input if the crate has a mass of 94 kg, the length of the plank is 5.1 m, the applied force is 490 N, and the platform is 1.2 m above the road.
   (c) What is the increase of potential energy of the crate from the road to the truck platform?
   (d) How much of the work that Howie does goes into overcoming friction?
   (e) What is the efficiency of the plank?

7-56. The pulley combination shown is used to lift a heavy block of weight W. Note that four strands of rope support the weight of the block.
   (a) If the block rises at constant speed, what is the tension in each strand?
   (b) How much pull is needed to raise the weight?
   (c) To raise the weight by a distance h, each of the four strands has to become shorter by a distance h. Through what distance must the pull be exerted to raise the weight by a distance h?
   (d) Write an expression for W_{output}.
   (e) Write an expression for W_{input}.
   (f) Ideally, how much pulling force should be required to lift a block of weight W?
   (g) Calculate the pull ideally required to lift a 240-N block.
   (h) If the actual pull required to lift the 240-N block is 130 N, determine the efficiency of the pulley.

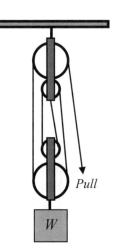

7-57. • The bucket shown hangs by a rope wrapped around an axle of radius r, attached to a larger wheel of radius R. The bucket is raised by pulling on a rope wrapped around the circumference of the wheel.

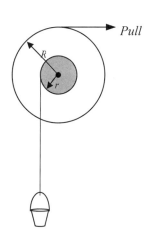

(a) If you pull on the rope and the wheel makes one complete turn, through what distance are you pulling? How much work will you be doing?
(b) Through what distance will the bucket be raised?
(c) How large is the output force compared to the input force?
(d) A wheel and axle system consists of a 3.0-cm radius axle mounted on an 18.0-cm radius wheel. How much larger is the output force than the input force?
(e) How is the wheel and axle system like a sort of circular lever?

7-58. Paul and Lil go to work out at the gym. They see two machines, A and B. For each machine, you pull down on the handle to lift the weight.

Lil goes to use machine A, which has 15 kg on it. Paul uses machine B, which has 20 kg on it.

(a) On which machine, A or B, do you have to apply more force to lift the weights?
(b) On which machine do you do more work if you pull down on the cable by 1 meter?

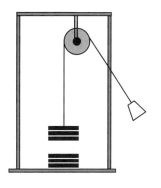
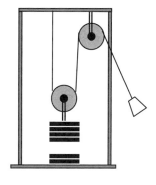

Machine A   Machine B

(c) On which machine do you do more work if you raise the weights by 1 meter?
(d) Lil, on machine A, pulls down her handle by ½ meter. How far will Paul, on machine B, have to pull down on his handle to do the same amount of work as Lil does?

Springs and Elastic-Potential-Energy Problems

7-59. In the lab you have a box of 1-kg cylinders that can be hooked on to one another. You suspend one cylinder from a strong spring and record the spring's stretch. You continue suspending additional cylinders, one at a time, keeping measurements of the stretch.
(a) How does the stretch change per added cylinder?
(b) Sketch what a graph of added weight vs. total stretch would look like for the spring.
(c) What does the slope of the graph indicate about the spring?

7-60. In the lab you find the spring constant k of a certain spring by applying a force F and measuring the resulting stretch, a distance x.
(a) Write an equation that relates the spring constant to F and x.
(b) Calculate the spring constant if the stretching force is 5.0 N and the stretch of the spring is 6.5 cm.

7-61. A force F compresses a spring that has a spring constant k.
  (a) How much is the resulting spring compression?
  (b) Calculate the compression distance if a 10-N force is used to compress a spring that has a spring constant 200 N/m.

7-62. A spring is compressed a distance x when a load of mass m is placed on it.
  (a) How much will the spring compress when supporting a load $4m$?
  (b) How much will it compress when supporting a load $8m$ (assuming that the spring doesn't reach its elastic limit)?
  (c) What is the spring constant k of the spring?
  (d) Calculate the answers to (a), (b), and (c) if $m = 2.00$ kg and x is 0.050 m.

7-63. • A truck of mass M hauls a trailer of mass m connected by a spring. The spring has a spring constant k. The truck accelerates from rest at a constant rate to speed v in time t while the stretched spring maintains a constant length.
  (a) What is the acceleration of the truck and trailer?
  (b) By how much is the spring stretched?
  (c) Does the mass of the truck matter? Defend your answer.
  (d) Calculate answers to (a) and (b) above for a 3200 kg truck pulling a 420 kg trailer from rest to 22 m/s in 15.0 seconds using a spring with $k = 3900$ N/m.

7-64. You hook one end of a spring with spring constant k to a wall. You stretch the other end a distance x and hook it to a frictionless cart of mass m. Then you release the cart.
  (a) What will be the initial acceleration of the cart?
  (b) As the cart accelerates, will the acceleration of the cart be constant, increasing, or decreasing? Explain.
  (c) What will be the *acceleration* of the cart when the spring has returned to its equilibrium, unstretched position?
  (d) Calculate the initial acceleration for a 0.50-kg cart connected to a spring with $k = 7.5$ N/m and that has been stretched by 40 cm.

7-65. An object of mass m is suspended from a spring of spring constant k that hangs from the ceiling of an elevator. How much does the spring stretch:
  (a) When the elevator is at rest?
  (b) When the elevator moves upward at a constant speed v?
  (c) When the elevator accelerates upward at a?
  (d) When the elevator accelerates downward at a?
  (e) When the elevator moves downward at a constant speed v?
  (f) Calculate the spring stretch in the case of a 4.5-kg object suspended from a spring having a spring constant $k = 750$ N/m in the elevator at rest.
  (g) Calculate the spring stretch for the same mass when the elevator accelerates upward at 0.50 m/s^2.
  (h) Calculate the spring stretch for the same mass when the elevator accelerates downward at 0.50 m/s^2.

7-66. A block of mass m is pulled across a frictionless horizontal surface by a spring having a spring constant k. The block undergoes an acceleration a.

  (a) What is the magnitude of the net force on the block?
  (b) By how much does the spring stretch?
  (c) Calculate the net force on the block and the stretch of the spring for a 1.5-kg block being pulled by a spring with $k = 15$ N/m with an acceleration of 0.30 m/s^2.

7-67. A block of mass m rests on a long horizontal air table (friction-free) and is attached to a practically massless horizontal spring. When the block is pulled horizontally from rest by the spring with the spring maintaining a constant stretch, the block reaches a speed v in time t.
   (a) How much force acts on the block?
   (b) By how much does the spring stretch?
   (c) Calculate the force on a 4.5-kg block and the stretch in the spring with $k = 84$ N/m if the block reaches a speed of 0.46 m/s from rest in 1.3 seconds.

7-68. • A spring of length L has a spring constant k. The spring is cut in half. A load of mass m is hung by one of the half-length vertical springs.
   (a) What is the tension in the spring of length L when supporting the load m?
   (b) What is the tension in a half-length spring supporting the same load? (*Hint*: It's okay that these questions are easy ones with easy answers.)
   (c) How much does the half-length spring stretch compared to the stretch of the whole-length spring supporting the same load?
   (d) What is the spring constant of each half-spring of length $L/2$?

7-69. A spring of spring constant k is stretched a distance x from its equilibrium position.
   (a) How much work is required to stretch the spring this distance?
   (b) Calculate the amount of work required if the spring constant is 45 N/m and the distance of stretch from the equilibrium position is 2.50 cm.

7-70. Timbo does a certain amount of work W to stretch a spring a distance x.
   (a) What is the spring constant?
   (b) Calculate the spring constant if it takes 4.0 J of work to stretch a spring 18 cm from its equilibrium position.

7-71. A spring with spring constant k is stretched a distance x and hooked onto a block of wood of mass m, which sits on the lab table. Then you release the block. The kinetic friction force between the block and the lab table is f.
   (a) What is the initial elastic PE stored in the stretched spring?
   (b) How much work does the friction force do on the block as the spring "relaxes" back to its equilibrium length?
   (c) How much KE will the block have when the spring returns to its equilibrium length?
   (d) What will be the speed of the block when the spring returns to its equilibrium length?
   (e) Calculate answers for (a) through (d) for a 0.44-kg block, a spring with $k = 12$ N/m, an initial stretch of 65 cm, and a constant kinetic friction force of 2.4 N.

7-72. Chandrika uses a spring to pull horizontally on a block of mass m that rests on a lab table. The spring constant k and the values of μ_k and μ_s are known quantities.
   (a) How much will the spring have to be stretched to get the block to start moving?
   (b) How much stretch is needed to keep the block moving at constant speed?
   (c) If the spring is then stretched by an amount X, more than the amount in part (b), find the acceleration of the moving block.
   (d) Evaluate each of the above for $k = 55$ N/m, $X = 20.0$ cm, $m = 3.2$ kg, $\mu_s = 0.50$, and $\mu_k = 0.30$.

7-73. • Bungee jumper Margaret of mass m steps off a bridge of height h above the ground. She is attached to a bungee cord of length L (less than h!) and spring constant k. One end of the cord is attached to the bridge and the other end to her.
   (a) What is the minimum value that k must have so that Margaret just avoids hitting the ground?
   (b) Calculate the minimum value of k for a 70.0-kg woman and a 32.0-m high bridge, with a 10.0-m long bungee cord.
   (c) Calculate Margaret's maximum acceleration.

7-74. • A carnival game consists of a spring of spring constant k that you compress to fire a small block of mass m up a frictionless ramp at an angle θ and length L. A small cup sits at ground level, directly against the base of the far edge of the ramp.

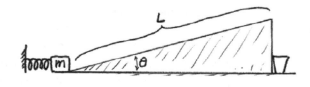

   (a) How far should you compress the spring so that the block will fall into the cup?
   (b) Calculate the amount of spring compression if $k = 10$ N/m, $m = 10$ grams, $L = 1.22$ m, and $\theta = 15°$.

7-75. • Place a block of mass m on top of a vertical spring of spring constant k mounted on the floor. Then push the block down a distance x from the spring's initial position.

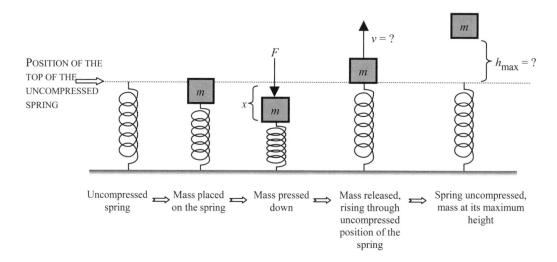

| Uncompressed spring | ⇒ Mass placed on the spring | ⇒ Mass pressed down | ⇒ Mass released, rising through uncompressed position of the spring | ⇒ Spring uncompressed, mass at its maximum height |

   (a) How much force must you exert to hold the block at rest on the compressed spring?
   (b) When you release the block, what is its speed when the spring has returned to its initial position?
   (c) How high above the initial position of the spring will the block rise?

Chapter 7 Energy

Show-That Problems for Work, Energy, and Power

7-76. Marshall applies a constant force of 20 N to a box and causes it to move at a constant speed of 4.0 m/s.
Show that the work he does in moving the box for 6.0 s is 480 J.

7-77. A 0.50-kg sphere at the top of an incline has a potential energy of 6.0 J relative to the base of the incline.
Show that when the sphere rolls halfway down the incline, its potential energy is 3.0 J.

7-78. A 20-N block falls freely from rest from a point 3.0 m above the surface of Earth.
Show that when the block has fallen 1.5 m, the gravitational potential energy of the block is 30 J relative to PE = 0 at the surface.

7-79. A 70-kg mountain climber with 35-kg of gear climbs from his base camp to the top of Mt. Everest, a vertical distance of approximately 3500 m.
Show that his change in PE is about 3.6×10^6 J (about the same as the amount of energy stored in three or four candy bars).

7-80. A cart moving at a constant speed of 25 m/s possesses 438 J of kinetic energy.
Show that the mass of the cart is 1.4 kg.

7-81. A steady force of friction acts on a 15-kg mass moving 10 m/s on a horizontal surface.
Show that if the mass is brought to rest over a distance of 12.5 m, the friction force is 60 N.

7-82. Mike's vintage sports car travels along a level road at 64 km/h, and skids to a stop with pre-antilock brakes.
Show that if its speed were 96 km/h, its skidding distance would be 2.25 times as much.

7-83. Mark pulls a 55-gram arrow back 71 cm on a bow that exerts an average force of 100 N on the arrow.
Show that the release speed is about 51 m/s.

7-84. A 1-kg projectile is fired at 10 m/s from a 10-kg launcher. The individual momenta of the projectile and the launcher have the same magnitude, but opposite directions.
Show that the KE of the projectile is ten times greater than the KE of the recoiling launcher.

7-85. The energy content of gasoline is about 3.5×10^7 J per liter, or 3.5×10^4 J per milliliter. Show that a perfectly efficient 1700-kg car facing no air resistance or friction would burn about 18 mL (less than an ounce) of gasoline in accelerating from rest to 27 m/s (about 60 mi/h).

7-86. Some of the newest hybrid cars have a regenerative braking scheme, so that when the brakes are applied, some of the KE of the car goes into charging batteries instead of into thermal energy in the brakes. Show that, with a 20% efficient system, a 1400-kg car moving at 24 m/s can provide 8.1×10^4 J of energy to the battery as the car slows to a stop.

7-87. A railroad boxcar traveling at speed v collides with and sticks to an identical boxcar initially at rest. The coupled cars then move together at speed $v/2$.
Show that the amount of energy converted to thermal energy in the collision is half the initial KE of the moving car.

7-88. A fisherman is fishing from a pier a height h above the water. He throws a fish that he caught back into the water, giving it an initial speed v_0.
    (a) Use conservation of energy to show that speed of the fish when it hits the water is
$$v_f = \sqrt{v_0^2 + 2gh}.$$
    (b) Why don't we have to worry about initial direction of the flying fish?

7-89. Romeo throws a package up to his love, Juliet, waiting on a balcony a height h above Romeo.
    (a) Show that the minimum speed at which Romeo can throw the package such that Juliet can catch it is $v_0 = \sqrt{2gh}$.
    (b) Show that if Romeo throws the package faster than this, it will have a speed $v_f = \sqrt{v_0^2 - 2gh}$ when Juliet catches it.

7-90. It takes work to compress the spring in a toy marble gun. The marble gun is laid on the floor, the spring is compressed though a distance x, and the marble is shot out and rolls up a ramp.

    (a) Show that the maximum height above the floor that the marble can reach is $h = \dfrac{kx^2}{2mg}$.
    (b) Show that the maximum height above the floor that the marble achieves isn't affected by the steepness of the ramp.

7-91. John pushes with a horizontal 120-N force on a 25-kg box to accelerate the box from rest along a horizontal surface. The friction force acting on the box is 90 N. Show that the box will have a speed of 5.4 m/s after it has been moved 12 m.

7-92. Sachi rolls a donut off the end of the kitchen counter. The donut leaves the counter with a speed of 2.5 m/s. Show that the donut has a speed of 5.1 m/s when it hits the floor 1.0 m below.

7-93. A pendulum is released from a distance h_0 above the tabletop. As it swings past its lowest point, it is a distance h_f above the tabletop. Show that the maximum speed of the pendulum is $\sqrt{2g(h_0 - h_f)}$.

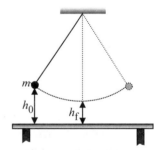

7-94. • A golf ball is hurled at and bounces from a very massive bowling ball that is initially at rest on a frictionless surface. Show that, after the collision, the golf ball has about the same kinetic energy, and that the bowling ball has nearly twice the initial momentum of the golf ball.

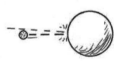

7-95. A kilowatt-hour is a unit of energy equivalent to work being done at the rate of 1 kilowatt for an hour. Show that 1 kW·h = 3,600,000 J.

7-96. An adult typically "burns" 2000 or more Calories a day, where 1 Calorie = 4.18 kJ. Show that this is an average power consumption of about 100 W.

7-97. One horsepower is equivalent to 746 watts. Show that a car engine that is purring along and putting out 40 hp is putting out 30 kW.

7-98. Dan finds that 700 watts of power is needed to keep his boat moving through the water at a constant speed of 10 m/s.
Show that the magnitude of the force exerted by the water on the boat is 70 N.

7-99. A rowboat oar is shown in the diagram.
    (a) Show that the oar produces an output force that is 1/3 as large as the input pull.
    (b) Explain why you'd want to use this device that *reduces* your input force.

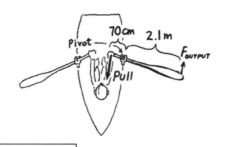

USE THE DIAGRAMS BELOW FOR PROBLEMS 7-100 THROUGH 7-104.

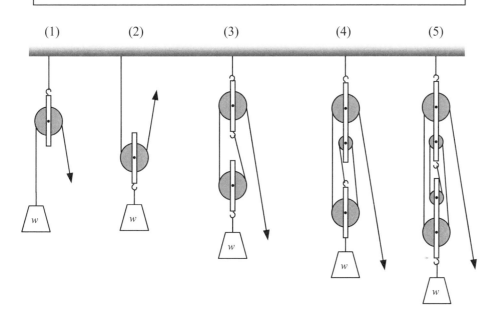

7-100. For pulley (1), show that the input and output forces are the same.

7-101. For pulley (2), show that you have to pull up 80 cm of rope to make the weight rise 40 cm.

7-102. For pulley (3), show that the output force is twice the input force, but the output distance is only half the input distance.

7-103. For pulley (4), suppose that the weight is 100 N and that a 40-N pull is required to lift the weight. Show that pulley has an efficiency of 0.83.

7-104. Suppose that pulley (5) has an efficiency of 0.70. Show that, in order to lift a 1400-N crate by 1.2 m, you'd have to pull down 4.8 m of rope with force of 500 N.

7-105. You hang a 550-gram mass on a spring and the spring stretches by 27 cm. Show that the spring constant is 20 N/m.

7-106. You have a spring with spring constant 25 N/m. Show that a 16-N force will stretch the spring 64 cm.

7-107. You decide you want to make a spring scale using a centimeter-ruler as for your scale, so a 1-N force would stretch the spring 1 cm, a 2-N force would stretch the spring 2 cm, etc. Show that you need a spring with a force constant $k = 100$ N/m.

7-108. You have to do 15 J of work to stretch a spring by 12 cm. Show that the spring constant is approximately 2100 N/m.

7-109. You have a spring with spring constant $k = 7.2$ N/m, and you stretch the spring by 23 cm. Show that the spring stores 0.19 J of elastic potential energy.

7-110. You stretch a spring with $k = 3.5$ N/m by 30 cm, attach it to a 400-g cart, and then release the cart. Show that the speed of the cart when the spring again reaches its equilibrium position is 89 cm/s.

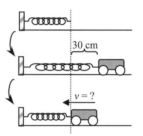

7-111. You pull a 350-gram block of wood across the table at constant speed with a spring of spring constant 4.0 N/m. The spring is stretched by 30 cm. Show that the coefficient of kinetic friction between the block and the table is 0.35.

7-112. Sue launches an 8.0-kg stone from a makeshift slingshot. The rubber bands making up the slingshot are stretched 2.2 m, and the boulder is launched at a near-horizontal speed of 12.0 m/s. Show that the spring constant of the combined rubber bands is 240 N/m.

8 Rotational Motion

In this chapter we distinguish between linear and rotational speeds. Linear speed v, as covered in Chapter 3, is motion measured in units of meters/second. When motion follows a curved path, we speak of tangential velocity, measured tangent or parallel to the curve at a particular point, which also has units meters/second. The curved paths treated here are either circles or arcs of circles.

Rotational speed ω is measured in RPM, rotations per minute. Or it can be rotations per second, rotations per month, or some number of complete rotations in a given time. Most physics texts treat it as some angle per unit of time.

Imagine that you are sitting on a rotating platform. The faster it spins, the greater distance you travel each second. Put another way, the greater its rotational speed, the greater your tangential speed:

$$v \sim \omega.$$

Sitting farther away from the axis of the rotating platform is another way to cover more distance in the same amount of time:

$$v \sim r\omega.$$

Combining both of these ideas, we say

$$v \sim r\omega.$$

We can use the exact equation $v = r\omega$ only if we use SI units for measuring ω. The SI unit for ω is radians/second, where 1 radian is an angle approximately equal to 57°. In keeping with the spirit of the textbook chapter, we've chosen not to include problems where the use of radians is necessary for a solution.

We say that an object following a circular path is accelerating because its velocity is changing continually—perhaps not in magnitude, but certainly in direction. We call the acceleration toward the center of a circle *centripetal* acceleration. It is defined by

$$a_{\text{centripetal}} = \frac{v^2}{r}.^*$$

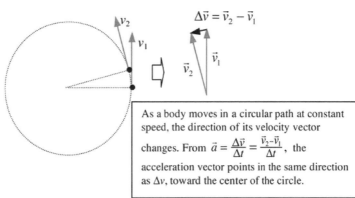

As a body moves in a circular path at constant speed, the direction of its velocity vector changes. From $\vec{a} = \frac{\Delta \vec{v}}{\Delta t} = \frac{\vec{v}_2 - \vec{v}_1}{\Delta t}$, the acceleration vector points in the same direction as Δv, toward the center of the circle.

The force producing the circular motion in any given situation is called the *centripetal force* and has magnitude $ma_{\text{centripetal}}$, or more specifically,

$$F_{\text{centripetal}} = m\frac{v^2}{r}.$$

Whenever something moves in a circular path there must be some force pushing or pulling it towards the center of the circle—otherwise it would fly off along a straight-line tangential path. In a problem involving rotational motion, first identify the physical force that pushes or pulls the rotating object towards the center, then set it equal in magnitude to mv^2/r, and solve. When there is an angle between the force and the radial direction (as in a conical swing, for example), equate the component of that force along the radial direction to mv^2/r and solve. Sometimes there is more than one force in the radial direction, in which case you set the *net* force to mv^2/r. Voila!

---

*Equivalently, $a_{\text{centripetal}} = r\omega^2$ (again, with ω in radians per second). In this book we stick with v^2/r.

© Paul G. Hewitt and Phillip R. Wolf

Sample Problem 1

Earth, with radius R_E, rotates once every 24 hours.

(a) Write an equation for the tangential speed of a point on Earth's equator.

Focus: $v = ?$ From $v = \dfrac{\text{distance}}{\text{time}} \Rightarrow \dfrac{2\pi R_E}{t}$.

(b) Calculate the tangential speed of a point on Earth's equator in km/h and m/s. ($R_E = 6370$ km.)

Solution: $v = \dfrac{2\pi R_E}{t} = \dfrac{2\pi(6370 \text{ km})}{24 \text{ h}} = 1670 \dfrac{\text{km}}{\text{h}} \times \dfrac{1000 \text{ m}}{1 \text{ km}} \times \dfrac{1 \text{ h}}{3600 \text{ s}} = 464 \dfrac{\text{m}}{\text{s}}$.

(c) Bob, mass $m = 81$ kg, stands at rest at the equator. Taking Earth's rotation into account, what is the net force on Bob?

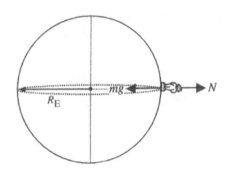

Focus: Since Bob moves in a circle, there is a centripetal force acting on him. From the force diagram at the right, we can see that the gravitational force on Bob must be a little bit larger than the normal force for there to be a net force pulling inward toward the center of his circular path:

$$F_{\text{net}} = mg - N = \dfrac{mv^2}{r} = \dfrac{m\left(\dfrac{2\pi R_E}{t}\right)^2}{R_E} = \dfrac{4\pi^2 m R_E}{t^2} = \dfrac{4\pi^2 (81 \text{ kg})\left(6370 \text{ km} \times \dfrac{1000 \text{ m}}{1 \text{ km}}\right)}{\left(24 \text{ h} \times \dfrac{3600 \text{ s}}{1 \text{ h}}\right)^2} = 2.7 \text{ N}.$$

Usually, problems set at Earth's surface are treated as an inertial (unaccelerated) frame of reference because the effects of rotation are usually small enough to be ignored. We can see that, compared to Bob's weight, the centripetal force is $\dfrac{2.7 \text{ N}}{(81 \text{ kg})(9.8 \text{ m/s}^2)} = 0.0034$, about a third of a percent.

(d) How short would the day have to be for Bob to lose contact with Earth's surface?

Answer: As Earth spins faster, the required centripetal force to keep Bob spinning with it gets larger and the normal force gets smaller. At some critical speed, the normal force goes to zero and Bob loses contact with Earth's surface. From

$$F_{\text{net}} = mg - N = \dfrac{mv^2}{r}, \text{ with } N = 0 \Rightarrow mg = \dfrac{mv^2}{r} = \dfrac{4\pi^2 m R_E}{t^2}$$

$$\Rightarrow t = 2\pi \sqrt{\dfrac{R_E}{g}} = 2\pi \sqrt{\dfrac{\left(6370 \text{ km} \times \dfrac{1000 \text{ m}}{1 \text{ km}}\right)}{9.8 \dfrac{\text{m}}{\text{s}^2}}} = 5056 \text{ s} \times \dfrac{1 \text{ h}}{3600 \text{ s}} = 1.41 \text{ h}.$$

This is $\dfrac{24 \text{ h}}{1.41 \text{ h}} = 17$ times faster than Earth's present rotation rate.

Sample Problem 2

Suppose you attach a ball of mass m to a string and whirl it overhead in a horizontal circular path. If you keep the ball moving with constant linear speed v and shorten the string, tension in the string increases. The string will break if you exceed a critical tension T.

(a) Derive an equation for the shortest length of string you can use so the string doesn't break. (Make the approximation that the string remains horizontal.)

Solution: Step 1. *Focus*: Length of string $L = ?$

Step 2. Since the ball is traveling in a circular path, there must be some force pulling the ball toward the center of the circle. String tension T provides the centripetal force. Recognizing that T acts as $F_{centripetal}$ and that the radius of the circle is L,

$$F_c = \frac{mv^2}{r} \text{ becomes } T = \frac{mv^2}{L}.$$

Step 3. From $T = \frac{mv^2}{L} \Rightarrow L = \frac{mv^2}{T}$.

If we plug in the *maximum* value T can have, this gives us the *minimum* value L can have for a given mass and speed.

(b) Calculate the shortest length of the string that would keep a 250-gram ball circling at a constant speed of 6.0 m/s if the string's breaking strength is 45 N.

Solution: First convert to consistent SI units: $250 \text{ g} \times \frac{1 \text{ kg}}{1000 \text{ g}} = 0.25 \text{ kg}$.

$$\text{Then } L = \frac{mv^2}{T} = \frac{(0.25 \text{ kg})(6.0 \frac{\text{m}}{\text{s}})^2}{45 \text{ N}} = 0.20 \frac{\text{kg} \cdot \frac{\text{m}^2}{\text{s}^2}}{\text{kg} \cdot \frac{\text{m}}{\text{s}}} = \mathbf{0.20 \text{ m}}.$$

(c) The string could remain horizontal if the ball were on a horizontal surface, but not if it's moving in a horizontal circle in the air. Why? Make a rough sketch to support your answer.

Answer: When a ball moves in a horizontal circle, the net force on the ball in the vertical direction must be zero, so mg must be countered by an equal upward force. On a horizontal tabletop, this would be the normal force. Without the table there is no normal force and the upward force must be supplied by a component of the string tension. Therefore, the string must make an angle with the horizontal as shown in the diagram.

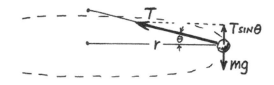

Sample Problem 3

A flyer can experience weightlessness at the top of a vertical loop-the-loop by flying at just the right speed.

(a) Derive an equation for the critical speed at the top of a vertical loop of radius *r* to make the pilot feel weightless.

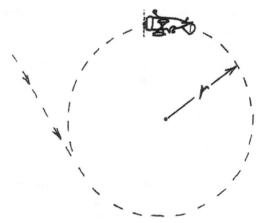

Solution: Step 1. *Focus*: $v = ?$

Step 2. The pilot is in circular motion, so this is a centripetal-force problem.

From $F = \dfrac{mv^2}{r} \Rightarrow v = \sqrt{\dfrac{Fr}{m}}$.

Two principal forces act on a flyer while executing a vertical loop—gravitational force mg and the normal force N from the seat pushing on the pilot's body. At the top of the loop, both forces are in the radial direction (downward, in this case). Together they provide the centripetal force.

The net force $F_{net} = mg + N = \dfrac{mv^2}{r}$.

When the pilot experiences weightlessness, the normal force is zero ($N = 0$), so the net force $F_{net} = mg$.

Step 3. $v = \sqrt{\dfrac{Fr}{m}} = \sqrt{\dfrac{mgr}{m}} = \sqrt{gr}$.

(b) Calculate the critical speed for achieving weightlessness at the top of a vertical loop of radius 200 m.

Solution: $v = \sqrt{gr} = \sqrt{\left(9.8 \dfrac{m}{s^2}\right)(200 \text{ m})}$ = **44 m/s**.

(c) If the plane makes the entire vertical loop at a constant speed, where along the loop will the pilot feel heaviest? Defend your answer with a sketch showing relative magnitudes of *N* and *mg*.

Answer: Normal force N is a measure of how heavy one feels. **The pilot feels heaviest at the bottom of the loop** because N there is a maximum. Whereas mg alone provides the centripetal force at the top when N is zero, the centripetal force at the bottom of the loop is provided by $N - mg$. The sketch indicates that N at the bottom must be $2\,mg$ if N at the top is zero and speed is the same in both locations. If the speed is greater at the bottom, N will be even larger. (At all points along the loop, only mg and N act on the pilot. At locations other than the top and bottom, the centripetal force is provided by the components of N and mg that lie along the radial direction—another story.)

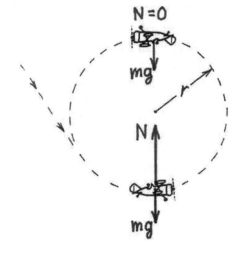

Torque

In Chapter 2 we learned the equilibrium rule: $\Sigma F = 0$. Now we complement the rule with the added condition $\Sigma \tau = 0$ (sum of the torques = 0), or equivalently $\Sigma \tau_{clockwise} = \Sigma \tau_{counterclockwise}$. (The sum of the clockwise torques = the sum of the counterclockwise torques.) In its simplest form, $\tau =$ lever arm × force. When the force and the lever arm are not mutually perpendicular, trigonometry comes into the picture.

Consider the following three cases involving a heavy, hinged beam. The hinge is attached to the wall. In each case, the weight of the bar exerts a torque, tending to make the bar rotate clockwise. A second force F is exerted on the end of the bar to attempt to balance the bar.

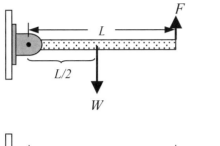

Case 1: *All* of the force F is exerted perpendicular to the lever arm. The torque due to F is FL.

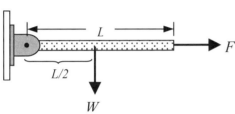

Case 2: *None* of the force F is exerted perpendicular to the lever arm. The torque due to F is zero. (In this case, the beam would start to rotate clockwise due only to W.)

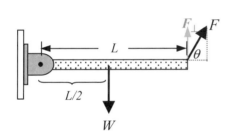

Case 3: The *component* of F exerted perpendicular to the lever arm will cause a torque on the beam. This component has a magnitude $F\sin\theta$, so the torque due to F is $(F\sin\theta)L$.

Notice that in Case 2, the force F doesn't tend to produce any rotation of the bar because none of F acts in a direction perpendicular to the bar. Notice that in Case 3, we take the component of F perpendicular to the lever arm because only this perpendicular component acts to produce rotation in the bar.[*]

So our more precise definition of torque is

$\tau =$ lever arm × component of force perpendicular to the lever arm.

---

[*] We would arrive at the same result if we instead multiplied F by the component of the lever arm that is perpendicular to the line of action of F. (Examples in the textbook illustrate this latter method.) Both methods are equivalent.

Sample Problem 4

A bar with mass m and length L is hinged at one end. A cable is attached to the other end of the bar and to the wall, making an angle θ with the wall as shown.

(a) Write an equation for the tension in the cable.

Focus: Cable tension $T = ?$

Since the bar is at rest, the net torque acting on it must be zero. Clockwise torque about the hinge is provided by the weight of the bar, which we assume acts at the center of the bar. Counter clockwise torque is provided by the tension in the cable. The two torques added algebraically equal zero.

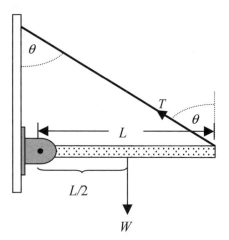

The clockwise torque due to the weight of the bar is simple enough—simply $mg \times (L/2)$. The counterclockwise torque due to cable tension acts at a distance L from the hinge and is provided by the component of T acting perpendicular to the bar, $T\cos\theta$.

So $T\cos\theta\, L - W\left(\dfrac{L}{2}\right) = 0 \Rightarrow T = \dfrac{mg}{2\cos\theta}$.

(b) Suppose the mass of the bar is 1.3 kg and the bar is 1.2 m long. Calculate the tension in the cable.

Solution: Cable tension $T = \dfrac{(1.3 \text{ kg})(9.8 \text{ m/s}^2)}{2\cos 58°} = 12$ N.

Sample Problem 5

A small space pod at the end of a tether line of length L moves at speed v_0 in a circular path about a massive space station.

(a) What will be the speed of the pod if the length of the line is reduced to $0.25\,L$?[*]

Focus: $v = ?$

The only force acting on the pod is the tension in the tether. The tension force is directed radially inward, toward the center of the pod's circular path. Hence the tension does not exert a torque on the pod, which means the angular momentum of the pod is conserved.

From (Angular momentum)$_\text{initial}$ = (Angular momentum)$_\text{final}$

$\Rightarrow mv_0 L = mv_\text{new} r_\text{new} = mv_\text{new}(0.25L) \Rightarrow v_\text{new} = \dfrac{v_0}{0.25} = 4v_0$.

(b) If the initial speed of the space pod is 1.0 m/s, what is the speed when pulled inward one-fourth of its original distance from the space station?

Solution: $v_\text{new} = 4(1.0 \text{ m/s}) = \mathbf{4.0}$ **m/s**. The pod gains four times as much speed.

---

[*] Strictly speaking, the rotating system is the space station + pod, where each of them revolves around the system's center of mass. For a station that normally is much more massive than the pod, pulling the pod inward toward the station does not appreciably move the station. For a tether line that is very long compared with the radius of the space station, the length of the tether is essentially the radius of the pod's path. Under these circumstances (massive station and long tether) it is reasonable to treat the pod as the only element in the system and to treat the length of the tether as the radius of the path. More complex solutions are generally beyond the aim of a *Conceptual Physics* course.

Problems involving trigonometry are toward the end of this chapter.

Circular Motion Problems

8-1. Little Megan rides on a horizontal rotating platform of radius r at an amusement park and feels the air passing through her hair as she moves at speed v. She rides at a location one-third the way from the center to the outer edge.
   (a) If the rotation rate of the platform remains constant, what will be her linear speed if she moves to the outer edge?
   (b) Calculate Megan's speed at the outer edge if she is moving at 1.0 m/s when she is one-third the radius from the center.

8-2. Manuel rides at speed v while in the seat of a rotating Ferris wheel. Horsing around, Manuel decides to climb down a spoke of the wheel. He climbs one-quarter the distance toward the center, with people below astounded.
   (a) What is his linear speed when located three-quarters the distance from the rotational axis?
   (b) What would be his linear speed at the central axis?

8-3. Joshua travels in his race car at a steady speed v around a circular track of radius r.
   (a) Write an equation for the time it takes to make a complete circle.
   (b) Calculate the time required to go around a track of radius 134 m at a speed of 60 m/s.

8-4. Chelcie races once around a circular track at a speed v in t seconds.
   (a) Write an equation for the radius of the track.
   (b) Calculate the radius for a speed of 5.5 m/s and a time of 73 s to complete one circle.

8-5. It takes a planet a time T to orbit the Sun once a year in an essentially circular orbit of radius R.
   (a) Write an equation for the planet's average orbital speed.
   (b) Earth takes 1 year to orbit the Sun at an average Earth-Sun distance of 1.50×10^{11} m. Calculate Earth's average orbital speed.

8-6. A car moves at speed v. Relative to the axle, the tangential speed of any point on the outer surface of the tire is also v.
   (a) Write an equation for the number of rotations per second made by a car tire of diameter D when the car moves at speed v.
   (b) Calculate the number of rotations per second for 60-cm diameter tires on a car driven at 108 km/h.

8-7. A ceiling fan rotates at n rpm (revolutions per minute). The tips of the fan are distance r from the axis.
   (a) Write an equation for the linear speed of the tips of the ceiling fan.
   (b) Calculate the linear speed of the tips if they are 0.70 m long and rotate at 15 rpm.

Rotational Inertia Problems

Rotational Inertias of Various Objects			
Point Mass	**Thin Hoop**	**Cylinder**	**Sphere**
$I = mr^2$	$I = mr^2$	$I = \frac{1}{2}mr^2$	$I = \frac{2}{5}mr^2$

8-8. Consider three objects: Object A is a hoop of mass m and radius r; Object B is a sphere of mass $2m$ and radius r; Object C is a disk of mass $3m$ and radius r.
(a) Which has the greatest rotational inertia?
(b) Which has the smallest rotational inertia?

8-9. The moment of inertia of every round object can be expressed as a constant $\times mr^2$. The acceleration of an object rolling down an incline is related to the magnitude of this constant. An object with a greater constant undergoes a smaller acceleration.
Consider the three objects of the previous problem, released at the same time from the top of an inclined plane.
(a) Which will get to the bottom in the shortest time?
(b) Which will take the longest time to get to the bottom?

8-10. When a block, initially at rest, slides down an incline of negligible friction, the speed at the bottom is $\sqrt{2gh}$ (as found when PE at the top transfers to KE at the bottom). All blocks starting from rest, regardless of mass, will have equal accelerations on an inclined, friction-free plane. Now consider a race between a block on a friction-free incline and a bowling ball on an incline that is identical, except that there is enough static friction to enable rolling. A ball rolling down an incline has both "translational" KE, covered in Chapter 7, and *rotational KE*, a topic not covered in the textbook.
(a) Without knowing the details of rotational KE, which will win the race: the block or the bowling ball?
(b) Which would win the race if both were on friction-free, same-angle planes (so that the ball would slide rather than roll)?

8-11. A playground carousel is free to rotate about its center on frictionless bearings. The carousel has a mass M, radius r, and is at rest. Assume that the carousel is simply a large disk.
(a) Write an equation for the rotational inertia of the carousel.
(b) Write an equation for the rotational inertia of a man of mass m sitting on the edge of the same rotating carousel.
(c) It so happens that the rotational inertias of bodies add when they share the same center of rotation. Write an equation for the combined rotational inertia of the man and carousel when the man sits at the edge of the carousel.
(d) Calculate the combined rotational inertia if the diameter of the carousel is 4.0 m, its mass is 50 kg, and the mass of the man is 70 kg.

Torque Problems

8-12. Mary Beth uses a torque feeler that consists of a meterstick held at the 0-cm end with forces applied to various positions along the stick. (See Figure 8.17 in your textbook.) In Trial A, force F pulls down at the 100-cm mark. In Trial B, force $2F$ pulls down at the 50-cm mark. In Trial C, $3F$ pulls down at the 40-cm mark. (Neglect the weight of the meterstick.)
(a) Which trial produces the greatest torque?
(b) Which trial produces the least torque?
(c) In which two trials are the torques equal?

8-13. James uses a wrench to turn a stubborn bolt.
(a) How much would the torque on the bolt increase if the applied force were doubled?
(b) How much would the torque increase if a pipe were inserted over the wrench to make the lever arm 3 times as long (assuming the same force)?

8-14. Jose pushes perpendicular to a door with force F at the door's far edge. Marie pushes the door from the opposite side and the door doesn't turn. Her push is perpendicular to the door, but at a point halfway from the hinge to the edge. (A top view of the door is seen in the sketch.)
(a) How hard does she push on the door?
(b) Calculate Marie's push if Jose pushes with a force of 40 N.

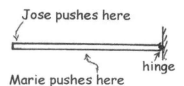

8-15. Marcus of mass M and Juanita of mass m balance each other on opposite sides of a seesaw. The fulcrum is located at the middle of the seesaw.
(a) If Marcus sits at distance D from the fulcrum, how far from the fulcrum does Juanita sit?
(b) If Marcus weighs 360 N and sits 1.25 m from the fulcrum, where does 315-N Juanita sit?

8-16. A boy and girl sit on a seesaw of length L, balanced at its center. The girl sits at the far end. The boy is twice as heavy as the girl, and therefore sits midway between his end and the center. Then the girl is given a bag of oranges weighing W, and the seesaw rotates out of balance. When the boy is given a bag of apples, balance is restored.
(a) What is the weight of the apples compared with the weight of the oranges?
(b) Calculate the weight of the bag of apples if the bag of oranges weighs 50 N.

8-17. Continue with the above problem with the girl of weight W_g, the boy with double her weight, and the bag of oranges with weight W.
(a) If the boy didn't have a bag of apples to increase his weight, how much farther from the center should he sit to restore balance?
(b) Calculate how far the boy must move for a 200-N girl given a 50-N bag of oranges on a seesaw 5.0 m long.

8-18. Karen wishes to determine the mass of a meterstick. She suspends the meterstick by a string tied around the 20-cm mark of the meterstick. She finds that it takes a mass m tied at the 5-cm mark to balance it.
(a) What is the mass of the meterstick?
(b) Calculate the meterstick's mass if it were balanced by a 140-gram mass.

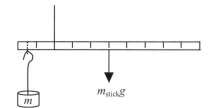

8-19. Muriel sees this mobile hanging from the ceiling. It is nicely constructed with string and two very stiff, lightweight wires.
(a) Compared with the central "Sun" of mass m, how do the masses m_1 and m_2 compare?
(b) How much tension is in the string holding the mobile to the ceiling?

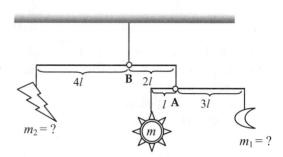

8-20. Consider two identical metersticks tied together with a loose piece of string attached to their ends. If you hold one stick horizontal, the other dangles vertically, as shown. You place your finger at an appropriate location beneath the horizontal stick so the two sticks will balance.
(a) Where along the horizontal stick should you place your finger so that the metersticks will balance?
(b) If you did the same with one horizontal meterstick and two vertical metersticks dangling from its one end, where would you place your finger to balance the system?

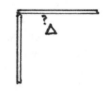

8-21. A uniform heavy plank of mass m lies on a raised horizontal surface, with practically half of the board hanging over the edge. Suppose you put your pet cat of mass $0.2\,m$ on the end of the overhung plank (with a safety net below). Of course the plank would topple, unless it is moved to the left.

(a) What percentage of the plank can overhang without toppling, with the cat at its end?
(b) If the cat were twice as heavy, what would be the maximum percentage of overhang distance?

8-22. Lydia has a mass m and stands at the end of a uniform plank of length L and mass M that overhangs the top of a building as shown. The maximum distance of overhang for tipping to not occur is one-eighth the length of the plank.
(a) How does Lydia's mass compare with the mass of the plank?
(b) Calculate the mass of the plank if Lydia's mass is 45 kg.

Centripetal Force Problems

8-23. Consider three cases of whirling a puck in a circular path at the end of a string on a friction-free air table. For Case A, the linear speed of the puck is v and the length of the string is L. For Case B, the linear speed of the puck is $0.5\,v$ and the length of the string is $2L$. For Case C, the linear speed of the puck is $1.5\,v$ and the length of the string is $1.5L$.
(a) In which case is tension in the string greatest?
(b) In which case is tension in the string least?
(c) Are there two cases where the tension is the same?

8-24. A toy electric train of mass m, moving at constant speed, rounds a circular track of diameter D once in time t.
(a) Derive an equation for the centripetal force provided by the tracks.
(b) Calculate the centripetal force acting on a 0.80-kg train rounding a 2.0-m diameter track every 12 s.

8-25. A centripetal force is needed to keep a block of mass m moving on a smooth floor in a circular path of radius r. The block makes a complete revolution in time t.
(a) Derive an equation for the centripetal force.
(b) Calculate the centripetal force needed if the mass is 2.0 kg, the radial distance is 1.0 m, and each revolution takes 5.0 s.

8-26. A model airplane of mass m flies in a horizontal circle of diameter D and completes each round-trip in time t.
(a) Write an equation for its speed.
(b) Calculate the speed if the plane has a mass of 0.80 kg, the diameter of the circular path is 5.0 m, and the time to make a revolution is 2.0 s.
(c) Calculate the centripetal acceleration of the plane.
(d) Calculate the centripetal force that acts on the plane.

8-27. An air puck of mass m is attached to a string of length L and moves in a horizontal circle.
(a) Derive an equation for the speed the puck must move so that the centripetal acceleration is equal to the acceleration of gravity g.
(b) Calculate the puck's speed for a 0.60-m long string and a 0.20-kg puck.

8-28. A ball revolves in a horizontal circle at speed v at the end of a string of length L. The tension in the string is T. (Assume the string is approximately horizontal.)
(a) Derive an equation for the mass of the ball.
(b) Calculate the mass when the string is 1.1 meters long, the speed is 7.9 m/s, and the tension is 25 N.

8-29. A ball of mass m is whirled in a horizontal circle by a rope attached to a spring having a spring constant k. The length of rope including the stretched spring is L and the speed of the ball is v.
(a) Write an equation for the magnitude of the centripetal force, assuming the string is practically horizontal.
(b) Derive an equation for the distance the spring is stretched.
(c) Calculate the magnitude of the centripetal force and the stretch in the spring with $k = 720$ N/m for a 0.36 kg ball being whirled at 7.0 m/s in a horizontal circle of radius 0.49 m.

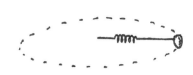

8-30. A car of mass m is driven at constant speed around a circular track of radius R, completing an entire trip around the track in time t.
   (a) What is providing the centripetal force keeping the car in a circular path?
   (b) Write an equation for the tangential speed of the car.
   (c) Write an equation for the magnitude of the friction force needed between the tires and the road.
   (d) Calculate the speed and the friction force acting on a 1400-kg car that makes a complete circuit of an 81-m radius track in 30 s.

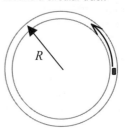
Top view of car driven around a circular track

8-31. At the equator (radius 6.37×10^6 m) you have a tangential speed due to Earth's daily rotation.
   (a) Suppose the spin rate increased. How large would your tangential speed have to be to lose contact with Earth's surface?
   (b) Why would an increased spin of Earth not throw people off Earth's poles?

8-32. A spherical asteroid of radius R has surface gravity g_A. A probe of mass m lands on the asteroid's surface.
   (a) Derive an equation for the rotational period of the asteroid such that a probe on the equator feels half the weight of a probe at the poles.
   (b) Calculate the period of a 10.0-km radius asteroid with surface gravitational acceleration of 0.016 m/s^2.

8-33. A pellet-filled tin can is tied to the end of a cable of length L. The tension in the cable is T when the can revolves around a center pole in a horizontal circle at a steady speed v.
   (a) Derive an equation for the mass of the can.
   (b) When the can moves at 15 m/s at the end of a 2.0-m long cable, the tension in the cable is 195 N. Calculate the mass of the can.

8-34. A centripetal force acts on an airplane of mass m flying at speed v in a horizontal circle of radius r.
   (a) Find the centripetal force divided by the mass. (This is the force per unit mass.)
   (b) Calculate the centripetal force per kilogram when the speed of the airplane is 200.0 m/s and the radius of its path is 12,500 m.

8-35. An electric force holds an electron of mass m in a circular orbit at a uniform speed v, and orbital radius r.
   (a) Write an equation for the electrical force holding the electron in orbit.
   (b) Calculate the force, given that the mass of an electron is 9.11×10^{-31} kg, its orbital radius is 5.3×10^{-11} m, and it whirls about the nucleus at speed 2.2×10^6 m/s.

8-36. Marshall swings a pail of water in a vertical circle of radius r.
   (a) Derive an equation for the minimum speed of the pail at the top of its path so that no water spills out.
   (b) Why does your answer not depend on the mass of the water?

8-37. • You are asked to design a space station of radius R that will spin fast enough that a resident standing within it at the outer circumference of the station would experience Earth-normal gravity. (See Figure 8.48 in the textbook.)
  (a) Derive an equation for the required rotational speed of the space station (in rotations-per-minute).
  (b) Calculate the rotational speed for a space station with a 1.0-km radius.

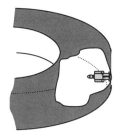

8-38. A crate of mass m rests on the bed of a flatbed truck that moves at speed v around an unbanked curve of radius r.
  (a) What coefficient of static friction between the crate and the flatbed is required to barely hold the crate from slipping off the flatbed?
  (b) Calculate the coefficient of friction needed to hold the 100-kg crate when the speed of the truck is a constant 14.0 m/s and the radius of the curve is 60.0 m.

8-39. • Investigators find that a drop of water of mass m will barely adhere to a piece of cloth in a washing machine drum when it rotates at 1200 rpm.
  (a) Find the force of adhesion between the drop and the cloth when circling at radius r.
  (b) Calculate the force of adhesion if the mass of the drop is 5×10^{-5} kg and the radial distance of the cloth from the axis of the drum is 0.40 m.

8-40. • A cotton-fiber string attaches a ball of mass m to a rotating device. The string can exert sufficient force to keep the ball circling at a radius r when whirled by the device. The string snaps if the motion is increased.
  (a) What is the minimum radius at which a ball of one-tenth mass m may be whirled by the same string at the same rotational speed in the same device without having the string break? (Assume that the string can be approximated as horizontal.)
  (b) Calculate the radius if it were initially 2.10 m for a mass of 2.0 kg.

8-41. A model airplane flies in a horizontal circle at the end of a string that, at most, can supply tension T. The mass of the airplane is m and its speed is v. Assume that the speed is great enough so that the string is nearly parallel to the ground.
  (a) Derive an equation for the smallest radius r at which the plane can be flown before the string breaks.
  (b) Calculate the radial distance if the plane's mass is 0.70 kg, its speed is 30 m/s, and maximum tension is 175 N.

8-42. A penny sits on the outer edge of a phonograph turntable of diameter D. The penny moves with a speed v.
  (a) Derive an equation for the minimum coefficient of static friction μ_s that will keep the penny from flying off the turntable.
  (b) Calculate μ_s if the speed of the penny is 0.75 m/s and the diameter of the turntable is 0.32 m.

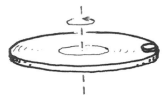

8-43. In a typical lab experiment, a rubber stopper of mass m is tied to a string and whirled around in a horizontal circle. The string is fed through a smooth pipe and tied to a hanging mass M. The whirling speed and radius are adjusted until the hanging mass remains stationary. It is found that it takes t seconds for the stopper to make n revolutions of radius r.

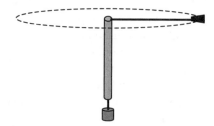

(a) Write an equation for the speed of the stopper.
(b) What force is supplying the centripetal force on the stopper?
(c) Derive an equation for the mass of the hanging mass.
(d) Calculate the hanging mass needed to keep a 14.6-gram stopper moving in a 53-cm radius circle if the stopper takes 62 seconds to make 100 revolutions.
(e) The stopper is going around in a circle at constant speed. Is it accelerating? Is there a net force on the stopper? Explain.

8-44. A small electric car of mass m moves at a constant speed v in a vertical circle on the track shown. The radius of the track is r.

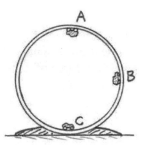

(a) Write an equation for the normal force acting on the car at A, the highest point.
(b) Write an equation for the normal force acting on the car at C, the lowest point.
(c) Write an equation for the normal force acting on the car at B, the 3 o'clock position halfway up the track.

8-45. A bowling ball of mass m is attached to the ceiling by a rope of length L. The ball is pulled to the side and released, swinging to and fro. At the bottom of the swing, the ball has speed v.

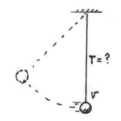

(a) Write an equation for the tension in the rope at the bottom of its swing.
(b) Calculate the rope tension at the bottom of the swing for a 7.3-kg ball, a 2.0-m long rope, with the speed of the ball at the bottom of 4.3 m/s.

Angular Momentum Problems

8-46. A stone of mass m is tied to a string and swung at speed v in a circle of radius r.
(a) Write the equation for the angular momentum of the stone.
(b) Calculate the angular momentum of a 0.40-kg stone swung at 4.5 m/s in a circle with radius 3.0 m.

8-47. A space pod circles a space station at the end of a long tether line.
(a) How will the linear speed of the pod compare with its original speed when the line has been pulled in to half its length?
(b) How will the linear speed compare when the line is pulled in to one-tenth its original length?

8-48. A small space telescope at the end of a tether line of length L moves at linear speed v about a central space station.
(a) What will be the linear speed of the telescope if the length of the line is reduced to $\frac{1}{3}L$?
(b) If the initial linear speed of the telescope is 1.0 m/s, what is its speed when pulled in to one-third its initial distance from the space station?

8-49. A ball of mass m at the end of a string moves at constant speed v_0 in a circle of radius r. Assume the speed is great enough so the string remains practically horizontal.
(a) If the ball is pulled into a circular path of half the radius, what will be its speed?
(b) By what factor will the tension have increased when the ball is at this new radius?
(c) Calculate the increased tension if the mass of the ball is 0.12 kg, initial speed is 18 m/s, and initial radius is 4.0 m.

8-50. • A mouse of mass m sits on the edge of a horizontal, spinning Lazy Susan, which can be considered as a cylinder of mass M and radius R. The Lazy Susan spins at an angular speed ω_0.
(a) What is the angular momentum of the system? (Use the idea angular momentum = $I\omega$ to guide your thinking.)
(b) The mouse moves so that now it is at a distance $R/2$ from the center. Write an equation for the new angular momentum of the system.
(c) What is the new angular speed of the system?
(d) Calculate the new angular speed for a 35-gram mouse on the edge of a 240-gram Lazy Susan, 60 cm in diameter, that initially spins at 0.75 rotations per second.

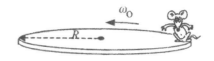

Problems with Some Trigonometry

8-51. Daniel pulls straight upward with force F on a wrench of length L as shown.
(a) Write an equation for the torque exerted on the bolt.
(b) Calculate the torque for a 72-N upward pull on a 22-cm-long wrench where the angle that the wrench makes with the horizontal is 30°.

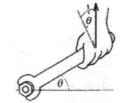

8-52. A ball rolls on an inclined plane because of an unbalanced torque. Suppose the plane is inclined at angle θ. (The pivot point is taken as the point of contact between the ball and the incline.)
(a) Write an equation for the torque on the ball.
(b) Will an incline of 2θ provide twice as much torque on the ball? Defend your answer.

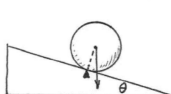

8-53. A hinged, square floor hatch door of mass m and side L is held open at an angle θ by a horizontal rope tied to a point on the door a distance $\frac{2}{3}L$ away from the door hinge.
(a) What is the tension in the rope?
(b) Calculate the tension for a 21-kg square door, 91 cm on a side, held open at a 57° angle.

8-54. • A lantern of weight W is suspended at the end of a horizontal bar of weight w and length L that is supported by a cable that makes an angle θ with the side of a vertical wall. Assume the weight of the bar is at its center.
(a) Derive an equation for the tension in the cable.
(b) Calculate the tension in the cable for a bar of weight 28 N and length 1.5 m, plus a lantern of weight 85 N, and the cable making a 37° angle to the vertical.

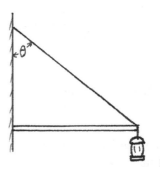

8-55. A popular lab activity that illustrates a conical pendulum is the toy Flying Pig. The sketch shows two forces acting on the Pig; tension T in a string of length L and weight mg.
(a) How does the centripetal force on the Pig compare with the horizontal component of T?
(b) Since the Pig isn't accelerating in the vertical direction, what is the vertical component of T?
(c) How does tension T compare with the weight of the Pig?
(d) What is the radius of the Pig's path (in terms of L)?
(e) Write an equation for the speed of the Pig in terms of L, θ, m, and g.
(f) Find the speed of a 180-gram Flying Pig at the end of a 1.50-m rope if the rope makes a 28° angle with the vertical.

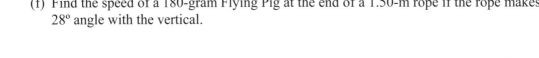

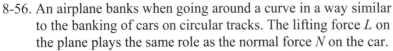

8-56. An airplane banks when going around a curve in a way similar to the banking of cars on circular tracks. The lifting force L on the plane plays the same role as the normal force N on the car.
(a) What angle should a plane traveling at speed v make with the horizontal when the radius of the horizontal turn is r?
(b) What is the actual angle of banking if the speed of the airplane is 210 m/s and the radius is 9700 m?
(c) How will this angle differ for an airplane of twice the mass, traveling at the same speed, and making the same radial turn?

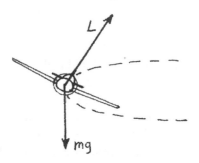

8-57. • At a popular carnival ride, one sits in a chair that is swung in a circular path by a cable as shown. The rider, in effect, is the bob of a conical pendulum. The mass of a rider and the chair is m, the length of the cable L, and the angle the cable makes with the vertical is θ.
  (a) Write an equation for the tension T in the cable.
  (b) Derive an equation for the occupant's speed.
  (c) Calculate the tension if the total mass of the rider and chair is 125 kg, the length of the cable is 15.0 m, and the angle the cable makes with the vertical is 35°.

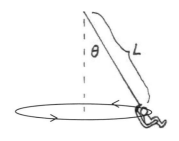

8-58. • Whereas friction provides the centripetal force to hold a car on a curved, non-banked track, a properly banked curve requires no friction at all. Consider a horizontal circular track of radius r banked at angle θ.
  (a) Inspect the sketch and compare the vertical component of N with the magnitude of mg. Why are they the same size?
  (b) What role does the horizontal component of N play?
  (c) Derive an equation for the speed of a car of mass m that will enable it to remain on the track without the need of friction.
  (d) Calculate the speed of the car if the angle is 30°, the radius is 125 m, and the mass of the vehicle is unknown.

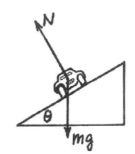

8-59. • A car rounding a curve at speed v can make a turn on a banked turn without the aid of friction. Consider a banked track of radius r. There are two principal forces acting on the car—gravity and the normal force.
  (a) Derive an equation for the angle the curve should be banked so that cars can travel safely at speed v without relying on friction.
  (b) Calculate the banking angle for a curve of radius 180 m designed for a driving speed of 24 m/s.

8-60. • Consider a ball rolling at constant speed on the inner surface of a cone. The surface of the cone makes an angle θ to the vertical as shown. Further assume friction is negligible and the ball rolls in a horizontal circle of radius r.
  (a) Only two forces act on the ball. What are they?
  (b) Is the normal force greater than, less than, or equal to the weight?
  (c) Derive an equation for the speed of the ball.
  (d) Calculate the speed of the ball if the radius of the ball's circular path is 0.96 m and the cone angle is 32°.

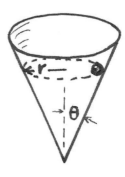

Show-That Problems for Rotational Motion

8-61. A child rides a flying horse at the outer edge of a merry-go-round. She is located 9.5 m from the central axis and is a bit frightened of the speed, so her parents place her on a horse nearer the center, 4.5 m from the axis.
  Show that the inner horse has a linear speed less than half the speed of the outer horse.

8-62. The space shuttle orbits Earth at 7800 m/s at an elevation of about 180 km above Earth's surface. Earth's radius is 6370 km.
Show that the shuttle completes an orbit once every 88 minutes.

8-63. Geosynchronous satellites orbit Earth with an orbital period of 1 day at a distance of 4.23×10^7 m from Earth's center.
Show that these satellites move at 3.08 km/s.

8-64. A 45-N force perpendicular to the 0.20-m-long handle of a wrench is successful in turning a stubborn bolt.
Show that the torque produced is 9.0 N·m, and that when the handle is extended with a length of pipe to 0.5 m, the torque is 2.5 times as much.

8-65. A boy weighs 388 N and sits 2.0 m from the fulcrum of a balanced seesaw. The girl, on the opposite side of the fulcrum, sits 3.0 m from the fulcrum.
Show that the girl weighs 259 N.

8-66. A 388-N boy and a 259-N girl balance on a seesaw. The boy sits 2.0 m from the fulcrum of the seesaw and the girl sits on the opposite side, 3.0 m from the fulcrum. The girl is handed a bag of apples weighing 75 N.
Show that she will have to move 68 cm closer to the fulcrum for the seesaw to remain in balance.

8-67. A 3.0-kg toy is tied to a string. Stephanie holds one end of the string above her head and whirls the toy so that it makes a horizontal circle of radius 1.2 m at waist level. The toy is effectively the bob of a conical pendulum, moving at 3.4 m/s with a string tension of 41 N.
Show that the angle of the string with the vertical is about 45°.

8-68. Consider a car rounding a horizontal, nonbanked curve of radius 55 m.
Show that the coefficient of static friction between the road and a car's tires must be at least 0.60 if the car is to round the curve at a speed of 18 m/s.

8-69. Because of Earth's spin, your weight at the equator is slightly less than on a nonrotating Earth.
Show that you would weigh 0.3% less at the equator than you would at the North Pole.

8-70. Gretchen moves at a speed of 3.2 m/s when sitting on the edge of a horizontal rotating platform of diameter 4.2 m. Her mass is 45 kg.
Show that the frictional force needed to prevent her from slipping off is about 220 N.

8-71. Suppose that the speed doubles for Gretchen in the previous problem, and she is able to hold onto the platform.
Show that her angular momentum about the platform's center will be about 600 kg m²/s.

8-72. A 0.60-kg puck revolves at 2.4 m/s at the end of a 0.90-m string on a frictionless air table.
Show that when the string is shortened to 0.60 m, the speed of the puck will be 3.6 m/s.

8-73. An astronaut on a space walk swings a 1.2-kg bucket of tools in a circular path at the end of a cable 3.6 m long. The astronaut then reels in the bucket to a 1.2-m distance.
Show that the speed of the bucket will be three times as great at the reduced distance.

8-74. An ice skater spins about a vertical axis with arms outstretched. She brings her arms in and reduces her rotational inertia to one-third her initial rotational inertia.
Show that she spins three times faster.

8-75. A skater rotates at 1.2 revolutions per second with arms at her side. When she raises her arms to a horizontal position, her speed decreases to 0.8 revolutions per second.
Show that her rotational inertia with outstretched arms has increased by 1.5 times.

8-76. A 610-gram squirrel stands on the end of a 2.5 m long branch that makes a 60° angle with the vertical trunk of the tree.
Show that the squirrel's weight exerts an additional 13 N·m torque on the branch.

8-77. Daniel does bicep exercises by keeping his elbows at his side while raising his forearms with a 15-kg dumbbell in each hand.
Show that 62 N·m of torque is exerted on each of Daniel's 45-cm long forearms when he holds his arm at a 20° angle above the horizontal.

8-78. A 6.1-kg uniform trap door set into the floor is a square 1.0 m on a side and has its handle set 0.80 m from its hinges. Joel leans over the hinges and grabs the handle, pulling at an angle of 33° to the floor.
Show that, to get the door to open, he must exert a force of 69 N.

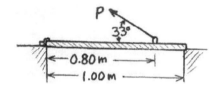

8-79. Garrett hinges a 3.0-m plank at the edge of his roof. The plank extends horizontally and is supported by a rope at the far end that reaches to a sturdy tree branch. The angle of the rope is 50° to the horizontal, Garrett's mass is 53 kg, and the mass of the plank is 12 kg.
(a) Show that the rope tension is slightly more than 300 N when Garrett stands 1.0 m from the hinge.
(b) Show that the tension is about 530 N when Garrett stands 2.0 m from the hinge.

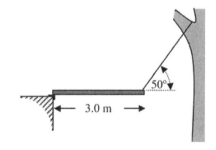

So we gotta distinguish between *velocity* and *tangential velocity* (both measured in m/s) — and between *tangential velocity* and *angular velocity* (measured in ω). Also we distinguish between *force* (measured in N) and *torque* (measured in N·m). And there's a difference between *momentum*, mv, and *angular momentum*, mvr (or Iω). There's a lot of distinctions to be made in this chapter, which for many students is the most difficult to master in mechanics. Good luck!

"What you learn is who you become."

9 Gravity

Just as beautiful music has a pattern that musicians of all nationalities comprehend, so a universal pattern underlies the cosmos. The stars, planets, and all other bodies in the universe dance in rhythm with this pattern—the universal law of gravitation. Isaac Newton discovered that every body in the universe is pulling on every other body with a force, $F = G\dfrac{m_1 m_2}{d^2}$. Here m_1 and m_2 are the masses of a pair of bodies and d is the distance between their centers. The force law as written applies to both particles and spherical bodies, as well as to nonspherical bodies sufficiently far apart.

The problems in this chapter are applications of this central rule of nature—as expressed in the equation for gravitational force.

Sample Problem 1

A group of students repeat Philipp von Jolly's experiment (discussed on page 153 of *Conceptual Physics, 11th Edition*). With m_2 not present, the mass m_1 is balanced. When m_2 is rolled into place, *additional* weights F must be added to the balance pan to balance out the additional downward force on m_1. The students record the following data: $m_1 = 5.00$ kg, $m_2 = 5775$ kg, $F =$ the weight of 0.589 milligram mass (the *extra* mass needed to maintain balance after the large sphere is rolled under the smaller one), and $d = 0.569$ m, the distance between the centers of m_1 and m_2.
(a) With these data, compute the universal gravitational constant, G.

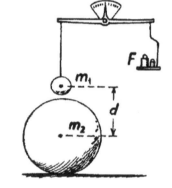

Focus: $G = ?$

From $F = G\dfrac{m_1 m_2}{d^2} \Rightarrow G = \dfrac{Fd^2}{m_1 m_2}$.

The difficulty in measuring G is that the gravitational force between ordinary bodies is so small! We can see just *how* small it is by recognizing that the pull between the 5.00 kg and the 5775 kg masses just over half a meter apart can be balanced out by Earth's pull on 0.589 milligrams (about the weight of the tiniest bit of aluminum foil you can tear from a roll of foil). Nevertheless, students in physics labs, using methods such as von Jolly's, can measure incredibly weak forces. The gravitational force between m_1 and m_2 effectively equals the weight of 0.589 milligrams of material. Call the mass of this bit of matter m, and its weight mg.

Then we get $G = \dfrac{Fd^2}{m_1 m_2} = \dfrac{mgd^2}{m_1 m_2} = \dfrac{\left(0.589 \text{ mg} \times \dfrac{1 \text{ g}}{1000 \text{ mg}} \times \dfrac{1 \text{ kg}}{1000 \text{ g}}\right)\left(9.8 \dfrac{\text{m}}{\text{s}^2}\right)(0.569 \text{ m})^2}{(5.00 \text{ kg})(5775 \text{ kg})}$

$= 6.47 \times 10^{-11} \dfrac{\text{N} \cdot \text{m}^2}{\text{kg}^2}$, about 3% less than the value of G accepted today.

© Paul G. Hewitt and Phillip R. Wolf

(b) If this experiment were performed on the Moon, which values would change and which wouldn't?

Answer: m_1, m_2, and d are independent of location so those values wouldn't change. Neither would the force between the two masses. But because g on the Moon is about six times smaller than it is on Earth, we'd have to add about six times as much extra mass onto the balance pan to get the same mg as occurs at Earth's surface.

Sample Problem 2

Sam, having read about a new "lose-weight-as-you-sleep" scheme, decides to move his bed upstairs, 3 m farther from the ground than he was before.
(a) How will his weight upstairs compare with his weight downstairs?

Focus: $\dfrac{W_{upstairs}}{W_{downstairs}} = ?$

Sam believes, correctly, that by moving his bed farther away from the center of Earth he will feel a smaller gravitational attraction to Earth—that he will "lose weight." The question is, how much weight will he lose?

His distance from Earth's center will increase from R_E (the radius of Earth) to $R_E + 3$ m.
We can calculate his weight upstairs and downstairs using the law of universal gravitation:

$$\frac{W_{upstairs}}{W_{downstairs}} = \frac{F_{\text{Earth and Sam upstairs}}}{F_{\text{Earth and Sam downstairs}}} = \frac{G\dfrac{M_E m_s}{(R_E + 3\,\text{m})^2}}{G\dfrac{M_E m_s}{R_E^2}} = \frac{R_E^2}{(R_E + 3\,\text{m})^2} = \left(\frac{R_E}{R_E + 3\,\text{m}}\right)^2$$

$$= \left(\frac{6{,}370{,}000\,\text{m}}{6{,}370{,}000\,\text{m} + 3\,\text{m}}\right)^2 = \mathbf{0.999999}$$

Sam's weight is reduced by a factor of $(1 - 0.999999)$, about one one-millionth. He could achieve a similar weight loss by sweating the equivalent of about two drops of water (which he likely lost when he moved his bed upstairs!).

(b) Sam weighs 940 N downstairs. How much will he weigh upstairs?

Answer: $W_{upstairs} = 0.999999\, W_{downstairs} = 0.999999 \times (940\,\text{N}) = \mathbf{940\,N}$—no noticeable change in weight.

Sample Problem 3

Your spaceship, mass m, is coasting at a point halfway between the Moon (mass m_M) and Earth (mass m_E). The distance between the centers of Earth and the Moon is d_{EM}.

(a) Find the magnitude and direction of the net force on your ship.

Focus: $F_{net} = ?$

Your ship, with its rocket engine off, is subject to two forces, one pulling your ship toward the Moon (to the left in the sketch) and another pulling your ship toward Earth (to the right in the sketch). The distance between your ship and the center of either body is $d_{EM}/2$. The *net* force on your ship will be the vector sum of these two forces acting in opposite directions.

If we call "toward Earth" the positive direction, then

$$F_{net} = F_{toward\ Earth} - F_{toward\ Moon} = G\frac{m_E m}{\left(\frac{d_{EM}}{2}\right)^2} - G\frac{m_M m}{\left(\frac{d_{EM}}{2}\right)^2} = G\frac{m}{\left(\frac{d_{EM}}{2}\right)^2}(m_E - m_M) = \frac{4Gm}{d_{EM}^2}(m_E - m_M).$$

Since you're the same distance from both Earth and the Moon, and Earth is the more massive of the two bodies, Earth's pull on your ship is greater than the Moon's. The net force points toward Earth.

(b) Find the magnitude and direction of the acceleration of your 35,000-kg ship halfway between Earth and the Moon.

$\left(G = 6.67 \times 10^{-11}\ \frac{N \cdot m^2}{kg^2},\ m_E = 5.98 \times 10^{24}\ kg,\ m_M = 7.36 \times 10^{22}\ kg,\ d_{EM} = 3.84 \times 10^8\ m\right)$

Focus: $a = ?$

$$a = \frac{F_{net}}{m} = \frac{\left(\frac{4Gm}{d_{EM}^2}(m_E - m_M)\right)}{m} = \frac{4G}{d_{EM}^2}(m_E - m_M)$$

$$= \frac{4\left(6.67 \times 10^{-11}\ \frac{N \cdot m^2}{kg^2}\right)}{(3.84 \times 10^8\ m)^2}(5.98 \times 10^{24}\ kg - 7.36 \times 10^{22}\ kg) = 0.0107\ \frac{m}{s^2} = 1.07\ \frac{cm}{s^2}.$$

Your ship accelerates slowly toward Earth.

(c) If your ship were 10 times more massive, what would its acceleration be?

Answer: $a = \frac{10\ times\ the\ force\ F}{10\ times\ the\ mass\ m}$ = the same acceleration as before = $1.07\ \frac{cm}{s^2}$.

A 10-times-more-massive ship means 10 times more gravitational force on it. So the acceleration of your ship would be the same. (Note in the solution to part (b) that the mass of the ship cancels out.) The same cancellation of mass occurs for all freely falling bodies. In accord with Newton's second law, the ratio of force to mass for all bodies in free fall—including this coasting spaceship—is the same, whatever the location. That ratio is the acceleration.

Gravity Problems

INFORMATION FOR SOLVING PROBLEMS IN THIS CHAPTER:

$G = 6.67 \times 10^{-11} \frac{\text{N} \cdot \text{m}^2}{\text{kg}^2}$.

	Earth	Moon	Sun
Mass	5.98×10^{24} kg	7.36×10^{22} kg	1.99×10^{30} kg
Radius	6.37×10^6 m	1.74×10^6 m	6.96×10^8 m
Center-to-Center distance to Earth		3.84×10^8 m	1.50×10^{11} m

9-1. The gravitational field (gravitational force per unit mass) at Earth's surface averages 9.80 N/kg. The field decreases with distance from Earth's center via the inverse-square law.
   (a) At what distance from Earth's center is the gravitational field 0.098 N/kg? (Express your answer in Earth radii.)
   (b) At what distance from Earth's *surface* is the gravitational field 0.098 N/kg?
   (c) Show with units that $9.80 \frac{\text{N}}{\text{kg}} = 9.80 \frac{\text{m}}{\text{s}^2}$.

9-2. An object of mass m is brought to the Moon, where the acceleration due to gravity at its surface is only one-sixth that of Earth.
   (a) What is the mass of the object on the Moon's surface?
   (b) What is the weight of the object on the Moon's surface?
   (c) If the object has a mass of 1.0 kg on Earth, what would be its mass and weight on the Moon?

9-3. Two asteroids with masses m and M are separated by a distance d and attract each other with a gravitational force F.
   (a) By how much does the force change if both masses are doubled, while the distance between the asteroids remains unchanged?
   (b) By how much does the force change if both masses double and the distance between them also doubles?
   (c) By how much does the force change if only one mass doubles, and the distance between them doubles?

9-4. The center-to-center distance between two objects changes by a factor k. (That is, the new distance is k times the old distance.)
   (a) How large is the new gravitational force compared to the original gravitational force?
   (b) Calculate the change in gravitational force if the distance between the objects is tripled ($k = 3$).
   (c) Calculate the change in gravitational force if the distance between the objects is halved ($k = \frac{1}{2}$).

9-5. Two bowling balls of mass m and radius R are touching.
   (a) Write an equation for the amount of force you would need to apply to overcome their gravitational attraction to each other.
   (b) Calculate the required force to separate two touching 5.2-kg bowling balls, each with radius 10.9 cm.

9-6. Suppose that the mass of a star increases by a factor of four while its diameter doubles.
   (a) By how much does the gravitational force on an object at the star's surface change?
   (b) Another star of the same mass collapses to one-tenth its normal radius with no mass change. How is the gravitational force on an object at the star's surface affected?

9-7. You, mass m_Y, see a charming someone, mass m_C, across the room at a distance x away.
   (a) Write an equation for the strength of your gravitational attraction to the charming person.
   (b) What is the charming person's attraction to you?
   (c) Calculate the strength of the attraction if you each have a mass of 75 kg and you stand 3.1 m apart.

9-8. A mass m is suspended from a vertical spring. (Recall Hooke's law: $F = kx$, where the stretch of a spring is proportional to the force on it.) The spring stretches a distance x_E on Earth. The same mass on the surface of another planet stretches the spring by a distance x_P.
   (a) Derive an equation for the value of g on the other planet.
   (b) Suppose that the planet were inflated in size but kept the same mass. At the new surface, would the spring stretch more, less, or the same amount as before? Defend your answer.

9-9. After you blast off in a rocket ship, the force of Earth's gravity on you decreases. Earth's radius is R and its mass is M.
   (a) At what distance, in Earth radii, from Earth's center would Earth's pull on you be one-quarter as much as it is at Earth's surface?
   (b) At what distance from Earth's center, in Earth radii, would gravitation on you be one-tenth that at Earth's surface?

9-10. • Earth of mass m_E and the Moon of mass m_M are separated by a distance R_{EM}. At some location between them, their gravitational forces on a third object will exactly cancel out.
   (a) Do you expect this location to be closer to Earth or to the Moon? Defend your answer.
   (b) Derive an equation for the location between the centers of Earth and the Moon where their net gravitational force on a third object will be zero.
   (c) Using the actual masses of Earth and the Moon and the actual distance between their centers, find how far, in meters, this point is from the center of Earth and from the center of the Moon.

9-11. The International Space Station (ISS) orbits the Earth at a distance h above Earth's surface. The radius of the Earth is R_E.
   (a) Write an equation for the acceleration due to Earth gravity at the location of the ISS.
   (b) Calculate the acceleration due to gravity on the ISS when it orbits at an altitude of 350 km.
   (c) In what direction would you have to propel a coin from the ISS so that the coin would fall straight down to Earth?

9-12. Pretend that the Sun (mass m_S and radius R_S) has contracted to become a white dwarf with a radius equal to that of Earth, R_E.
   (a) Derive an equation for the acceleration due to gravity at the surface of the contracted Sun.
   (b) If the Sun's radius contracted still further, to half R_E, what would g be at its surface?

9-13. Neil Armstrong tosses a ball straight up at an initial speed v from the surface of the Moon of mass m_M and radius R_M.
   (a) Write a formula for g at the Moon's surface.
   (b) What maximum height will the ball reach given this initial speed?
   (c) How much time passes after the ball leaves Neil's hand before he can catch the ball again?
   (d) Calculate the maximum height and total travel time for the ball to go up and back if it is tossed upward at 9.0 m/s.

9-14. You, mass m, stand on a bathroom scale on the surface of a planet of radius R and find that your weight is W.
   (a) Derive an equation for the mass of the planet from this information.
   (b) Calculate the planet's mass if your weight on the planet is 469 N, your mass is 67.0 kg, and the radius of the planet is 5.97×10^6 meters.

9-15. Planet X has a mass that is x times the mass of Earth and a radius that is y times the radius of Earth. Your weight on the surface of Earth is W_E.
   (a) Write an expression for the ratio of your weight on Planet X compared with your weight on Earth.
   (b) How much would your weight be on the surface of the same planet if its diameter shrunk to half its original size with no change in its mass?

9-16. On Earth's surface you have weight W. Pretend that somehow Earth shrank in size, keeping the same mass, with no effects due to Earth's rotation.
   (a) With what force would gravity pull on you at the surface of a contracted Earth if it shrinks to 79% of its present radius?
   (b) With what force would gravity pull on you if Earth shrank to 10% its present radius?
   (c) Suppose instead that Earth stayed its normal size, but that you could burrow to Earth's center. What would your weight be there?

9-17. A blob of water on Earth is gravitationally attracted to the Sun and also to the Moon. Call the mass of the blob m_b.
   (a) Determine the ratio between the Sun's pull and the Moon's pull on the blob. (You should come up with an actual number.)
   (b) How do you reconcile your answers with the fact that the Moon is more responsible for ocean tides on Earth than the Sun?

9-18. The Sun has a mass 333,000 times that of Earth. While you stand on Earth, the average distance between you and the center of the Sun is 23,500 times the distance between you and Earth's center. Both bodies exert a gravitational pull on you.
   Calculate the ratio of the two pulls on you. Which pulls harder on you, Earth or the Sun?

9-19. Suppose that you are standing on a very tall ladder so that you are two Earth radii above Earth's surface. You give a rock a gentle toss upward.
   (a) Calculate the ratio: $\dfrac{\text{the maximum distance the rock travels upward for a toss at the top of the ladder}}{\text{the distance it would have risen for the same toss at Earth's surface}}$.
   (b) Derive an equation for the maximum rise for a rock tossed upward from the top of this ladder, with initial speed v_0.
   (c) Calculate the maximum rise for a rock tossed upward at 1.8 m/s from the top of this ladder.

9-20. • Four identical asteroids of mass M occupy the corners of a large square in space. The sides of the square are L.
(a) Determine the net gravitational force (magnitude and direction) on one of the asteroids due to the other three. (*Hint*: These forces combine as vectors.)
(b) What is the net gravitational force due to the four asteroids acting on a dust speck of mass M located at the center of the "square"?

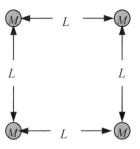

Show-That Problems for Gravity

9-21. Equate your weight mg to Newton's equation for gravitational force, $G\frac{mM}{R^2}$, where M is the mass of Earth and R is its radius.
Show that $g = \frac{GM}{R^2}$.

9-22. The symbol g can mean acceleration due to gravity or gravitational field strength.
Show that its units can be m/s² or N/kg.

9-23. Aileen drops a dense mass m from a high tower and it undergoes an acceleration g.
Show that if she dropped a dense mass $2m$, it would have the same acceleration.

9-24. Consider two identical twins, each of mass 40 kg, located 1 meter apart.
Show that the gravitational attraction between them is about the same as the weight of a 10-microgram speck of dust. (1 microgram = 10^{-6} grams.)

9-25. Suppose that the mass of a star somehow doubles without any change in its radius.
Show that the acceleration due to gravity at its surface also doubles.

9-26. Suppose that the diameter of a star doubles, without a change in mass.
Show that the acceleration due to gravity at its surface reduces to one-fourth its previous value.

9-27. In a high-flying plane you are farther from Earth's center than at the surface.
Show that at a 10-km altitude, gravity's pull on you is ~0.3% less than it is on the ground.

9-28. Geosynchronous satellites orbit Earth with an orbital period of 1 day at a distance of 35,900 km from Earth's surface.
Show that at this distance, Earth's gravitational force on the satellite is 2.3% of the value at Earth's surface.

9-29. Venus has a mass 81.5% that of Earth, and its radius is 94.9% that of Earth.
(a) Show that the value of g on the surface of Venus is 8.9 m/s². (Refer to Problem 9-21.)
(b) Show that the weight of a 1.0-kg stone on the surface of Venus is 8.9 N.

9-30. The mass of Mars is 0.11 times that of Earth and its radius is 0.53 times that of Earth. Show that your weight on the surface of Mars would be about 39% of your weight on the surface of Earth.

9-31. In November of 2005, the Japanese space probe *Hayabusa* visited an asteroid with a mass of about 3.5×10^{10} kg. The acceleration due to gravity on the surface of the asteroid is estimated to be about 0.0001 m/s^2 and the escape speed for the asteroid is estimated to be about 0.2 m/s.
(a) Show that to experience the same acceleration due to gravity, you'd have to be about 300 Earth radii from Earth's center.
(b) Show that the speed that would let you escape the asteroid would raise you to a maximum height of 2 mm on Earth.

9-32. The Hubble Space Telescope is 570 km above Earth's surface.
(a) Show that the force of Earth's gravity on a 73-kg astronaut on Earth's surface is about 720 N.
(b) Show that the force of Earth's gravity of the same astronaut doing repairs on the Hubble Telescope is about 600 N.

9-33. Jack climbs a beanstalk so high that his weight at the top is half that at Earth's surface.
Show that Jack is about 9000 km from Earth's center.

9-34. The Moon is being pulled by both nearby Earth and the faraway Sun.
Show that the Sun pulls on the Moon with more than twice as much force as Earth does.

9-35. Show that gravitation between Earth and the Sun is 177 times stronger than gravitation between Earth and the Moon.

9-36. Ocean tides are caused more by the Moon than the Sun.
(a) Show that the difference between the Sun's pull on 1 kg on the nearer side of Earth and 1 kg on the far side of Earth is 1.00×10^{-6} N.
(b) Show that the difference between the Moon's pull on 1 kg on the nearer side of Earth and 1 kg on the far side of Earth is 2.21×10^{-6} N.
(c) What do the above answers tell you about the relative influence of the Moon's gravity and the Sun's gravity on Earth's tides?

9-37. A 1.0-kg melon weighs 9.8 N at Earth's surface. If the melon were located two Earth radii above Earth, its weight would be less.
Show that the gravitational force on the melon would be 1.1 N at this distance.

9-38. Consider a mountain of mass 6.0×10^{11} kg, 1.0 km away from you.
Show that the mountain exerts more gravitational force on you than the Moon does.

9-39. A newborn baby of mass 3.5 kg is held in the arms of her mother, who has a mass 75 kg. Assume the distance between their centers of gravity is 0.30 m.
Show that the force of gravity between mother and child is more than 8 times as strong as the gravitational force exerted by Mars on the child.
(The mass of Mars is 6.4×10^{23} kg and its closest distance to Earth is 8.0×10^{10} m.)

10 Projectile & Satellite Motion

The central idea in solving projectile motion problems is recognizing that the horizontal and vertical components of the motion are independent. In the absence of air resistance, the only force acting on a projectile is gravitational, so the only acceleration of the projectile is downward. The horizontal motion proceeds as though the vertical motion wasn't there, and the vertical motion proceeds as if the horizontal motion wasn't there.

A satellite in a circular Earth orbit is a projectile in free fall. The satellite moves fast enough so that the curvature of its path parallels the curvature of Earth's surface and the satellite doesn't hit the ground. For such a satellite orbiting a distance r from Earth's center, gravity provides the centripetal force. Kepler's third law tells us that the square of the satellite's orbital period is related to the orbital radius cubed, or $T^2 \sim r^3$.

Sample Problem 1

Zephram leans over the edge of a cliff of height h and drops a rock. It falls straight down and hits the ground below.

(a) Derive an equation for the time it takes the rock to hit the ground. Ignore air drag.

Focus: $t = ?$

This is a straightforward kinematics problem.

Let's call downward the positive direction. From $d = v_0 t + \frac{1}{2}at^2 \Rightarrow d = \frac{1}{2}gt^2 \Rightarrow t = \sqrt{\frac{2h}{g}}$.

(b) Suppose that the rock is instead thrown horizontally from the cliff at a speed v_x. How long is the rock in the air before it hits the ground?

Focus: $t = ?$

Since the horizontal motion of the rock doesn't affect its vertical motion, it will still hit the ground in the same time $t = \sqrt{\frac{2h}{g}}$.

(c) Write an equation for how far from the base of the cliff the rock lands.

Focus: $x = ?$

The fact that the rock is falling doesn't affect its horizontal velocity. $\Rightarrow x = v_x t = v_x \sqrt{\frac{2h}{g}}$.

(d) Calculate how far from the base of the cliff the rock lands when it is thrown horizontally at 12 m/s. The cliff is 33 m high.

Solution: $x = v_x \sqrt{\frac{2h}{g}} = 12 \frac{\text{m}}{\text{s}} \sqrt{\frac{2(33 \text{ m})}{\left(9.8 \frac{\text{m}}{\text{s}^2}\right)}} = 31 \text{ m}$.

© Paul G. Hewitt and Phillip R. Wolf

Sample Problem 2

A can of soup of mass m is projected at a speed v_0 from the ground at an angle θ with the ground. Ignore air drag.

(a) What are the horizontal and vertical components of the can's initial velocity?

From the diagram, $\cos\theta = \dfrac{v_{0x}}{v_0} \Rightarrow v_{0x} = v_0 \cos\theta$.

Similarly, $\sin\theta = \dfrac{v_{0y}}{v_0} \Rightarrow v_{0y} = v_0 \sin\theta$.

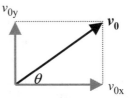

(b) Derive an equation for the maximum height reached by the soup can.

Focus: h = ?

Since we're not given the time, we'll apply the time-independent kinematics equation $v_f^2 - v_0^2 = 2ad$ to the vertical direction. Calling upward the positive direction, $v_{0y} = v_0 \sin\theta$ and $a = -g$. At the maximum height $d = h$ and $v_{fy} = 0$.

So $v_f^2 - v_0^2 = 2ad \Rightarrow -(v_0 \sin\theta)^2 = 2(-g)h \Rightarrow \boxed{h = \dfrac{(v_0 \sin\theta)^2}{2g}}$.

(c) Derive an equation for the time the can of soup is in the air. Assume that the can's launch height and landing height are the same.

Focus: t = ?

From $a = \dfrac{\Delta v}{\Delta t} = \dfrac{v_f - v_0}{t} \Rightarrow t = \dfrac{v_f - v_0}{a}$.

The can's trajectory is symmetric. The can hits the ground with the same speed at which it was launched. So the y-component of the can's velocity when hitting the ground is $-v_0 \sin\theta$.

So $t = \dfrac{v_f - v_0}{a} = \dfrac{-v_0 \sin\theta - v_0 \sin\theta}{-g} = \dfrac{2v_0 \sin\theta}{g}$.

Another approach is to find how long the soup can takes to reach its maximum height, and then double that time to find the time for the whole trip.

From $v_f = v_0 + at \Rightarrow 0 = v_0 \sin\theta + (-g)t \Rightarrow t = \dfrac{v_0 \sin\theta}{g}$. This is the time for the can to reach its maximum height. The total time in the air is twice this $\Rightarrow \boxed{t = \dfrac{2v_0 \sin\theta}{g}}$.

(d) Derive an equation for how far from the launch point the can hits the ground (that is, its range).

Focus: x = ?

$x = v_x t = (v_0 \cos\theta)t$. The vertical motion of the can doesn't affect its horizontal motion. All we need here is the time in the air, which we found in part (c). Then

$x = v_x t = (v_0 \cos\theta)\dfrac{2v_0 \sin\theta}{g} = \boxed{\dfrac{2v_0^2 \sin\theta \cos\theta}{g}}$.

(e) For a given initial velocity v_0, what throwing angle will give the maximum range for the soup can?

Focus: $\theta = ?$

We want to find the angle θ that results in the maximum value of x for given values of v_0 and g.

From $x = \dfrac{2v_0^2 \sin\theta \cos\theta}{g}$ we can see that x will be maximum when the product $\sin\theta \times \cos\theta$ is maximum. This table of various launching angles shows that the maximum range occurs when **the launch angle is 45°**. (Note that the range is the same for 30° and 60°, or any pair of complementary angles, as discussed in the textbook.)

θ	$\sin\theta$	$\cos\theta$	$\sin\theta \times \cos\theta$
15°	0.259	0.966	0.250
30°	0.500	0.866	0.433
40°	0.643	0.766	0.492
45°	0.707	0.707	0.500
50°	0.766	0.643	0.492
60°	0.866	0.500	0.433
75°	0.966	0.259	0.250

(f) Calculate the maximum height, time in the air, and horizontal range for a 0.40-kg soup can thrown from the ground at 21 m/s at an angle of 35° above the horizontal.

Solution:

$$h = \dfrac{(v_0 \sin\theta)^2}{2g} = \dfrac{\left(21\tfrac{m}{s} \sin 35°\right)^2}{2\left(9.8\tfrac{m}{s^2}\right)} = 7.4 \text{ m}.$$

$$t = \dfrac{2v_0 \sin\theta}{g} = \dfrac{2\left(21\tfrac{m}{s}\right)\sin 35°}{9.8\tfrac{m}{s^2}} = 2.5 \text{ s}.$$

$$x = \dfrac{2v_0^2 \sin\theta \cos\theta}{g} = \dfrac{2\left(21\tfrac{m}{s}\right)^2 \sin 35° \cos 35°}{9.8\tfrac{m}{s^2}} = 42 \text{ m}.$$

Sample Problem 3

A horizontally moving tennis ball barely clears the net, a distance y above the surface of the court. To land within the court, the ball must not be moving too fast.

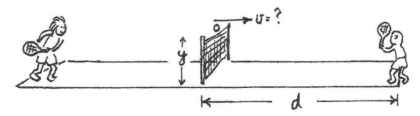

(a) Ignoring air drag, find the maximum speed the ball can have as it clears the net to land within the tennis court's border, a distance d from the bottom of the net.

Focus: $v = ?$

We model the tennis ball as a projectile moving in the absence of air drag, so the horizontal and vertical components of the ball's velocity are independent. We're asked to find its horizontal speed, so we write,

$$v = v_x = \frac{d}{t}$$

where d is horizontal distance traveled in time t. We're given d, but we're not given the time. So this isn't a simple one-step problem. Importantly, *the equation directs us to consider time*, which we might not have considered otherwise!

Now some important physics: The time t it takes for the ball to hit the court will be the same as if we had just dropped it from the top of the net from rest, a vertical distance y. We say "from rest" because the initial *vertical* component of velocity is zero—at the highest point in its path, the ball moves horizontally over the net.

From $y = \frac{1}{2}gt^2 \Rightarrow t^2 = \frac{2y}{g} \Rightarrow t = \sqrt{\frac{2y}{g}}$.

So $v = v_x = \frac{d}{t} = \frac{d}{\sqrt{\frac{2y}{g}}}$.

(b) Calculate the maximum speed a horizontally moving tennis ball can have as it clears the net of height 1.00 m and still strike within the court's border, 12.0 m distant from the net.

Solution: $v = \dfrac{d}{\sqrt{\dfrac{2y}{g}}} = \dfrac{12.0 \text{ m}}{\sqrt{\dfrac{2(1.00 \text{ m})}{9.80 \frac{\text{m}}{\text{s}^2}}}} = 26.6 \frac{\text{m}}{\text{s}}$.

Sample Problem 4

A watermelon rolls off the horizontal edge of a building at speed v_0.

(a) Find its speed after time t as it sails through the air. Ignore air drag.

Focus: $v = ?$

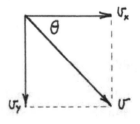

At any given time, there are two components of the watermelon's velocity—a vertical component v_y, and a horizontal component v_x. The speed at any moment is the magnitude of the vector sum of these two components: $speed = \sqrt{v_x^2 + v_y^2}$. The equation tells us to find expressions for v_x and v_y.

The horizontal component $v_x = v_0$, since no horizontal forces act on the watermelon to produce horizontal accelerations. The vertical component at any time t is $v_y = gt$.

So $speed = \sqrt{v_x^2 + v_y^2} = \sqrt{v_0^2 + (gt)^2}$

(b) What angle does the watermelon's velocity vector make with the horizontal at time t?

Focus: $\theta = ?$

Look at the diagram. v_x, v_y, and v form a right triangle, with v_y as the side opposite θ, v_x as the side adjacent to θ, with v as the hypotenuse. We can see that

$$\tan\theta = \frac{v_y}{v_x} \Rightarrow \theta = \tan^{-1}\left(\frac{v_y}{v_x}\right) = \tan^{-1}\left(\frac{gt}{v_0}\right).$$

(c) Calculate the velocity of the watermelon 1.0 second after rolling off the edge at 5.0 m/s.

Solution:

$$\text{speed} = \sqrt{v_0^2 + (gt)^2} = \sqrt{\left(5.0\frac{m}{s}\right)^2 + \left(9.8\frac{m}{s^2}(1.0s)\right)^2} = 11\frac{m}{s};$$

$$\theta = \tan^{-1}\left(\frac{v_y}{v_x}\right) = \tan^{-1}\left(\frac{gt}{v_0}\right) = \tan^{-1}\left(\frac{9.8\frac{m}{s^2}(1.0s)}{5.0\frac{m}{s}}\right) = 63°.$$

So the velocity of the watermelon at $t = 1.0$ s is **11 m/s at an angle of 63° below the horizontal**.

Sample Problem 5

Ramon fires a potato at speed v_0 at an initial angle θ with the horizontal ground. The potato strikes a balloon when it reaches the top of its trajectory.

(a) Ignoring air drag, what is the speed of the potato when it hits balloon?

Focus: $v = ?$

At the potato's maximum height, the vertical component of its velocity is zero, so the magnitude of v will be the magnitude of the x-component of v_0

$$\Rightarrow v = v_{0x} = v_0 \cos\theta.$$

(b) Derive an equation that shows the height of the balloon.

Focus: $h = ?$

The vertical component of the potato's initial velocity $v_{0y} = v_0 \sin\theta$. Since we're not given time, we consider the time-independent kinematics equation $v_f^2 - v_0^2 = 2ad$ applied to *vertical* motion, where $a = -g$, $v_f = 0$, and $d = h$.

From $v_f^2 - v_0^2 = 2(-g)h \Rightarrow h = \frac{-v_{0y}^2}{2(-g)} = \frac{(v_0 \sin\theta)^2}{2g}.$

(c) Assume that the projectile was launched at 100 m/s at a projection angle of 60°. Calculate the speed of the projectile when it hits the balloon and the height of the balloon.

$$v = v_0 \cos\theta = \left(100\frac{m}{s}\right)\cos 60° = 50\frac{m}{s}; \quad h = \frac{(v_0 \sin\theta)^2}{2g} = \frac{\left(100\frac{m}{s}\sin 60°\right)^2}{2\left(9.8\frac{m}{s^2}\right)} = 380 \text{ m}.$$

© Paul G. Hewitt and Phillip R. Wolf

(d) Suppose that the Sun is directly overhead, so that as the potato sails through the air, it casts a shadow on the ground. How fast does the shadow move across the ground?

Answer: The shadow across the ground has the same speed as the horizontal component of the projectile's velocity. So $v = v_x = v_0 \cos\theta = \left(100 \frac{\text{m}}{\text{s}}\right) \cos 60° = 50 \frac{\text{m}}{\text{s}}$.

Sample Problem 6

A satellite of mass m circles Earth at a constant altitude h above the Earth's surface.

(a) Derive an equation for the speed of the satellite.

Focus: $v = ?$

Since the satellite travels in a circular orbit, the gravitational force must provide a centripetal force mv^2/r. Math-wise,

$$G \frac{M_E m_{sat}}{(d_{\text{Earth to sat}})^2} = \frac{m_{sat} v^2}{r_{orbit}}.$$

There are three things to notice before we go any further:
- The term $d_{\text{Earth to sat}}$ is the distance from the *center* of Earth to the satellite, equal to $R_{\text{Earth}} + h$.
- The radius of the orbit (r in the mv^2/r term) is also measured from the center of Earth, and is exactly the same as $d_{\text{Earth to sat}}$.
- We don't need to know the mass of the satellite (because it cancels out of the equation).

From $G \dfrac{M_E m_{sat}}{(d_{\text{Earth to sat}})^2} = \dfrac{m_{sat} v^2}{r_{orbit}}$ $\Rightarrow G \dfrac{M_E}{(R_E + h)^2} = \dfrac{v^2}{(R_E + h)}$ $\Rightarrow v = \sqrt{\dfrac{GM_E}{(R_E + h)}}$.

(b) The International Space Station orbits Earth in a nearly circular orbit, averaging a height of 350 km above Earth's surface. Find the speed of the ISS as it circles Earth.

From above, $v = \sqrt{\dfrac{GM_E}{(R_E + h)}}$.

There are two things to notice before we go any further:
- G has units of N·m²/kg², so all distances we use should be expressed in meters.
- The mass and radius of Earth are listed at the top of page 120.

First let's convert the units:

height $h = 350$ km $= 3.5 \times 10^5$ m.

$R_E + h = 6.37 \times 10^6$ m $+ 3.5 \times 10^5$ m $= 6.72 \times 10^6$ m.

Then $v = \sqrt{\dfrac{6.67 \times 10^{-11} \frac{\text{N} \cdot \text{m}^2}{\text{kg}^2} (5.98 \times 10^{24} \text{kg})}{6.72 \times 10^6 \text{m}}} = 7700 \frac{\text{m}}{\text{s}}$ (which is a brisk $27,700 \frac{\text{km}}{\text{h}}$!).

The space station moves about 25 to 30 times faster than a normal commercial jet. This high speed ensures that it falls around Earth rather than into Earth's surface! (As a matter of interest, even at this altitude, a tiny bit of air resistance exists, which means the station needs small rocket boosts from time to time to keep its orbit from decaying!)

(c) **The Moon orbits Earth once each 27.3 days. Use this information to calculate the time T that the International Space Station takes to complete an orbit.**

Focus: $T = ?$

Both the space station and the Moon orbit Earth. From Kepler's third law, the square of the satellite's orbital period is proportional to the cube of its orbital radius, or $T_{orbit}^2 \sim r_{orbit}^3$.

So $\left(\dfrac{T_{ISS}}{T_{Moon}}\right)^2 = \left(\dfrac{r_{ISS\,orbit}}{r_{Moon\,orbit}}\right)^3 \Rightarrow T_{ISS} = T_{Moon}\sqrt{\left(\dfrac{r_{ISS}}{r_{Moon}}\right)^3} = 27.3 \text{ days}\sqrt{\left(\dfrac{6.74\times10^6 \text{ m}}{3.84\times10^8 \text{ m}}\right)^3}$

$= 0.0635 \text{ days} \times \dfrac{24\,\text{h}}{1\,\text{day}} = \mathbf{1.52\ h}$.

We could have calculated the answer directly from our answer to part (b):

From $v = \dfrac{d}{t} \Rightarrow T = \dfrac{2\pi r}{v} = \dfrac{2\pi(6.72\times10^6\,\text{m})}{7700\,\frac{\text{m}}{\text{s}}} = 5484 \text{ s} \times \dfrac{1\,\text{h}}{3600\,\text{s}} = \mathbf{1.52\ h}$.

Note that we get the same answer either way!

Problems for Projectile Motion

(Air drag significantly affects the motion of such things as fast baseballs and even satellites in near-vacuum territory. To keep the physics simple, however, all problems in this chapter ignore the effects of air resistance.)

10-1. Three balls are tossed off the roof of a building. Ball A is thrown straight downward with speed v. Ball B is thrown upward with speed v. Ball C is simply dropped from rest.
   (a) Which ball hits the ground first? Last?
   (b) Which ball hits the ground with the greatest speed? The least speed?
   (c) Which ball spends the most time in the air? The least time?

10-2. Three ballplayers—Abe, Bob, and Cal—throw balls straight up. Abe throws at speed v, Bob throws at speed $1.5v$, and Cal throws at speed $2v$. The balls are caught at the same height from which they were thrown.
   (a) Which player's ball is in the air longest? Shortest?
   (b) Which player's ball goes highest? Lowest?
   (c) Which player's ball has the greatest acceleration in flight? Which has the least acceleration?

10-3. Three projectiles are launched with the same initial velocity v. Projectile A is launched at an angle of 30° from the horizontal, Projectile B at 45°, and Projectile C at 60°. Each projectile lands at the same elevation from which it was launched.
   (a) Which projectile has the greatest horizontal velocity? The least?
   (b) Which projectile has the greatest initial vertical velocity? The least?
   (c) Which projectile attains the greatest height? The least?
   (d) Which projectile has the greatest range? The least?
   (e) Which projectile has the greatest time in the air? The least?

10-4. A lemming waddles horizontally off a small cliff of height h with a speed v.
  (a) Derive an equation for how far from the base of the cliff the lemming lands.
  (b) Derive an equation for the lemming's speed upon landing.
  (c) Calculate answers to parts (a) and (b) if the height of the cliff is 4.9 m, the initial speed of the lemming is 0.50 m/s, and its mass is 0.4 kg.

10-5. That strange classmate of yours slides your physics textbook of mass m off the lab table with a speed v. It meets the floor in time t.
  (a) Derive an equation for the height of the tabletop above the floor.
  (b) Write an equation for the horizontal distance the book lands from the edge of the table.
  (c) Calculate the answers to parts (a) and (b) if the book leaves the table at 1.2 m/s, the time in the air is 0.43 s, and you don't know the mass of the book.

10-6. Chelsea gives a horizontal speed v to a ball that then rolls off a lab bench y meters high.
  (a) Derive an equation for the time it takes for the ball to reach the floor.
  (b) Derive an equation for the distance the ball lands from a point on the floor directly below the edge of the bench.
  (c) Calculate how long and how far for $v = 1.5$ m/s and a bench height of 1.2 m.

10-7. Chuck's ball rolls horizontally at speed v off the edge of a platform and lands a horizontal distance x beyond the edge of the platform.
  (a) Derive an equation for the time that the ball is airborne.
  (b) Derive an equation for the height of the platform.
  (c) Calculate the height of the platform if the ball rolls off the edge at 15 m/s and lands 18 m away.

10-8. Manuel throws a ball horizontally with speed v from the top of a building of height h.
  (a) Derive an equation for how far from the bottom of the building the ball strikes the ground.
  (b) Calculate the horizontal distance from the bottom of the building that the ball will land for a horizontal speed 15 m/s and a building 32 m tall.
  (c) Imagine that the above situation were to take place on the surface of the Moon, where the acceleration due to gravity is only $1/6\ g$. Calculate the horizontal distance from the base of the lunar building that the ball lands.

10-9. Caillen stands atop a cliff and tosses a stone horizontally into the ocean below. The stone has a mass m, and it hits the water in time t after leaving her hand.
  (a) Derive an equation for the height of the cliff.
  (b) Calculate the height of the cliff if the stone's mass is 0.80 kg and it hits the water 2.4 seconds after leaving Caillen's hand.

10-10. A shoe is thrown horizontally with velocity v from the top of a tall building of height h. At the same time, a similar shoe is simply dropped from the same height and takes time t to reach the ground below.
  (a) Write an equation for how far from the bottom of the building the thrown shoe lands.
  (b) Calculate the range if the thrown shoe's initial velocity was 12 m/s and the time for the other shoe to drop is 2.0 s.

10-11. An emergency package of mass m is dropped from a horizontally moving airplane traveling at constant speed v when its altitude is h.
 (a) Derive an equation for the time it takes for the package to reach the ground.
 (b) Derive an equation for how far forward from its dropping point the package lands.
 (c) At the moment the package hits the ground, what is its distance from the airplane?
 (d) Calculate answers to these questions for an airplane flying at 115 m/s at an altitude of 440 m and dropping a 22-kg package.

10-12. On a baseball field, the distance between the pitcher's mound and home plate is x. Pedro pitches a fastball that moves horizontally as it leaves his hand and takes 0.45 s to reach the home plate.
 (a) Write an equation for the initial speed of the pitch.
 (b) Calculate this initial speed for a distance of 18.4 m between the pitcher's mound and home plate.
 (c) Calculate the vertical drop of this ball on its way from the pitcher's mound to home plate.
 (d) What is the vertical drop of the ball when it is halfway to the plate?

10-13. • Mala, practicing with her bow and arrow, fires one that strikes horizontally into a third-floor barn window shutter. The arrow travels a horizontal distance x and a vertical distance y in time t. (Note again: The arrow strikes horizontally into the window shutter. What does this tell you about the vertical component of the arrow's velocity when it strikes?)
 (a) Write an equation for the speed of Mala's arrow when it hits the shutter.
 (b) Derive an equation for the speed of Mala's arrow when it left her bow.
 (c) Calculate the speed of Mala's arrow when it left her bow if the horizontal distance traveled is 32 m, the vertical distance is 9.5 m, and time of flight is 1.39 s.
 (d) Could you have solved this problem if the time of flight was not given? Explain.

10-14. Firemen shoot a stream of water at a burning building. Water leaves the high-pressure hose at speed v at an angle θ to the horizontal.
 (a) Write an equation for the speed of the water at the highest point in its trajectory. Assume that the water reaches its highest point before hitting the building.
 (b) What is the acceleration of the water at its highest point?

10-15. A pellet is fired from a rifle at a speed v at an angle θ above the horizontal.
 (a) Write an equation for the pellet's maximum height.
 (b) Derive an equation for the time it takes to reach its maximum height.
 (c) Derive an equation for the horizontal range of the pellet.
 (d) Calculate answers to the above questions for an initial speed of 150 m/s and an angle of 37° above the horizontal.

10-16. Fred throws a ball at an angle θ above the horizontal with speed v.
 (a) Write an equation for the ball's speed when it reaches its highest point.
 (b) Derive an equation for the speed of the ball 1.0 s after reaching its highest point, assuming it is still in the air.
 (c) Calculate answers to parts (a) and (b) if the initial speed is 30 m/s and the angle is 60° above the horizontal.

10-17. In a new twist on birthday party games, a piñata hangs from the branch of a very tall tree. Alex aims and fires a small coconut from the ground at an angle θ above the horizontal. He fires it at velocity v_0. The coconut strikes the piñata just as it reaches the top of its trajectory.
   (a) Write an equation for the speed of the coconut as it hits the piñata.
   (b) Calculate the speed of the coconut when it hits the piñata. Assume an initial speed of 18 m/s and a projection angle of 59°.
   (c) How high is the piñata?
   (d) How long will it take the candies that spill from the piñata to hit the ground?

10-18. Stranded on a tropical island, Isabella loads her slingshot and fires a compact seashell at a coconut. The seashell leaves the slingshot at ground level at a speed v, making an angle θ with the horizontal. The coconut is hit at the top of the shell's trajectory.
   (a) Write an equation for the speed of the seashell as it hits the coconut.
   (b) Calculate the speed of the seashell when it hits the coconut. Assume an initial speed of 24 m/s and a projection angle of 52°.
   (c) How high is the coconut?

10-19. Stan the stuntman steps out from a helicopter that is flying at a horizontal velocity v.
   (a) Write an equation for Stan's velocity (speed and direction) at the instant he steps out from the helicopter.
   (b) Derive an equation for Stan's velocity (speed and direction) t seconds later.
   (c) Derive an equation for the time it takes for Stan to hit the safety net a vertical distance y below.
   (d) Find Stan's velocity when hitting the net and his time of fall, assuming the helicopter travels at 13 m/s and is 26 m above the safety net when Stan begins his drop.

10-20. • Stan the stuntman again steps out from a helicopter that this time is flying at velocity v, slightly downward at an angle θ with the horizontal.
   (a) Write an equation for Stan's velocity (speed and direction) at the instant he steps out from the helicopter.
   (b) Derive an equation for Stan's velocity (speed and direction) t seconds later.
   (c) Derive an equation for the time it takes Stan to hit the safety net a vertical distance y below.
   (d) Calculate Stan's velocity when he lands on the net and his time in the air if the helicopter's velocity is 12 m/s at an angle of 15° below the horizontal, and the helicopter is 26 m above the safety net when Stan begins his drop.

10-21. • Stephanie is lying on the ground outside, playing with a high-pressure water hose, squirting water in all directions. When she's done playing, she finds that the ground is wet up to a distance d away from where she was holding the hose.
   (a) Derive an equation for the speed of the water coming out of the hose.
   (b) Calculate the speed of the water if the radius of the wet circle is 45 m.

10-22. • A projectile of mass m is fired at an angle θ above the horizontal with a speed v.
(a) Write equations for the horizontal and vertical components of the projectile's initial momentum.
(b) What will be the horizontal and vertical components of the projectile's momentum when it reaches its highest point?
(c) The projectile explodes into two equally massive fragments at the top of its trajectory. One fragment, the first, falls vertically to the ground below. Derive an equation for the speed of the second fragment immediately after the explosion.
(d) How much farther away from the launch point will the second fragment land compared with where the projectile would have landed if it didn't explode?

Problems for Satellite Motion

(*As before, assume no air resistance for these problems.*)

10-23. The speed of a satellite in circular orbit about Earth depends on its distance from Earth's center. The force of gravity provides the centripetal force necessary for circular orbit.
(a) Equate centripetal force to gravitational force and produce an equation for the speed of a satellite circling at a distance r from Earth's center.
(b) Show that this speed in close orbit, a distance just a bit greater than Earth's radius, is about 8 km/s.
(c) Show that the time it takes for the satellite to complete one orbit at this distance will be a little less than 90 minutes.

10-24. The speed of a spaceship in circular orbit about the Moon depends on its distance above the Moon's surface. (Let the Moon's radius be R and its mass M.)
(a) Derive an equation for the speed for a satellite in a circular orbit at altitude h above the Moon's surface (that's a total radial distance $R + h$).
(b) Derive an equation for the time the satellite takes to complete a lunar orbit at this altitude.
(c) Calculate the speed and period of the spaceship if its distance above the Moon's surface is 25.0 km. ($M_M = 7.36 \times 10^{22}$ kg. $R_M = 1.74 \times 10^6$ m.)
(d) Why can we assume there is no air resistance with this problem?

10-25. Robert is calculating the orbit for a weather satellite so that it will circle the Earth with an orbital period T.
(a) Write an equation for the radius of this orbit in terms of T.
(b) How high above Earth's surface will the satellite orbit?
(c) How fast will the satellite move in its orbit?
(d) Calculate the altitude and speed of a weather satellite that circles Earth every 150 minutes.

10-26. A rocket is fired from Earth's surface to a height equal to Earth's radius.
(a) Write an equation comparing the force of gravity on the rocket at this height with the force of gravity on the rocket back at Earth's surface.
(b) Derive an equation for the orbital speed needed for a circular orbit at this height.
(c) Calculate the orbital speed of a rocket in a circular orbit at this height.

10-27. Satellites that transmit TV signals are in geosynchronous orbit (above Earth's equator with an orbital period of 24 hours).
(a) How high above the equator are these satellites?
(b) Why is it especially nice to have a satellite with a period of 24 hours? (Think about how satellite dishes are oriented when installing satellite TV.)

10-28. A satellite at distance r from the center of Earth completes one circular orbit in time T. Newton's law of gravitation gives the pull of Earth's gravity on the satellite. This same pull is given by the equation for centripetal force.
(a) Equate gravitational force to centripetal force and solve for the mass M of Earth in terms of r, T, and the gravitational constant G.
(b) Show that Kepler's third law of planetary motion follows from this relationship among M, r, T, and G.

10-29. The radius of Earth's orbit around the Sun is 1.5×10^{11} m. The period of Earth's orbit around the Sun is 365.25 days.
(a) Use this information and Kepler's third law to compute the mass of the Sun.
(b) Why does this information and Kepler's third law not allow you to calculate the mass of Earth?
(c) Why can the masses of planetary bodies with satellites be easily found, but not the masses of bodies without satellites?

10-30. Io, a satellite of Jupiter, has a mean distance from Jupiter of 4.22×10^8 m and a period of 1.77 Earth days.
(a) Use this information and Kepler's third law to compute the mass of Jupiter.
(b) Why won't this information and Kepler's third law allow you to calculate the mass of Io?

10-31. You are exploring an as-yet uncharted solar system and you discover a small planet orbiting its parent star in a circular orbit of radius r. The planet has an orbital period T.
(a) Derive an equation for the mass of this star.
(b) The star HD 38801 has a recently discovered planet that orbits the star once every 696 days at a distance of 1.71 AU. (1 AU = 1 Astronomical Unit = the average Earth-Sun distance = 1.50×10^8 km.) Calculate the mass of the star.

10-32. An astronomical unit (AU) is defined as the average distance between the centers of Earth and the Sun. Jupiter is on average 5.19 AU from the Sun.
(a) What is Jupiter's orbital period around the Sun?
(b) How would Jupiter's orbital period be affected if Jupiter were closer to the Sun?

Show-That Problems for Projectile Motion

10-33. The speed of a projectile anywhere along its path is given by $V = \sqrt{v_x^2 + v_y^2}$.
Show that the speed of a projectile at the peak of its trajectory is $V\cos\theta$, where θ is the angle V makes with the horizontal.

10-34. A waterfall flowing over the horizontal edge of a 28-m high cliff lands 5.0 meters from the base of the cliff.
Show that the speed of the water at the top of the cliff is 2.1 m/s.

10-35. Suppose that a flight attendant on a commercial jet moving 840 km/h pours a cup of coffee from a pot held 28 cm above the coffee cup.
Show that the coffee travels a horizontal distance of 56 meters relative to Earth between the time it leaves the pot and the time it meets the cup.

10-36. Tenny's compound archery bow can fire an arrow at 55 m/s (or more). Assume the arrow is shot horizontally at 55 m/s, pointed directly at the bull's-eye of a target 22.6 m away.
Show that the arrow hits the target 83 cm below the bull's-eye.

10-37. Suppose Tenny uses the same bow and the same arrow as in the preceding problem and aims the arrow 2.1° above the horizontal.
Show that she will hit the bull's eye.

10-38. • Diane can throw something up at a particular angle from the ground so that its maximum height and its horizontal distance traveled will be the same.
Show that this angle is 76°.

10-39. On February 6, 1971, Apollo 14 astronaut Alan Shepard hit a golf ball on the Moon with an improvised golf club he is rumored to have smuggled aboard the spaceship. The acceleration due to gravity on the surface of the Moon is 1.62 m/s².
Show that the ball would land about 480 meters away if he gave it an initial velocity of 32 m/s at an angle of 25° with the horizontal ground.

10-40. A World War II British artillery gun fired its shell at 550 m/s.
Show that if air resistance is neglected, such a gun would have a maximum range of almost 31 km.

10-41. • Show that the maximum height h of a projectile following a parabolic path is half the height H of a isosceles triangle whose base is equal to the range, with the angle of the two sides equal to the projectile's launch angle.

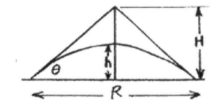

Show-That Problems for Satellite Motion

10-42. Equate the force of gravity to centripetal force where M is the mass of Earth and G is the gravitational constant.
Show that the speed v of a satellite in circular Earth orbit of radius r is $v = \sqrt{\dfrac{GM}{r}}$.

10-43. The distance covered in one complete circular orbit is simply the circumference of a circle.
Show that the speed of a satellite in circular orbit is $\dfrac{2\pi r}{T}$ where r is the radius of the satellite's orbit and T its orbital period.

10-44. Combine the results of the previous two problems (10-42 and 10-43).

Show that the period $T = 2\pi\sqrt{\dfrac{r^3}{GM}}$ for a satellite in circular orbit about Earth.

10-45. Titan is a moon of Saturn that orbits Saturn at an average distance of 1.22 million km with an orbital period of 15.9 days.

Show that the mass of Saturn is 5.69×10^{26} kg.

10-46. Mars has two moons, Phobos and Deimos. The radius of Mars is 3397 km. Phobos orbits Mars at a distance of 5980 km above Mars's surface with a period of 0.319 days. Deimos orbits Mars with a period of 1.262 days.

(a) Draw and label a diagram showing Mars, Phobos, and Deimos with the appropriate distances included.

(b) Show that Deimos is approximately 20,100 km above the surface of Mars.

10-47. Consider the data in the previous question.

Show that Phobos moves 1.58 times faster than Deimos if their orbits are approximated as circles.

10-48. Continue with the data in Problem 10-46.

Show that the mass of Mars is 6.42×10^{23} kg.

10-49. A particular satellite circles Earth at a distance of 775 km above the surface.

Show that the orbital speed of the satellite is 7470 m/s.

10-50. Another satellite orbits Earth at 4,990 m/s in a circular orbit.

Show that its distance from Earth's center is 16,000 km.

10-51. Make a simple force-vector diagram of a satellite in a circular orbit.

Show that Earth's gravitational force does not work on the satellite.

10.52. Consider a binary star system consisting of two identical stars of mass m orbiting about their common center of mass located exactly halfway between the two stars. The radius of their orbit is R.

Show that each star moves with speed $v = \sqrt{\dfrac{Gm}{4R}}$.

10.53. *Io*, the innermost of Jupiter's four largest moons, orbits Jupiter with an average orbital radius of 4.22×10^5 km and with a period of 1.77 days. *Europa* orbits Jupiter with an average orbital radius of 6.71×10^5 km.

Show that the period of *Europa's* orbit is 3.55 days.

10-54. Use the data in the previous problem to show that Jupiter's mass is 1.90×10^{27} kg.

12 Solids

Although scaling is the highlight of this chapter, the concept of density is introduced here and is then covered in greater depth in Chapter 13. Whereas springs and Hooke's law are treated in Chapter 12 of the textbook, a mathematical treatment of springs is included in Chapter 7 of this Problems book.

Sample Problem 1

A movie depicts two pirates carrying a chest full of pure gold, which has a density of 19,300 kg/m³.
(a) If the chest has dimensions 0.60 m × 0.35 m × 0.30 m, find the mass of the gold.

Focus: $m = ?$ From the dimensions of the chest, we can get its volume. From the density we can figure out the mass of the gold's volume.

From $\rho = \dfrac{m}{V} \Rightarrow m = \rho V = \rho(l \times w \times h) = 19{,}300 \dfrac{\text{kg}}{\text{m}^3}(0.60\,\text{m} \times 0.35\,\text{m} \times 0.30\,\text{m}) = \mathbf{1200\ kg}$.

(b) Find the weight of the gold.

$W = mg = (1200\,\text{kg})\left(9.8\,\dfrac{\text{m}}{\text{s}^2}\right) \approx \mathbf{12{,}000\ N}$.

(c) Is it likely that the pirates could carry such a chest?

No! This is over 2600 lbs, about the weight of a small car. Even pirates who have been working out would have a difficult time lifting this.

Sample Problem 2

Consider a telephone pole of length L and average diameter D.
(a) What is its volume?

Answer: The pole is a cylinder, so its volume = (circular area) × length =

$$\pi R^2 L = \pi\left(\dfrac{D}{2}\right)^2 L = \dfrac{\pi D^2 L}{4}.$$

(b) What is its surface area?

Answer: For a cylinder the surface area is the area of the two circular ends, $2 \times \pi R^2 = 2 \times \pi\left(\dfrac{D}{2}\right)^2 = \pi\dfrac{D^2}{2}$, plus the area of the side. We can imagine that if we were to remove the ends and unroll the side, we'd have a rectangle with a height of πd (the circumference of the circular end) and length L. So the surface area is

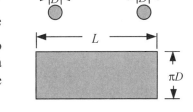

$$\dfrac{\pi D^2}{2} + \pi D L = \pi D\left(\dfrac{D}{2} + L\right).$$

(c) When scaled up by 2 (twice as long with twice the diameter), by how much will the cylinder's total surface area increase?

Solution: $A_{new} = \pi(2D)\left(\dfrac{(2D)}{2} + (2L)\right) = 4\pi D\left(\dfrac{D}{2} + L\right) = 4\,A_{initial}$.

A rule of scaling states that when a body is scaled up by a factor x, the surface area increases by x^2. So scaling up by 2 yields an increase in area of 2^2. The surface area **increases by a factor of 4**.

(d) If the cylinder is scaled up by a factor of 2, by how much will its volume increase?

Solution: $V_{new} = \dfrac{\pi(2D)^2(2L)}{4} = 8\left(\dfrac{\pi D^2 L}{4}\right) = 8\,V_{initial}$.

A rule of scaling states that when a body is scaled up by a factor x, the volume increases by x^3. So the volume of the cylinder increases by a factor of **8**.

(e) What is the effect on the surface-area-to-volume ratio for the pole when it is scaled up by a factor of 2?

Answer: Since the surface area increases by a factor of 4 and the volume increases by a factor of 8, the

surface-area-to-volume ratio, $\dfrac{\text{surface area}}{\text{volume}} = \dfrac{4 \times A_{initial}}{8 \times V_{initial}} = \dfrac{1}{2}\left(\dfrac{A_{initial}}{V_{initial}}\right) =$ **half the original surface-area-to-volume ratio**. The same ratio applies to the cross section of the pole compared with its weight. A pole twice as long and twice as wide will have four times the cross-sectional area (so four times the stiffness) but eight times the weight, so it will have half the strength compared with its weight. This accounts for the greater sag of the larger pole when supported horizontally at its ends. What would happen to this ratio if the pole's length and diameter were tripled?

••Sample Problem 3

A gold prospector finds a solid rock that is composed solely of quartz and gold. The density of quartz is ρ_{quartz} and the density of ρ_{gold} is ρ_{gold}.
(a) Find the mass of gold contained in the rock of mass M with a volume V.

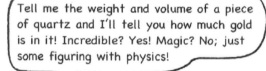

Focus: $m_{gold} = ?$

This is a difficult problem to solve because it involves two unknowns—the amount of gold and the amount of quartz. The "trick" is recognizing that the masses of the gold and quartz have to add to M, and that their volumes have to add to V.

(*Equation 1*): $m_{gold} + m_{quartz} = M$.

(*Equation 2*): $V_{gold} + V_{quartz} = V$.

What ties mass and volume together is density ρ: $\Rightarrow \rho_{gold} = \dfrac{m_{gold}}{V_{gold}}$ and $\rho_{quartz} = \dfrac{m_{quartz}}{V_{quartz}}$.

Strategy: We'll write Equation 1 in terms of m_{gold} and let the terms guide us to a solution.

$$m_{gold} = M - m_{quartz}.$$

$m_{gold} = M - \rho_{quartz} V_{quartz}$, and we know from Equation 2 that $V_{quartz} = V - V_{gold}$. So

$m_{gold} = M - \rho_{quartz}(V - V_{gold})$, and we know that $V_{gold} = \dfrac{m_{gold}}{\rho_{gold}}$. So

$$m_{gold} = M - \rho_{quartz}\left(V - \dfrac{m_{gold}}{\rho_{gold}}\right) = M - \rho_{quartz} V + m_{gold}\dfrac{\rho_{quartz}}{\rho_{gold}}.$$

Rearrange the equation so all of the m_{gold}'s are on the left side and we get

$m_{gold} - m_{gold}\dfrac{\rho_{quartz}}{\rho_{gold}} = M - \rho_{quartz} V$. The left side can be expressed as

$$m_{gold}\left(1 - \dfrac{\rho_{quartz}}{\rho_{gold}}\right) = m_{gold}\left(\dfrac{\rho_{gold} - \rho_{quartz}}{\rho_{gold}}\right).$$

Then from $m_{gold}\left(\dfrac{\rho_{gold} - \rho_{quartz}}{\rho_{gold}}\right) = M - \rho_{quartz} V$

$$\Rightarrow m_{gold} = \dfrac{\rho_{gold}(M - \rho_{quartz} V)}{(\rho_{gold} - \rho_{quartz})}.$$

How wonderful! The mass of the gold in the rock can be determined by knowing only the mass and volume of the rock, plus the densities of gold and quartz!

(b) Calculate the mass of gold contained in the rock if the rock's mass is 10.00 kg and its volume is 3.50×10^{-3} m^3. The density of quartz is 2650 kg/m^3 and the density of gold is 19,300 kg/m^3.

Solution:

$$m_{gold} = \dfrac{\rho_{gold}(M - \rho_{quartz} V)}{(\rho_{gold} - \rho_{quartz})} = \dfrac{19{,}300\,\frac{kg}{m^3}\left[10.00\ kg - 2650\,\frac{kg}{m^3}(3.50\times 10^{-3} m^3)\right]}{19{,}300\,\frac{kg}{m^3} - 2650\,\frac{kg}{m^3}} = 0.840\ kg.$$

This is 8.4% of the total mass of the rock.

(c) How much of the rock's volume is quartz?

Focus: $V_{quartz} = ?$

Since we know how much of the rock's mass is gold (0.840 kg), we know that the rest of the rock's mass (10.00 kg – 0.84 kg = 9.16 kg) must be quartz. Since we know the mass and density of the quartz, we can find its volume.

From $\rho = \dfrac{m}{V} \Rightarrow V_q = \dfrac{m_q}{\rho_q} = \dfrac{9.16\ kg}{2650\,\frac{kg}{m^3}} = 3.46\times 10^{-3}\ m^3$.

Percentage-wise, the quartz makes up

$\dfrac{3.46\times 10^{-3}\ m^3}{3.50\times 10^{-3}\ m^3} \times 100\% = \mathbf{98.9\ \%}$ **of the total volume of the rock.**

Solids Problems

12-1. A cube of metal with sides of length L is said to be pure platinum. Platinum has a density 2.14×10^4 kg/m^3, 21.4 times the density of water.
  (a) What is the volume of the platinum cube?
  (b) Suppose the cube is 0.35 m on each side. Calculate the mass of the cube.
  (c) Calculate the weight of the cube in newtons, and then in pounds. (Recall that 1 N = 0.22 lb.) Could you lift it?

12-2. Linsey has a cube of plastic with sides of length L and mass m.
  (a) What is its density?
  (b) Calculate its density if its mass is 15 kg and the length of its sides is 0.15 m.

12-3. A block of metal has dimensions l, w, h, and it has a mass m.
  (a) What is the density of the metal block?
  (b) Calculate the density of the block if its dimensions are $l = 0.20$ m, $w = 0.30$ m, $h = 0.15$ m, and its mass is 25 kg.

12-4. A circular slab of ice has a diameter D, thickness d, and density ρ.
  (a) What is the mass of the ice?
  (b) Calculate the mass if the diameter of the slab is 5.00 m, its thickness is 0.020 m, and its density is 917 kg/m^3.

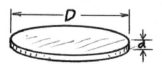

12-5. A liter of a certain liquid has a mass m.
  (a) What is its density?
  (b) Calculate its density if its mass is 13.6 kg. (Recall that 1 L = 10^{-3} m.)

12-6. A sphere of a certain material has a mass m and radius r. The volume of a sphere is $\frac{4}{3}\pi r^3$.
  (a) What is its density?
  (b) Calculate the sphere's density if its mass is 35 kg and its radius is 0.22 m.

12-7. Neutron stars have incredibly high densities.
  (a) Calculate the density of a spherical neutron star of mass 3.0×10^{28} kg and radius 1.3×10^3 m.
  (b) What is the mass of 1 teaspoon of this neutron star? 1 teaspoon is about 5 cubic centimeters.

12-8. A uniform lead sphere and a uniform aluminum sphere have the same volumes. (The density of lead is 11,344 kg/m^3 and the density of aluminum is 2700 kg/m^3.)
  (a) What is the ratio of densities for the lead and aluminum spheres?
  (b) What is the ratio of masses?

12-9. The density of water is 1000 kg/m^3 and the density of ice is 917 kg/m^3.
  (a) When a given mass of water changes phase to form ice, by what factor does its volume increase?
  (b) Which weighs more, a liter of water or a liter of ice? (Recall that 1 L = 1000 cm^3.)

12-10. A uniform lead sphere and a uniform aluminum sphere have the same mass. (The density of lead is 11,344 kg/m^3 and the density of aluminum is 2700 kg/m^3.)
(a) What is the ratio of radii for the lead and aluminum spheres?
(b) What is the ratio of their diameters?

12-11. • A silver prospector finds a solid rock that is composed solely of quartz and silver. (The density of quartz is 2650 kg/m^3 and the density of silver is 10,500 kg/m^3.)
(a) Calculate the mass of silver contained in the rock of mass 5.50 kg with a volume of 2.00×10^{-3} m^3.
(b) How much of the rock volume is quartz?

12-12. Silver can be pounded into extremely thin sheets. The density of silver is 10,500 kg/m^3.
(a) If a mass m of silver can be pounded into a sheet of thickness t, what is the area A of the sheet?
(b) Calculate the area of such a sheet 3.00×10^{-7} m thick made from 0.500 kg of silver.

12-13. In the lab you measure a rectangular piece of aluminum foil to have width w, length l, and mass m. The density of aluminum is ρ_{Al}.
(a) What is the thickness of the aluminum foil?
(b) Calculate the thickness of a piece of aluminum foil that is 9.4 cm wide, 79 cm long, and has a mass of 4.21 grams. (The density of aluminum is 2.70 g/cm^3.)

12-14. Consider a rectangular beam of length L and cross-sectional area A.

(a) What is its volume?
(b) If its linear dimensions are scaled up by a factor of 3, how will its volume increase?
(c) When scaled up by 3, by how much will its total surface area increase?
(d) What is the effect of scaling up the linear dimensions by 3 on the surface-area-to-volume ratio for the beam?
(e) Compared to the original beam, will the scaled-up beam be more or less likely to sag when supported at its ends?

12-15. A plastic cube has sides of length L.
(a) What is the area of one of its sides?
(b) What is the total surface area of the cube?
(c) What is the volume of the cube?
(d) What is the ratio of total surface area to volume?
(e) For a larger cube, is the total-surface-area-to-volume ratio more or less?

12-16. A cube of cheese with sides L is sliced into smaller cubes with sides $L/2$.
   (a) How many cubes result?
   (b) What is the area of each side of the smaller cube?
   (c) What is the total surface area of each smaller cube?
   (d) What is the sum of the total surface area of the smaller cubes, and how does it compare with the total surface area of the single cube with sides L?
   (e) What is the volume of each smaller cube?
   (f) What is the surface-area-to-volume ratio of the cube with side L?
   (g) What is the surface-area-to-volume ratio for each of the smaller cubes with side $L/2$?
   (h) How does the surface-area-to-volume ratio change as the cube is cut into smaller and smaller cubes?

12-17. • A spherical blob of mercury breaks into two identical smaller spherical blobs. (The surface area of a sphere is $4\pi r^2$ and the volume is $\frac{4}{3}\pi r^3$.)
   (a) How does the volume of the larger blob compare with that of each smaller blob?
   (b) How does the surface area of the larger blob compare with that of each smaller blob?
   (c) What is the surface-area-to-volume ratio for the larger blob?
   (d) What is the surface-area-to-volume ratio for each of the smaller blobs?
   (e) If each of the blobs is forced into a cube shape, how is the surface-area-to-volume ratio affected?
   (f) If the blobs break into even smaller ones, how is the surface-area-to-volume ratio affected?

12-18. Four little spheres of mercury, each with a radius r, coalesce to form a single sphere.
   (a) What will be the radius of the coalesced sphere?
   (b) How does the surface area of the coalesced sphere compare with the total surface areas of the four smaller ones?
   (c) Has the area/volume ratio increased or decreased for the coalesced sphere?

12-19. A storage tank is doing a good job supplying water to a population of people. The design of a new storage tank is being considered in a location that serves half the population. A rookie designer proposes that the new tank have half the height and half the diameter of the present tank.
   (a) How much smaller is the volume of the new tank compared with that of the present tank?
   (b) For painting purposes, how much smaller is the overall surface area of the new tank?
   (c) Why was the rookie's proposal rejected?

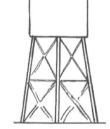

12-20. Gillian plans to decorate her party with helium balloons. The balloons come in two different diameters, x and y.
   (a) How many times more helium will she need if she gets the larger y-diameter balloons instead of the smaller x-diameter balloons?
   (b) Calculate how much more helium Gillian will need if she uses the same number of 9-inch diameter balloons instead of 7-inch diameter balloons.

Show-That Problems

12-21. A 2.900-kg solid cylinder that is 0.1200 m tall has a radius of 0.0300 m.
Show that the density of the cylinder is about 8550 kg/m^3.

12-22. A "stick" of butter (113 g) measures 8.1 cm long, 3.8 cm wide, and 4.0 cm tall.
(a) Show that the density of butter is about 0.92 g/cm^3.
(b) Show that the volume of 5 pounds (about 2.2 kg) of body fat (essentially butter) is about 2.4 liters.

12-23. Cork has a density of 300 kg/m^3.
Show that a cubic meter of cork would be too heavy for a person to lift.

12-24. A solid iron sphere ($\rho = 7874$ kg/m^3) has a diameter of 0.040 m.
Show that its mass is 0.26 kg.

12-25. A queen orders a 0.2000-kg crown made from pure gold. When it arrives from the crown maker, the volume of the crown is found to be 1.036×10^{-5} m^3. (Recall that $\rho_{gold} = 19{,}300$ kg/m^3.)
Show that the crown has the volume you'd expect if it were, indeed, pure gold.

12-26. A standard queen-sized waterbed mattress is 5 feet wide, 7 feet long, and 9 inches thick. Show that the water in such a mattress weighs more than 1600 pounds. (The density of water is 62.4 lbs/ft^3.)

12-27. A solid copper sphere of radius 0.0150 m is suspended from a vertical rubber band. The density of copper is 8960 kg/m^3.
Show that the tension in the rubber band is 1.24 N.

12-28. An aluminum can containing a half-liter of water has a mass of 0.525 kg.
Show that the volume of aluminum required to manufacture the can is 9.3×10^{-6} m^3. (The density of aluminum is 2700 kg/m³ and the density of water is 1.00 kg/L.)

12-29. Earth's mass is 5.98×10^{24} kg and Earth's radius is 6.37×10^6 m.
Show that the average density of Earth is approximately 5500 kg/m^3.

12-30. The Sun's mass is 1.99×10^{30} kg and Earth's radius is 6.96×10^8 m.
Show that the average density of the Sun is approximately one-fourth that of Earth's density.

12-31. Some neutron stars have the same mass as the Sun but much smaller radii.
Show that the density of a neutron star having the mass of the Sun but a radius of 20.0 km is 5.94×10^{16} kg/m^3.

12-32. A spherical tank has a 5-m^3 capacity.
Show that a 10-m^3 spherical tank would be 1.26 times larger in diameter than the original tank.

12-33. Judy decides to reduce both the length and width of her lawn by 25%.
Show that this smaller lawn will require only 56% of the water that the original lawn required.

12-34. A crystal cube has a mass of 12.0 kg.
Show that the same cube scaled down to half size would have a mass of 1.5 kg.

12-35. A bust of a famous president cast in solid brass has a mass of 12.0 kg.
Show that the same bust scaled down to half size would have a mass of 1.5 kg (using the logic of the previous problem).

12-36. A model steel bridge is 1/20 the exact scale of a bridge that is to be built.
Show that if the model bridge weighs 50 N, the real bridge will weigh about 400,000 N.

12-37. Toothpicks are made from a log that is 100 times longer and 100 times thicker than an individual toothpick.
Show that the weight of the log is a million times greater than the weight of one of the toothpicks (assuming no wastage in the production process).

12-38. When supported horizontally at both ends, the toothpick of the previous problem shows no noticeable sag. But the log from which it was made shows noticeable sag when similarly supported.
Show that each unit of cross-sectional area of the log supports 100 times more weight than each unit of cross section of the toothpick.

12-39. Two similarly proportioned people at the beach, one twice as heavy as the other, apply suntan lotion to their bodies.
Show that the amount of lotion needed by the twice-as-heavy sunbather is about 1.6 times as much.

12-40. Standard gas-fired tank water heaters typically have 1/2-inch-diameter gas lines and have 3-inch diameter exhaust vent pipes. Newer design on-demand water heaters typically require 3/4-inch-diameter gas lines and have 4-inch diameter exhaust vent pipes.
(a) Show that the 3/4-inch gas lines can carry twice as much gas as the 1/2-inch lines can.
(b) Show that the 4-inch vent pipes have almost twice the capacity of the 3-inch vent pipes.

12-41. A baker facing hard times decides that she can save some dough if she uses 10% less dough in each of her cookies. She rolls the dough to the same thickness as before but now makes smaller diameter cookies.
Show that her formerly 3.2-inch diameter cookies will now be 3.0 inches in diameter.

12-42. • Iron has a density of 7870 kg/m^3. This is equivalent to 8.48×10^{28} iron atoms in 1 cubic meter of iron. The atomic nucleus accounts for virtually all of the atoms mass. The iron nucleus has a radius of about 4×10^{-15} m.
(a) Show that in 1 cubic meter of iron, all of the iron nuclei taken together occupies only 2.27×10^{-14} m^3.
(b) Show that this would form a cube 28 μm to a side (1 μm = 10^{-6} m).
(c) Show that the density of the atomic nucleus is 3.5×10^{17} kg/m^3.

12-43. Ethanol has a mass density of 0.79 kg/liter and an energy density of 28 MJ/kg. (1 MJ = 10^6 J). Gasoline has a density is 0.73 kg/liter and an energy density of 44 MJ/kg.
Show that you would need 1.45 L of ethanol to provide the energy contained in 1 L of gasoline.

13 Liquids

The chapter begins with pressure, defined as the force per unit area. The SI unit for pressure is the *pascal* (Pa). 1 Pa = 1 N/m². 1 pascal is a small pressure. Standard atmospheric pressure is 1.01×10^5 Pa.

The key concept in this chapter is Archimedes's principle, which states, "An immersed body is buoyed up by a force equal to the weight of the fluid it displaces." If an object placed into a fluid displaces say, 2 N of fluid, the object will experience a 2-N upward force from the fluid. We call this upward force a *buoyant force*.

We conclude the chapter with problems that illustrate Pascal's principle, which states: "A change in pressure at any point in an enclosed fluid at rest is transmitted undiminished to all points in the fluid."

It will be useful to know that the density of water is 1000 kg/m³ = 1 kg/L = 1 g/cm³.

Sample Problem 1

A swimming pool has a uniform depth h and cross-sectional area A.
(a) What is the pressure due to water at the bottom of the pool?

Focus: $p_{water} = \dfrac{F}{A}$. A is given. What is F?

Taking only the water into account, the pressure at the bottom of the pool is due to the weight of the water spread over the area of the bottom of the pool.

$p_{water} = \dfrac{m_{water} g}{A}$. From $\rho = \dfrac{m}{V} \Rightarrow m = \rho V$.

So $p_{water} = \dfrac{m_{water} g}{A} = \dfrac{(\rho_{water} V_{water})g}{A} = \dfrac{\rho_{water}(Ah)g}{A} = \rho_{water} gh.$

Notice that the pressure is independent of the actual area of the pool. It only depends upon the depth of the water and its density. Let's confirm that ρgh yields units of pressure:

$$\rho\left(\dfrac{kg}{m^3}\right) g\left(\dfrac{m}{s^2}\right) h(m) = \dfrac{\left(kg \cdot \dfrac{m}{s^2}\right) \cdot m}{m^3} = \dfrac{N}{m^2} = Pa.$$

(b) What is the total pressure at the bottom of the pool if atmospheric pressure is P?

Focus: $p = ?$

The total pressure at the bottom of the pool will be the pressure due to the atmosphere plus the pressure due to the water.

$p = P + p_{water} = \mathbf{P + \rho_{water} gh}$

(c) Calculate the pressure at the bottom of the pool if it's 2.55 m deep, has a cross-sectional area of 50.0 m², and atmospheric pressure is 1.01×10^5 Pa.

Solution: $p = P + \rho_{water} gh = 1.01 \times 10^5 \text{ Pa} + \left(1000 \dfrac{kg}{m^3}\right)\left(9.8 \dfrac{m}{s^2}\right)(2.55 m) = \mathbf{1.26 \times 10^5}$ **Pa**.

© Paul G. Hewitt and Phillip R. Wolf

Sample Problem 2

A rectangular block of wood of mass m floats in water, with $x\%$ of the block submerged.
(a) What weight of water does the block displace?

Analysis: The floating block is at rest, so the net force on it is zero ($\Sigma F = 0$), which means that the upward buoyant force on the block must exactly balance the downward gravitational force on it. From Archimedes's principle, the buoyant force (BF) on the block must equal the weight of the water that the block displaces, so from

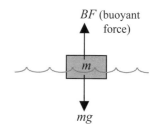

$BF = mg$ and $BF =$ weight of water displaced

$\Rightarrow$ the weight of water displaced = $\boldsymbol{mg}$.

A corollary to Archimedes's principle is the *principle of flotation*, which states that the BF on a floating body is equal to the body's weight.

(b) What is the density of the wood?

Focus: $\rho_{wood} = \dfrac{m}{V_{block}}$. We know m. What is V_{block}? We are given that

$V_{submerged} = \dfrac{x}{100} V_{block} \Rightarrow V_{block} = \dfrac{100}{x} V_{submerged}$. So now we need to find $V_{submerged}$.

(Notice how the equations guide our thinking about the problem!) As in part (a)

$BF = m_{water\,displaced}\, g = \rho_{water} V_{submerged}\, g$ and $BF = m_{block}\, g$, so

$\rho_{water} V_{submerged}\, g = m_{block}\, g \Rightarrow V_{submerged} = \dfrac{m}{\rho_{water}}$.

Now we can put all of this together:

$$\rho_{wood} = \dfrac{m}{V_{block}} = \dfrac{m}{\left(\dfrac{100}{x} V_{submerged}\right)} = \dfrac{m}{\left(\dfrac{100}{x}\dfrac{m}{\rho_{water}}\right)} = \dfrac{x}{100}\rho_{water}.$$

(c) Calculate the density of a balsa wood block that floats with 14% of its volume under water.

Solution: $\rho_{wood} = \dfrac{x}{100}\rho_{water} = \dfrac{14}{100}\left(1000\dfrac{\text{kg}}{\text{m}^3}\right) = 140\dfrac{\text{kg}}{\text{m}^3}$.

Sample Problem 3

A chunk of ore suspended by a light string has weight W in air. When the chunk is totally immersed in a container of water, tension in the string is w.
(a) How large is the buoyant force acting on the immersed chunk of ore?

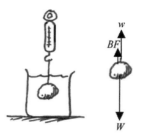

Analysis: The chunk is in equilibrium, so $\Sigma F = 0$. The upward forces on the ore are the tension w in the string and the buoyant force; the downward force is the ore's weight.

$\Sigma F = \text{Tension} + BF - \text{Weight} = 0 \Rightarrow w + BF - W = 0 \Rightarrow \boldsymbol{BF = W - w}$.

(b) Derive an equation for the density of the ore.

Focus: $\rho_{ore} = ?$ We know that $\rho_{ore} = \dfrac{m_{ore}}{V_{ore}}$. We also know that $m_{ore} = \dfrac{W}{g}$, so $\rho_{ore} = \dfrac{W}{V_{ore}g}$.

So now we need to find V_{ore}.

Since the chunk is totally submerged,

$$V_{ore} = V_{displaced} = \dfrac{m_{water\,displaced}}{\rho_{water}} = \dfrac{\left(\dfrac{\text{weight of water displaced}}{g}\right)}{\rho_{water}} = \dfrac{\left(\dfrac{BF}{g}\right)}{\rho_{water}} = \dfrac{W-w}{\rho_{water}\,g}.$$

Putting this all together: $\rho_{ore} = \dfrac{W}{gV_{ore}} = \dfrac{W}{g\left(\dfrac{W-w}{\rho_{water}\,g}\right)} = \left(\dfrac{W}{W-w}\right)\rho_{water}.$

Notice how the definition of density guides our way through solving the problem.

(c) Suppose the chunk is then totally immersed in another liquid of unknown density, and the tension in the string is 0.9w. What is the density of the unknown liquid? Give your answer in terms of W, w, and ρ_{water}.

Analysis: The tension in the string is less than before, so the new liquid must be exerting a greater buoyant force than the water does. The same *volume* of liquid is being displaced. This means that the displaced volume of the new liquid must weigh more than the same volume of water. The new liquid therefore has a higher density than water does. If the liquid were twice as dense, the weight of the liquid displaced by the ore would be twice as much, so the buoyant force would be twice as much.

We are given that $\dfrac{BF_{new}}{BF_0} = \dfrac{W-0.9w}{W-w}$ and we have figured out that $\dfrac{BF_{new}}{BF_0} = \dfrac{\rho_{new}}{\rho_{water}}$.

Putting this together:

$$\dfrac{W-0.9w}{W-w} = \dfrac{\rho_{new}}{\rho_{water}} \Rightarrow \rho_{new} = \left(\dfrac{W-0.9w}{W-w}\right)\rho_{water}.$$

(d) A chunk of ore is suspended from a spring scale that reads in grams. The scale reads 324 grams when the ore is in air, 220 grams when the ore is in water, and 198 grams (0.9 × 220 g) in glycerin. What are the densities of both the ore and the glycerin?

Solution: Notice that "weights" are given in grams, units of mass. Although weights are properly expressed in newtons, the direct proportion between weight and mass allows us to use grams in buoyancy problems. The g's in mg's cancel out, so we can use the mass readings directly. When doing buoyancy problems in the lab, it is common to talk of "gram-weights" or "kilogram-weights" as units of force.

$$\rho_{ore} = \left(\dfrac{W}{W-w}\right)\rho_{water} = \left(\dfrac{324\text{ g}}{324\text{ g}-220\text{ g}}\right)1000\dfrac{\text{kg}}{\text{m}^3} = 3120\dfrac{\text{kg}}{\text{m}^3}.$$

$$\rho_{glycerin} = \left(\dfrac{W-0.9w}{W-w}\right)\rho_{water} = \left(\dfrac{324\text{ g}-198\text{ g}}{324\text{ g}-220\text{ g}}\right)1000\dfrac{\text{kg}}{\text{m}^3} = 1210\dfrac{\text{kg}}{\text{m}^3}.$$

© Paul G. Hewitt and Phillip R. Wolf

Sample Problem 4

A pie pan with negligible mass floats on the surface of water in a small aquarium tank. When a piece of iron of density ρ and mass m is placed in the pie pan (which remains floating), the water level rises.

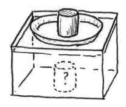

(a) What weight of water rises?

Answer: The pan floats at rest, so the buoyant force on the pan is equal to the weight of the piece of iron. By Archimedes's principle, the weight of the water displaced = the weight of the piece of iron = $m_{iron}\, g$.

(b) What volume of water rises?

Focus: $V_w = ?$ From $\rho = \dfrac{m}{V} \Rightarrow V_w = \dfrac{m_w}{\rho_w}$. From part (a), the weight of the water displaced is equal to the weight of the iron. So the mass of the water displaced is equal to the mass of the iron: $\Rightarrow V_w = \dfrac{m_{iron}}{\rho_w}$.

(c) What volume of water would rise if the iron were not floating, but at the bottom of the tank instead?

Focus: $V_w = ?$

When the iron was in the floating pie pan, the weight of water displaced was equal to the weight of the iron. This time the iron is totally submerged, so the *volume* of water displaced is equal to the *volume* of the iron. Therefore $V_w = V_{iron} = \dfrac{m_{iron}}{\rho_{iron}}$.

Since iron is almost 8 times denser than water, the volume of water displaced when the iron is submerged is about one-eighth the volume displaced when the iron is floating in the pie pan.

(d) When the iron "falls overboard" from the pie pan and sinks in the tank, does the water level of the tank rise, fall, or remain unchanged?

Answer: The water level **falls** compared with the water level when the iron floats in the pan. As mentioned in (c), less water is displaced when the iron is submerged than when it is in the floating pan.

Sample Problem 5

In order to lift a heavy load, a hydraulic machine injects oil into a closed cylinder with a pressure $p + P$, where P is atmospheric pressure and p is the so-called "gauge pressure"— the pressure beyond atmospheric. The plunger in the cylinder has a radius r, and has negligible mass compared with the loads to be lifted.

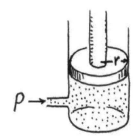

(a) What force is exerted on the plunger?

Focus: $F = ?$

The upward pressure on the plunger from the oil below it is $p + P$, and the downward pressure from the air above it is P, so the net upward pressure is p, which results in an upward force on the plunger. (We assume that changes of pressure within the closed cylinder caused by gravity can be neglected relative to the high pressure $p + P$. In other words, we take the pressure $p + P$ to be uniform within the cylinder.)

From $Pressure = \dfrac{F}{A} \Rightarrow F = pA = p\pi r^2$.

(b) Calculate the upward force exerted on the plunger if it has a radius 0.120 m and the gauge pressure input is 3.25×10^6 N/m².

$$F = p\pi r^2 = \left(3.25 \times 10^6 \, \frac{\text{N}}{\text{m}^2}\right)\pi(0.12\,\text{m})^2 = 1.5 \times 10^5 \, \text{N}.$$

(c) How many kilograms of mass can be lifted by this machine?

Focus: $m = ?$ If the machine is lifting at its maximum capacity, the lifting force of the plunger will be exactly balanced by the weight of the mass being lifted.

$$F_{\text{machine}} = mg \Rightarrow m = \frac{F_{\text{machine}}}{g} = \frac{1.5 \times 10^5 \, \text{N}}{9.8 \, \frac{\text{m}}{\text{s}^2}} = 1.5 \times 10^4 \, \text{kg}.$$

(d) Would the machine be able to lift a cube of granite, 1.70 m on a side? (The density of granite is 2750 kg/m³.)

Answer: **Yes**. The mass of such a cube = $\rho V = \left(2750 \, \frac{\text{kg}}{\text{m}^3}\right)(1.70\,\text{m})^3 = 1.35 \times 10^4$ kg, which is less than 1.5×10^4 kg.

Problems for Liquids

13-1. When you stand with two feet on a bathroom scale, your weight is given by the pointer or digital readout on the scale.
  (a) What happens to the pressure on the scale when you stand on one foot?
  (b) Does the scale show this increase in pressure? Defend your answer.

13-2. A petite girl wishes to test the strength of her boyfriend's stomach muscles. She stands on his stomach. Her weight is W and the total area of the bottom of her feet is A.
  (a) What is the pressure that her feet exert on his stomach?
  (b) Calculate the pressure she exerts if her weight is 400 N and the area of her two feet is 0.020 m².
  (c) Calculate the pressure she exerts if she's wearing spike heels and stands on only the heels of combined area 0.0002 m³.
  (d) Why are spike heels not a very good idea on a new linoleum floor (or on your stomach)?

13-3. A brick of mass m has length l, width w, and height h, with $l > w > h$.
  (a) What pressure does the brick exert on the floor when the brick is lying on its largest side?
  (b) Its smallest side?
  (c) Calculate answers for (a) and (b) for a 2.00-kg brick that is 20.0 cm × 8.9 cm × 5.8 cm.

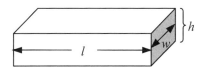

13-4. The circular glass plate on a scuba diver's mask has a radius r. Above the surface of a lake, the only force on the outside of the glass plate is that due to the pressure of the atmosphere. (Atmospheric pressure is normally 1.01×10^5 Pa.)
  (a) Write an equation for the force on the outside of the glass plate due to atmospheric pressure. Explain why this is not a net force.
  (b) The diver then dives to a depth D in the lake. What is the force on the mask due only to water?
  (c) Calculate answers to parts (a) and (b) if the radius of the mask is 0.075 m, and water depth D is 10.3 m.

13-5. A canoe of mass m floats on a freshwater lake.
  (a) What is the buoyant force acting on the canoe?
  (b) Find the volume of water V displaced by the canoe.
  (c) Calculate answers to parts (a) and (b) for a 33-kg canoe.

13-6. A rectangular block of length l, width w, and height h floats in water as shown.
  (a) What is the weight of the block?
  (b) What is the density of the block?
  (c) Calculate the weight and density of a block 25.0 cm long, 15.0 cm wide, and 11.0 cm high that floats with 9.0 cm of the block submerged.

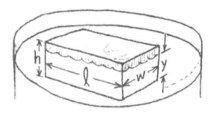

13-7. A ball of volume V and mass m, which normally floats, is held fully submerged in a fluid of density ρ.
  (a) Draw a force diagram for the ball when it is held underwater by a downward push P.
  (b) Write an equation for the buoyant force acting on the ball.
  (c) Write an equation for the gravitational pull downward on the ball.
  (d) How much downward push P would you have to exert on the ball to keep it totally submerged?
  (e) Calculate answers for a 0.71-kg, 4.4-liter ball in a fluid of density 975 kg/m^3.

13-8. The density of ice is 0.92 times the density of water. A slab of ice of volume V_{ice} floats on a freshwater lake.
  (a) Derive an equation for the minimum volume of the slab so that a 30-kg youngster could stand on it without getting her feet wet.
  (b) If the slab has thickness t, what must be the cross-sectional area A?
  (c) Calculate the area of the slab if the slab is 0.11 m thick.

13-9. A coal barge of mass m floats in fresh water. When empty, the waterline is shown by the dashed line in the sketch.

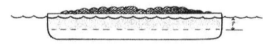

When loaded, the line is beneath the water surface. The area of the bottom of the barge is A, and the mass of coal is M.
  (a) How large is the buoyant force keeping the loaded barge afloat?
  (b) Find the total volume of water displaced.
  (c) How much deeper into the water does the barge descend when loaded with coal?
  (d) Calculate numerical values for the above for a 5000-kg barge with a 10,000-kg load of coal and a bottom area of 30 m².

13-10. A submarine can rise or sink by controlling its overall density. The submarine has tanks that can be filled with water or air. To descend, water is allowed to flow in, increasing the sub's overall density. To rise, the water is forced out by compressed air that lowers the sub's overall density.

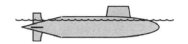

(a) Suppose that a submarine of overall volume V floats on the ocean's surface with $x\%$ of its volume above the surface. Find the minimum volume of seawater that the sub must take on to totally submerge beneath the surface.
(b) Calculate the minimum volume of seawater intake needed to submerge a 25,000-m³ sub initially floating with 12% of its volume above water. (The density of seawater is 1025 kg/m³.)

13-11. A rock suspended by a spring scale in air weighs W. When suspended while submerged in water, it weighs w.*

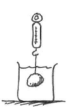

(a) Find the volume of the rock.
(b) Find the density of the rock.
(c) Calculate these values if the rock weighs 16.5 N in air and 10.3 N in water.

13-12. A piece of metal suspended by a spring scale in air weighs W and has density ρ_m.
(a) What will the scale read if the metal is totally submerged in water?
(b) Calculate the weight in water of a 0.65-N piece of aluminum. ($\rho_{Al} = 2.70$ g/cm³.)

13-13. A metal cylinder of weight w is suspended in a container of water of weight W, fully immersed without touching the bottom. The container is placed on a weighing scale.

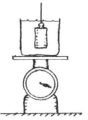

(a) Will the reading on the scale be $w + W$, less than $w + W$, or more than $w + W$? Defend your answer.
(b) What will be the reading if the cylinder is released and allowed to rest on the bottom of the container of water?

13-14. In the lab you find that a piece of metal has weight W when you hang it from a spring scale in air, but when the metal is suspended in water, the scale reads weight w.
(a) Find the density of the metal.
(b) Calculate the density if the scale reads 66 g when the metal is in air, but only 54 g when the metal is suspended in water.

---

* "Weight in water" means "This is what the scale reads when the rock is submerged in the water."

13-15. A wooden block of mass m is placed on a scale next to a beaker of water, which has a mass M.
  (a) What is the reading on the scale in newtons?
  (b) If the block is placed in the beaker and floats, what will be the reading on the scale?
  (c) If the block is instead a solid cube of iron of mass $8m$, what will be the reading on the scale when the cube is beside the beaker?
  (d) What will be the reading if the iron cube is sitting on the bottom of the beaker?

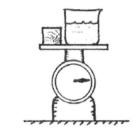

13-16. A wooden ball of mass m and volume V is held beneath the surface of water in a swimming pool.
  (a) What is the buoyant force acting on the ball?
  (b) How much force must be exerted to hold the ball down?
  (c) What would be the tension in a cord attached to the bottom of the pool, holding the ball motionless beneath the surface?
  (d) Are your answers to (b) and (c) the same or different? Defend your answer.
  (e) Calculate the tension in the cord when a wooden ball of mass 0.056 kg and volume 7.2×10^{-5} m^3 is attached to the bottom of a water-filled pool.

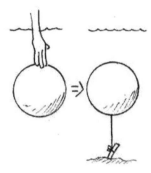

13-17. A lump of clay has twice the density of water. When it is put in water, it sinks.
  (a) When the clay has sunk, is the volume of water displaced less than, the same as, or greater than the clay's volume?
  (b) The clay is removed and reformed into the shape of a bowl or a boat, which floats on water. Is the volume of water displaced by the clay less than, the same as, or greater than the volume of clay itself?
  (c) Calculate the buoyant force on the clay of mass 1.00 kg when submerged in water.
  (d) Calculate the buoyant force on it when it is shaped like a boat and floats.

13-18. A lobster walks onto a bathroom scale on the ocean bottom. The scale, calibrated for normal weight readings in air, shows a weight w.
  (a) If the density of the lobster is twice the density of seawater, what is the mass of the lobster? (The density of seawater is 1025 kg/m^3.)
  (b) Find the volume of the lobster.
  (c) If the scale reads 1.0 N when on the ocean bottom, find the mass and volume of the lobster.

13-19. A fish of weight w is in the middle of a container of water of weight W. The container with the fish inside is placed on a weighing scale that reads in newtons.
  (a) What is the average density of the fish? Defend your answer.
  (b) What is the volume of the fish?
  (c) What is the reading on the weighing scale?
  (d) What will be the scale reading if the fish expands its air sac and floats on the water?

13-20. • A copper sphere of radius R is going to be partially hollowed out so that it will just barely float when submerged in 20°C water.

$(\rho_{w,\,20°C} = 998.2$ kg/m^3 $\rho_{copper} = 8960$ kg/m^3.)

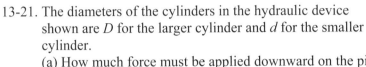

(a) What must be the inner radius r of the copper sphere? (Ignore the weight of the air inside the sphere.)
(b) Find the thickness of the walls of the sphere.
(c) Calculate the thickness of the walls of a copper sphere of outer radius 4.000 cm that will just barely float in 20°C water.
(d) When the water and copper are heated together, the water expands more than the copper does. If you heat the water (and therefore the copper) slightly, will the sphere rise or sink? Defend your answer.

13-21. The diameters of the cylinders in the hydraulic device shown are D for the larger cylinder and d for the smaller cylinder.

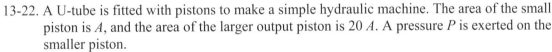

(a) How much force must be applied downward on the piston of the smaller cylinder to exert an upward force F on the piston of the larger cylinder?
(b) How far down must the smaller piston be pushed to lift the larger piston by a distance H?
(c) Calculate the force required on a 5.0-cm diameter piston to raise a 1500-kg load mounted on a 25-cm diameter piston. Calculate also the distance moved by the smaller piston to raise the larger piston by 64 cm.

13-22. A U-tube is fitted with pistons to make a simple hydraulic machine. The area of the small piston is A, and the area of the larger output piston is $20\,A$. A pressure P is exerted on the smaller piston.
(a) What is the pressure exerted on the larger piston?
(b) If force F acts on the smaller piston to produce pressure P, how much force acts on the larger piston?
(c) Calculate the force acting on the larger piston if a 450-N force acts on the smaller piston, whose area is 12 cm^2.

13-23. The output piston of a hydraulic press has a cross-sectional area A.
(a) How much fluid pressure is needed to produce a force F on the output piston?
(b) If the area of the input piston is 0.20A, what force is needed on the input piston to produce a force F on the output piston?
(c) Calculate the input force needed to produce an output force of 1.5×10^6 N for a 0.25 m^2 output piston.

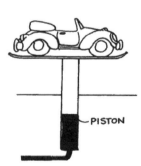

13-24. In an automotive service station, a car is hoisted with a hydraulic press. Pressurized fluid is pumped into a cylinder below a piston that is raised, and which raises the car.
   (a) If the fluid is pressurized to a gauge pressure P^*, and the area of the piston that raises the car is A, what weight can the piston support?
   (b) Calculate the weight that the piston can support if input gauge pressure is 3.5×10^6 Pa, the area of the piston is 0.40 m^2, and the weight of the hydraulic fluid is neglected. (Recall that 1 Pa = 1 N/m^2.)

13-25. A backhoe uses a hydraulic cylinder to lift heavy loads. A pump injects hydraulic oil into a cylinder at a gauge pressure P and drives an output plunger, which has a cross-section A.
   (a) What force can the output plunger exert?
   (b) Calculate this force if the input gauge pressure is 3.3×10^6 Pa and the cross section of the piston is 0.070 m^2.

Show-That Problems for Liquids

13-26. A sea wall in Holland springs a leak through a hole of area 1 cm^2 at a depth of 2 meters below the ocean's surface.
   Show that in order for a boy to plug the leak with his thumb, he must apply a 2-N force.

13-27. A flat-bottomed swimming pool has an area of 30.0 m × 10.0 m and a depth of 2.0 m. Show that the total weight of water in the pool when full is about 5.9×10^6 N, and that the pressure due to this weight of water on the pool bottom is nearly 20,000 Pa (recall that 1 Pa = 1 N/m^2).

13-28. A water tower supplies water to The Villages in Florida. The tower is 50 m high.
   Show that the pressure at the base of the tower due to the water is nearly 500,000 Pa.

13-29. Water next to the Hoover Dam is 221 m deep.
   Show that the pressure at the base of the dam due to the water is about 2.2 million Pa.

13-30. The vertical face of a reservoir dam is 105 m wide and 12.0 m high.
   (a) Show that when water reaches the crest of the dam, the average water pressure on the face is about 60,000 Pa.
   (b) Show that the outward force on the dam is over 7×10^7 N.

13-31. The Mariana Trench off the coast of the Philippines is about 11,000 m deep. The average density of seawater is 1025 kg/m^3.
   Show that the pressure due to the water on the window of a submersible at that depth is more than 1000 times atmospheric pressure.

---

* *Gauge pressure* is the difference between the actual absolute pressure and the atmospheric pressure. The fluid on the inside of the cylinder pushes the gauge in one direction and atmospheric pressure pushes it in the opposite direction. The gauge reads the *difference* between those two pressures. Here the gauge pressure is the *net* pressure acting upward on the piston.

13-32. A wooden block 3.8 cm tall, 7.6 cm wide, and 13 cm long floats in water. When 91 g of metal weights are evenly distributed on the top of the block, it becomes barely submerged.
Show that the density of the block is 0.76 g/cm³.

13-33. An air mattress floating in a pool has dimensions 2.0 m × 0.60 m × 0.20 m, and has a mass of 1.0 kg (including the air within it).
Show that it can support (barely) a group of children with a total mass of 230 kg, but that if another child with a mass of 30 kg joins the party, it will sink.

13-34. The density of seawater is 1.03×10^3 kg/m³. (The density of ice is 0.92×10^3 kg/m³.)
Show that 11% of the total volume of an iceberg extends above water level.

13-35. An aluminum cube 0.15 m on each side is submerged in water.
Show that the buoyant force that acts on the cube is 33 N.

13-36. A sample of pure gold of mass 1.20 kg is placed in a container of water.
Show that the volume of water displaced is 6.22×10^{-5} m³. (The density of gold is 19.3 times that of water.)

13-37. • In the lab, you place the edge of a pan balance slightly over the edge of the lab table and then hang an aluminum cylinder from the pan. The pan balance is level when adjusted to read 57.2 grams. When you submerge the hanging aluminum cylinder in a can of water already filled to the brim, water overflows from the can into a beaker. With the aluminum suspended in the water, the pan balance levels again when the balance weights are adjusted to read 36.0 grams.

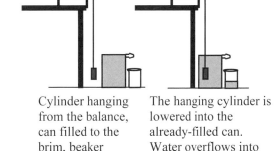

Cylinder hanging from the balance, can filled to the brim, beaker empty

The hanging cylinder is lowered into the already-filled can. Water overflows into the beaker.

(a) Show that 21.2 grams of water overflows into the beaker.
(b) Show that the density of the aluminum is 2.7 g/cm³.

13-38. A log floats in water with one-fourth of its volume above the surface.
Show that the density of the log is 750 kg/m³.

13-39. A ferry boat is 4.0 m wide and 6.0 m long. When Big Freddy steps on board, the boat sinks 0.0060 m deeper in the water.
Show that Big Freddy weighs about 1400 N.

13-40. A rectangular air mattress of dimensions 2.0 m × 0.60 m × 0.10 m floats in a swimming pool.
Show that, including its own mass, the maximum mass it can support before being completely submerged is 120 kg.

13-41. A novice weightlifter interested in aerobics can just barely lift a 30-kg iron barbell in the gym. (The density of iron is 7874 kg/m^3.)
Show that when the iron barbell is completely underwater, the weightlifter can lift nearly 34 kg.

13-42. While at a birthday party, Phil notices some apples floating in a half-barrel of water. He thinks of an idea to measure an apple's density. He carefully marks the part of the apple sticking above water, and then cuts it off and weighs it. The entire apple was 119 g, and the part above the water was 20 g.
Show that the average density of the apple is 0.83 times that of water.

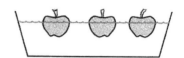

13-43. Martin decides to make a raft from a 34-kg sheet of plywood. His plan is to glue empty 2-L soda bottles to the bottom of the plywood sheet so that he (80 kg) can sit on the raft and read in the middle of his swimming pool without the plywood getting wet. Assume that the weight of the bottles and glue can be ignored.
Show that Martin will need to use at least 57 bottles.

13-44. A submarine has a weight of 10,000 tons. When floating partially submerged, it displaces 10,000 tons of water.
Show that, using reasoning, appreciably more than 10,000 tons of water must be displaced when the sub is in equilibrium beneath the water's surface.

13-45. • A 1967 Kennedy half-dollar is a mixture of silver and copper and has a mass of 0.01150 kg. When submerged in water, the coin weighs 0.1011 N.
Show that the coin contains nearly equal volumes of silver and copper. (The density of silver is 10,490 kg/m^3; the density of copper is 8,920 kg/m^3.)

13-46. A U-tube with pistons fitted at both ends comprises a hydraulic machine.
Show that the ratio of input to output piston areas A_{input}/A_{output} is equal to the ratio of input to output forces F_{input}/F_{output}.

13-47. A force of 100 N is applied to the small cylinder of the hydraulic device shown. The output cylinder has a diameter 2.3 times greater than the input cylinder.
Show that the output force is 530 N.

13-48. The piston in the small cylinder of the previous problem moves 6.9 cm downward when it does work on the system.
Show that the output cylinder moves a distance of 1.3 cm.

13-49. A machine uses a hydraulic cylinder to lift a heavy load. Hydraulic oil is injected into the cylinder at a gauge pressure of 4.06×10^6 Pa, which drives the output plunger. The output plunger has a radius of 0.120 m.
Show that the weight the device can lift is 1.84×10^5 N.

14 Gases

Much of the physics of liquids applies to gases, for both are fluids—gases being much less dense than liquids. Another big difference between liquids and gases is that, in many applications, the density of a liquid remains essentially constant over a huge range of pressures, whereas the density of a gas is quite sensitive to changes in pressure. At constant temperature, gas density and pressure are related by *Boyle's law*, $P_1/\rho_1 = P_2/\rho_2$, which can also be written $P_1 V_1 = P_2 V_2$, where P is pressure, ρ is density, and V is volume. Gas pressures are usually measured in units of pascals (N/m²) or in atmospheres (atm). Standard atmospheric pressure is 1 atm = 1.01×10^5 Pa.

The physics of fluids in motion is described by Bernoulli's principle, $P + \frac{1}{2}\rho v^2 = $ constant in steady horizontal flow. This means that for water flow in a horizontal pipe, $P + \frac{1}{2}\rho v^2$ in one section of the pipe is equal to $P + \frac{1}{2}\rho v^2$ in another. So as speed increases, pressure decreases (and vice versa).

Sample Problem 1

Air has weight. The density of 20°C air at sea level is 1.20 kg/m³.
(a) Calculate the weight of air in a typical school gym with rectangular dimensions 60.0 m × 30.0 m × 10.0 m.

Focus: $W = ?$

From the dimensions of the room, we can figure out the volume of the air, and from the volume and density, we can determine the mass.

From $\rho = \dfrac{m}{V} \Rightarrow m = \rho V.$ $W = mg = \rho V g = \rho(l \times w \times h)g$

$= 1.20 \dfrac{\text{kg}}{\text{m}^3}(60.0\,\text{m} \times 30.0\,\text{m} \times 10.0\,\text{m})\left(9.8\dfrac{\text{m}}{\text{s}^2}\right) = \mathbf{212{,}000\,N}.$

(b) What would this weight of air be if it were compressed?

Answer: Compression of air changes volume but not weight. It would still weigh **212,000 N**.

(c) Could you lift this weight of air if it were compressed?

Answer: **No**. This weight of air is over 20 tons!

Sample Problem 2

A SCUBA tank with an internal volume of 11.0 liters is slowly filled with air to bring its internal pressure up to 200 atmospheres.
(a) What volume of 20°C air at normal 1.00 atm pressure will fill the tank at 200 atm?

Focus: $V_{1\text{-atm}} = ?$ According to Boyle's law, since the gas pressure in the tank has increased 200 times, the gas volume must have decreased 200 times. It will take 200 × 11.0 L = 2220 L of room air to fill the SCUBA tank. More formally, from $P_1 V_1 = P_2 V_2 \Rightarrow V_1 = \dfrac{P_2}{P_1} V_2 = \left(\dfrac{200\,\text{atm}}{1\,\text{atm}}\right)(11.0\,\text{L}) = \mathbf{2200\,L}.$

© Paul G. Hewitt and Phillip R. Wolf

(b) How much more does the filled tank weigh compared to the "empty" one?

Focus: $mg = ?$ The "empty" tank originally had air in it at a pressure of 1 atm. We had to add 199 times more air molecules (that's 199 × 11.0 L of 1-atm air) to bring the pressure in the tank up to 200 atm.

From $\rho = \dfrac{m}{V}$ $\Rightarrow m_{\text{air added}} = \rho_{\text{air}} V_{\text{1-atm air}} = 1.20 \dfrac{\text{kg}}{\text{m}^3}\left(199 \times 11.0 \text{ L} \times \dfrac{10^{-3} \text{ m}^3}{1 \text{ L}}\right) = 2.63 \text{ kg}.$

The weight of this added air is $mg = 2.63 \text{ kg}\left(9.8 \dfrac{\text{m}}{\text{s}^2}\right) =$ **26 N**, about 6 lbs heavier.

Sample Problem 3

A buoyant force acts on us at all times. It also acts on inflated balloons.
(a) Lillian has a mass of 47 kg. What approximate volume of air of density 1.20 kg/m³ matches her mass?

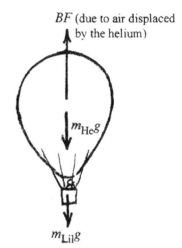

Focus: $V = ?$ From $\rho = \dfrac{m}{V}$ $\Rightarrow V = \dfrac{m}{\rho} = \dfrac{47 \text{ kg}}{1.20 \dfrac{\text{kg}}{\text{m}^3}} =$ **39 m³**.

(b) How big would a helium-filled balloon need to be to lift Lillian? Neglect the buoyancy on Lillian herself. Assume the mass of the balloon itself is negligible and that the density of helium is 0.178 kg/m³.

Focus: $V = ?$ The buoyant force on the balloon has to be just enough to lift the weight of the helium in the balloon and Lillian's weight.

From $BF = m_{\text{air displaced}}\, g$ and $BF = m_{\text{He}}\, g + m_{\text{Lil}}\, g$ $\Rightarrow m_{\text{air displaced}} = m_{\text{He}} + m_{\text{Lil}}.$

We can get V into our equation from the relation $\rho = \dfrac{m}{V}$ $\Rightarrow m = \rho V.$

The volume of helium and the volume of air displaced are the same, so

$m_{\text{air displaced}} = m_{\text{Lil}} + m_{\text{He}}$ $\Rightarrow \rho_{\text{air}} V = m_{\text{Lil}} + \rho_{\text{He}} V$ $\Rightarrow (\rho_{\text{air}} - \rho_{\text{He}})V = m_{\text{Lil}}$

$\Rightarrow V = \dfrac{m_{\text{Lil}}}{\rho_{\text{air}} - \rho_{\text{He}}} = \dfrac{47 \text{ kg}}{1.20 \dfrac{\text{kg}}{\text{m}^3} - 0.178 \dfrac{\text{kg}}{\text{m}^3}} =$ **46 m³**.

This is a balloon with a diameter of about 4.5 meters.

Sample Problem 4

Water flows through a horizontal pipe into a horizontal narrower pipe. Water pressure is P where the speed is v_1 through the larger cross-sectional area A_1. The flow continues through the smaller section of cross-sectional area A_2.

(a) Find the speed where the area of the pipe A_2 is one-fourth the area of A_1.

Analysis: Let's consider a "chunk" of fluid flowing through the pipe. We see the chunk at some time t_1 where the pipe has a cross-sectional area A_1, and the same chunk a little while later at time t_2 where the pipe has a cross-sectional area A_2. If the fluid is *incompressible* (as with water, where its density remains constant even as the pressure changes), then the volume of the fluid at these two times must be the same, so $A_1 \Delta x_1 = A_2 \Delta x_2$.

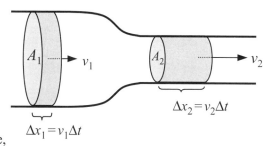

Referring to the diagram: It takes the same amount of time for the fluid to travel a distance Δx_1 in the wider section of pipe as it does for that same volume of fluid to travel a distance Δx_2 in the narrower section of pipe. Again, fluid doesn't bunch up. This means that $A_1 \Delta x_1 = A_2 \Delta x_2$ becomes $A_1(v_1 \Delta t) = A_2(v_2 \Delta t) \Rightarrow A_1 v_1 = A_2 v_2$. This last expression can be thought of as a restatement of the law of conservation of mass for an incompressible moving fluid, and is called the *equation of continuity*.

For our particular problem we have $A_2 = A_1/4$, and we're looking for v_2.

From $A_1 v_1 = A_2 v_2 \Rightarrow v_2 = \dfrac{A_1}{A_2} v_1 = \dfrac{A}{\left(\frac{A}{4}\right)} v_1 = 4v_1$.

(b) If the pressure in the wide part of the pipe is P_1, what is the pressure in the narrower section of pipe?

Analysis:

Bernoulli's principle relates pressure changes to speed changes. In steady horizontal flow,

$P_1 + \dfrac{1}{2}\rho v_1^2 = P_2 + \dfrac{1}{2}\rho v_2^2$, so it follows that the pressure in the narrower pipe section is

$P_2 = P_1 - \dfrac{1}{2}\rho(v_2^2 - v_1^2)$. In this example, $v_2 = 4v_1$, so

$P_2 = P_1 - \dfrac{1}{2}\rho\left((4v_1)^2 - v_1^2\right) = P_1 - 7.5\rho v_1^2$.

(c) If the speed of the water in the wider pipe is 4.0 m/s, what is the change of pressure from the wider to the narrower pipe? How many atmospheres of pressure change is this? Use $\rho_{water} = 1000$ kg/m^3.

Solution:

$P_2 - P_1 = -7.5\rho v_1^2 = -7.5\left(1000\dfrac{\text{kg}}{\text{m}^3}\right)\left(4.0\dfrac{\text{m}}{\text{s}}\right)^2 = -120,000$ Pa.

Divide this by atmospheric pressure, 101,000 Pa, to get a pressure **decrease of about 1.2 atmospheres**.

Problems for Gases

14-1. The dimensions of a typical living room are 4.0 m × 5.0 m × 2.5 m.
   (a) What is the weight of air in the room? Use $\rho_{air} = 1.20$ kg/m^3.
   (b) What are the mass and weight of an equal volume of water?
   (c) Would a standard floor be able to support this weight of water? Defend your answer, with the knowledge that floors are usually designed to support a static load of 1900 N/m^2 or so.

14-2. A Cartesian diver consists of a partially filled eyedropper submerged in a water-filled plastic bottle. When you squeeze the bottle, the pressure inside increases, which compresses the air inside the dropper. The water moving into the dropper makes the dropper denser—and it sinks. When you stop squeezing the bottle, a decrease in pressure results, density of the dropper decreases, and the eyedropper moves upward.
   (a) When the eyedropper is motionless, how does buoyant force on it compare with its weight (including the water in it)?
   (b) Can the eyedropper remain motionless if even more water gets into it? Defend your answer.
   (c) Calculate the buoyant force on the dropper when it is motionless if its mass with water inside is m.

14-3. An air-filled party balloon is squeezed to one-third its volume (without a change in its temperature).
   (a) Compared with the pressure before being squeezed, what is the new pressure of air in the balloon?
   (b) Compared with its mass before being squeezed, what is the mass of air after being squeezed?
   (c) Compared with the density before being squeezed, what is the density of air in the balloon after being squeezed?

14-4. Air in a cylinder is compressed to one-fifth its original volume with no change in temperature.
   (a) Find the pressure compared with the original pressure.
   (b) Find the density of the compressed gas compared with the original density.
   (c) Calculate the new pressure and density if the original pressure was 1.1×10^5 Pa and the original density was 1.3 kg/m^3.

14-5. A mercury barometer reads standard atmospheric pressure at sea level. When it is carried to an altitude of 5.6 km, the mercury column is half of its original height.
   (a) What is the air pressure at this altitude relative to sea-level pressure?
   (b) If the barometer is taken up another 5.6 km to an altitude of 11.2 km, will the height of its mercury column fall to zero? Defend your answer.

14-6. • Mala decides to make a barometer filled with oil of density ρ_{oil} rather than with mercury.
   (a) Derive an equation for the height of the oil column required to indicate standard atmospheric pressure.
   (b) Calculate the necessary height of the oil column if she uses oil with a density of 920 kg/m^3.
   (c) Would a small change in atmospheric pressure cause a bigger change in the height of the mercury-filled barometer or the oil-filled barometer? Explain. (For comparison, $\rho_{mercury}$ = 13,600 kg/m^3.)

14-7. Air pressure is nicely demonstrated with Magdeburg hemispheres. When the hemispheres are joined and then evacuated, the atmospheric pressure P holds the two halves together. The cross-sectional area of the circle where the hemispheres meet is A. (Interestingly, the net force over the curved surface of a hemisphere is the same as over a flat disk of area A.)
   (a) When the hemispheres are fully evacuated, what force is needed to pull them apart?
   (b) The air between the hemispheres is initially at atmospheric pressure. What force is needed to separate them when 80% of the air molecules have been evacuated from the hemispheres?
   (c) Calculate these values when atmospheric pressure is 1.01×10^5 Pa and the cross-sectional area A is 0.200 m^2.

14-8. Professor John tosses a rubber mat onto the top of a stool. With a hook secured to the top of the mat, he lifts the mat and stool together.
   (a) What is your explanation for the lifting of the stool?
   (b) If the area of the rubber mat is 0.90 m^2, what is the maximum weight it can support in this fashion (assuming the object being lifted is in contact with the full area of the mat)?

14-9. A popular classroom demo showing that air occupies space is dunking an empty drinking glass mouth downward in a container of water. Air in the glass prevents water from filling it.
   (a) Once submerged, how does the force needed to hold the cup in place change as the cup is pushed deeper into the water?
   (b) To what depth must the cup be submerged for the air in the glass to be compressed to half its initial volume?

14-10. A vacuum pump removes air from a cylinder of cross-sectional area A. A mass m is supported by a lightweight piston in the cylinder. Atmospheric pressure pushing up on the piston provides a force to balance the downward pull on the piston by the hanging mass.
   (a) Derive an equation for the largest mass that can be held up by the piston.
   (b) Calculate the mass that can be supported by a 29-cm^2 piston.

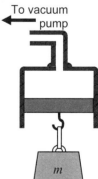

14-11. Pretend that the atmosphere of Earth consisted entirely of water vapor instead of oxygen and nitrogen, with the same atmospheric pressure that Earth now experiences.
  (a) If the water vapor condensed, what would be the depth of the liquid atmosphere? (*Hint*: Think of a water barometer.)
  (b) If our present atmosphere of nitrogen and oxygen were to condense, would it be less deep or deeper than the depth of a liquid water atmosphere? Nitrogen and oxygen in their liquid phases are respectively 0.8 and 0.9 times denser than water.

14-12. Professor John places a bit of dry ice in an uninflated balloon. He ties the balloon shut and sets it on a digital balance. As the dry ice sublimes (that is, changes into gas) and the balloon inflates, the reading on the scale goes down.
  (a) What is your explanation for the decreased scale reading?
  (b) If the scale reading decreases by about 2.4 grams, to what volume does the balloon inflate?

14-13. Professor John places a wooden shingle on the lecture table so it overhangs the edge a bit. He covers the shingle with a flattened sheet of newspaper, and then strikes the overhanging part of the shingle with a "karate chop" and breaks the shingle.
  (a) The dimensions of the newspaper are 70 cm × 56 cm. What "weight of air" bears down on the paper?
  (b) What role does inertia play in this popular classroom demonstration?

14-14. A balloon is weighted so that it's just barely submerged in water. You poke it a few centimeters deeper beneath the surface of the water.
  (a) Does the balloon sink, remain motionless, or return to the surface?
  (b) What happens to the volume of the balloon at this lower depth?
  (c) Supposing that the balloon sinks, how does depth affect the buoyant force on the balloon as sinking progresses?
  (d) How far would it have to sink so the buoyant force on the balloon is half what it was at the surface?

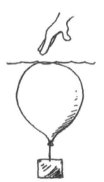

14-15. On a windless morning, Sung hovers motionless at low altitude in a hot-air balloon. The total weight of the balloon, including its load and the hot air, is W.
  (a) What is the weight of the displaced air?
  (b) What buoyant force acts on the balloon?
  (c) What is the buoyant force if the total weight of the balloon is 2700 N? (Why do you not need a calculator?)

14-16. A gas-filled balloon rises in air when it weighs less than the air it displaces. It is common to use a gas less dense than air to hold the sides of the balloon out so that it will displace a great amount of air. When filled with helium, the balloon's mass is m and its volume is V.
  (a) Write an equation for the buoyant force on the motionless balloon.
  (b) Derive an equation for the tension in the string that holds the balloon in place.

14-17. An empty balloon has a mass m. You wish to fill it with enough helium for it to be able to lift a small package of mass M. The density of helium in the balloon is 0.180 kg/m^3.
   (a) Derive an equation for the minimum volume that the balloon needs to lift the load.
   (b) Calculate the volume if the mass of the balloon is 150 kg and the package to be lifted has a mass of 1000 kg.

14-18. The Hindenburg was a German dirigible (also called a zeppelin) in the late 1930s, filled with bags containing hydrogen. The density of air at 0°C is 1.29 kg/m^3 and the density of hydrogen is 0.090 kg/m^3. The weight of the empty dirigible was 130 tons (1.18×10^6 N). Its volume, when filled, was 2.01×10^5 m^3.
   (a) What was the buoyant force on the filled dirigible?
   (b) How much force would be needed to hold it down if it were filled with a hypothetically massless gas?
   (c) What mass of hydrogen would completely fill the interior of the dirigible?
   (d) If completely filled with hydrogen gas, what would have been the net force on the dirigible?

14-19. Two identical, unfilled Mylar balloons are tied onto either side of a platform balance. The balloon on the right side is slowly filled with carbon dioxide (CO_2), which has a density of 1.84 kg/m^3 (= 1.84 g/L) at 20°C. Use $\rho_{air} = 1.20$ kg/m^3.
   (a) Which side of the balance goes down as gas is added to the balloon? Explain.
   (b) When a 1.00-gram mass is added to the balance (to which side?) to bring the balance level again, how much gas is in the balloon?
   (c) If, instead of carbon dioxide, you added a little bit of air to one of the Mylar balloons, would the platform balance move? Explain.

14-20. An airplane has a mass m and a total wing area A. During level flight, the pressure on the lower surface of the wings is P.
   (a) Draw a force diagram for the airplane. Then derive an equation for the pressure on the upper surface of the wings.
   (b) Will the air pressure on the lower surface of the wings increase, decrease, or remain the same when the angle of attack (that is, the angle that the wing makes with the oncoming air) is slightly increased?
   (c) How do the pressure differences on the upper and lower wing surfaces compare when the airplane is descending at a steady rate?

14-21. A home shower has a flow rate of V liters per minute through a pipe of internal diameter D.
   (a) Derive an equation for the speed of the water in the pipes.
   (b) Calculate the speed of water flowing through a 12.7-mm diameter pipe to a shower head that allows a flow of 10 liters per minute.

14-22. Water flows at speed v through a fire hose of cross-section A. The fire hose ends in a nozzle of smaller cross-section B.
   (a) Derive an equation for speed of water that exits the nozzle.
   (d) Calculate the exit speed of the water if the speed of water in the main hose is 4.0 m/s and the cross-section at the nozzle is one-fifth the area of the hose.

14-23. Air flows through a horizontal pipe, flowing first through a section of area A and then through a narrower section of area $0.40\,A$.
   (a) Compare the relative speeds of air in the wide and narrow sections of the pipe, assuming that the air's density doesn't change appreciably.
   (b) In which section of the pipe is air pressure greater?

14-24. A volume V of gasoline flows from a gas pump into a car's gas tank in a time t through a pipe of diameter D.
   (a) Derive an equation for the speed of the gasoline as it flows through the pipe.
   (b) Calculate the speed of the gasoline if 1 liter flows through a 2.0-cm diameter pipe into the car's gas tank every 8 seconds.

14-25. • A carburetor employs a *venturi tube*, which is a tube with a narrow constriction. Air flows at speed v through the air intake of diameter D and then speeds up as it travels through the narrow venturi neck of diameter d. Since the air is moving faster through the neck, the pressure in the neck is reduced. The difference between this lower pressure and the outside atmospheric pressure forces the fuel into the carburetor.

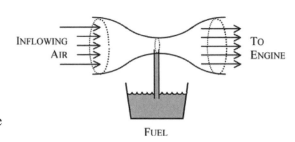

   (a) Starting with Bernoulli's principle, derive an equation that gives the pressure difference between the neck and the air-intake.
   (b) Calculate the pressure drop as 3.0 m/s air with density 1.20 kg/m³ moves from a 7.5-cm diameter intake tube to a 1.5-cm diameter venturi neck.

14-26. • As part of a lecture demonstration of Bernoulli's principle, Paul blows compressed air from a tank across the opening of a hollow straw of area A to lift a metal cylinder of mass m that sits inside the straw.

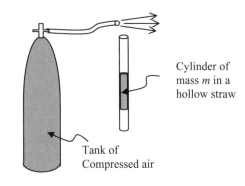

Cylinder of mass m in a hollow straw

   (a) Derive an equation for the minimum air speed that will raise the cylinder in the straw.
   (b) Determine the minimum air speed that will lift a 0.17-gram cylinder in a straw with $A = 0.30$ cm².

Show-That Problems for Gases

14-27. We usually don't notice atmospheric pressure, even though it's quite enormous.
   Show that the total weight of air pressing down on a sheet of newspaper of area 0.200 m² is 20,200 N.

14-28. A party balloon is squeezed to 2/3 of its initial volume.
   Show that the pressure in the balloon increases 1.5 times.

14-29. Suppose you have a syringe that contains air but has no needle. You put your thumb over the opening and then slowly squeeze the plunger of the syringe from the 15.0 mL mark to the 6.0 mL mark. Show that the pressure in the syringe will be 250 kPa.

14-30. A suction cup on a ceiling is able to support a 72-kg student who dangles from it. Show that the contact area between the cup and the ceiling must be a minimum of 0.0068 m^2.

14-31. You make a barometer by attaching a tube with a 0.040 cm^2 cross section to an airtight container. You place a small blob of mercury in the tube such that the volume of the trapped air is 250.0 cm^3. Now suppose that atmospheric pressure drops by 0.20%.
(a) Show that the volume of air expands to 250.5 cm^3.
(b) Show that the mercury blob will move 12.5 cm along the tube. (This simple device is a very sensitive way to measure changes in atmospheric pressure!)

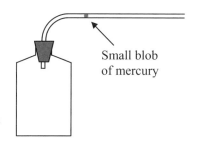

14-32. The air in the bicycle pump cylinder shown in the sketch is at atmospheric pressure ($\approx$100 kPa) while the air in the bicycle tire is at 340 kPa. (A one-way valve prevents air from flowing out of the tire.) The cylinder is 2.6 cm in diameter and the initial height of air in the pump cylinder is 55 cm.
(a) Show that you must push the pump handle downward by 39 cm before air will flow into the tire.
(b) Show that at that point you will be exerting a 130 N downward force on the pump handle.
(c) Which (if either) of your answers would be different if the area of the pump cylinder was half as much?

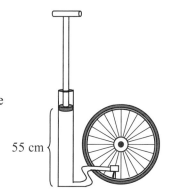

14-33. A spherical 5.0-kg balloon of volume 490 m^3 is filled with helium (density 0.179 kg/m^3). The balloon carries a load and floats motionless in air (density 1.29 kg/m^3). Show that the mass of the load is 540 kg.

14-34. In 1982 Larry Walters ascended from his home in Long Beach, California, to an altitude of 4900 m (16,000 ft) after tying 42 helium-filled, 1.9-m diameter weather balloons to his patio chair. Show that the buoyant force on these balloons at sea level would be about 1800 N.

14-35. The air inside a hot-air balloon is typically around 100°C. At this temperature the density of air is around 0.89 kg/m^3. The mass of fuel, balloon fabric, passengers, burner, and basket totals to about 820 kg.
(a) Show that 2500 m^3 of hot air inside the balloon would be more than sufficient to get the balloon off the ground on a morning where the outdoor air temperature is 10°C (so ρ_{air} = 1.25 kg/m^3). (It is safe to assume that the volume occupied by the balloon fabric, passengers, etc. is negligible compared to the volume of hot air.)
(b) Show that later in the day, when it is 20°C outside and ρ_{air} = 1.20 kg/m^3, this same volume of hot air would *not* be sufficient to get the balloon off the ground.

14-36. Air is a liquid at –196°C. A liter of liquid air has a mass of about 0.87 kg. The density of gaseous air at room temperature is $\rho_{air} = 1.20$ kg/m^3.
Show that when the liquid air changes to the gaseous phase at room temperature, its volume will expand over 700-fold.

14-37. Average atmospheric pressure at Earth's surface is 1.01×10^5 N/m^2. Earth's radius is 6.37×10^6 m.
Show that the total weight of Earth's atmosphere is about 5.15×10^{19} N.

14-38. In Paul Doherty's physics classes at the Exploratorium, atmospheric pressure is nicely illustrated with a 1.31-m (4.3-foot) bar of steel with a cross section of 1 square inch. The bar weighs 14.7 pounds, and when it is placed upright on the palm of your hand you can appreciate the magnitude of atmospheric pressure. (Atmospheric pressure is 14.7 lb/in^2 at sea level.)
Show that to simulate atmospheric pressure on Mars, only a wafer 0.8 cm thick does the trick. (Atmospheric pressure on Mars is 0.006 that of Earth's.)

14-39. Continuing the previous problem, atmospheric pressure on the surface of Venus, on the other hand, is much greater: 90 times that of Earth's. Simulating this pressure requires a much longer bar than the 1.31-m bar for Earth.
Show that the length of a bar to simulate the atmospheric pressure of Venus would need to be 118 m tall (which couldn't be held by your hand!).

14-40. Referring to the previous problem, instead of using such a long bar to simulate atmospheric pressure on Venus, the cross section of the Earth bar at the very end can be reduced to show the same greater pressure (which, on the palm of your hand, would be dangerous to try to hold).
Show that the area at the end of the "Earth bar" should be 1/90th of a square inch.

14-41. Oil flows at a speed of 1.20 m/s through a pipeline with a radius of 0.30 m.
Show that about 30,000 cubic meters of oil flow in one day.

14-42. Water in a pipe flows into a section that has half the cross-sectional area of the initial pipe.
Show that the speed of the water doubles.

14-43. The speed of blood in a normal aorta is 0.40 m/s.
Show that if the cross-sectional area of the aorta is enlarged by a factor of 1.6, the same volume of blood will flow at a speed of 0.25 m/s.

14-44. A 2% difference in atmospheric pressure exists between the underside and top of a roof during strong winds.
Show that the resulting lifting force on a roof that has a projected horizontal area of 2000 m^2 is more than 4 million N.

14-45. Relative to an airplane, the speed of air over the plane's wing is 63 m/s and under its wing is 60 m/s. Suppose that the plane has a wing area of 24 m^2 and that the density of air at the airplane's flying altitude is 1.13 kg/m^3.
Show that the lifting force is about 5000 N.

14-46. A 1000-Megawatt nuclear power plant can require 60 m^3 of cooling water each second. Show that in order to get this flow through an intake pipe at an intake speed of approximately 2 m/s, the intake pipe would have to be 6.2 m in diameter.

15 Temperature, Heat, and Expansion

Simply put, temperature is a measure of the average translational KE of particles in a material. Raising the temperature of a material means increasing the average speed of the particles within it.

The amount of energy involved in a temperature change depends upon three factors:

1. The nature of the material involved.
   All else being equal, some materials require more energy per gram than others to raise their temperature by 1°C. This "amount of energy to raise the temperature of 1 gram of something by 1°C" is called the *specific heat capacity* of the material, and is given the symbol c. It has units of J/g·°C or kJ/kg·°C.[*]

2. The mass of the sample, m.
   (It takes more energy to raise the temperature of more particles of the same substance.)

3. The number of degrees of temperature change, ΔT.
   (A larger change in temperature requires more energy than a smaller temperature change.)

All of the above apply equally well to objects cooling or objects warming.

We can summarize the above factors as follows:

Energy involved in a temperature change = specific heat capacity of the substance × mass of material changing temperature × number of degrees change in temperature

or

$Q = cm\Delta T$. (Q is the conventional symbol for "quantity" of heat.)

As materials warm, their particles jiggle faster and move farther apart, on average. (An exception is rubber.) The result is an expansion of the material. The amount of expansion depends on the type of material and the temperature change. For the same 1-degree temperature change, some materials will expand more than others. The fractional change in length ($\Delta L/L$) per degree (ΔT) is called the *coefficient of linear expansion* and is given the symbol α, with units 1/°C.

So $\alpha = \dfrac{(\Delta L/L)}{\Delta T}$.

The change in length of an object depends on the material of which it is composed (which determines α), its initial length, and the number of degrees it is warmed or cooled. That is,

$\Delta L = \alpha L \Delta T$.

---

[*] Historically, people took equal masses of hot material (for example 10 grams of water, iron, lead, etc. at 100°C) and placed them on a chunk of ice. Each hot sample melted a different amount of ice. Whichever sample melted the most ice had the largest "heat capacity." *Specific* heat capacity is "heat capacity per gram." A specific heat capacity of 1 calorie/g·°C was assigned to water, and all of the other materials were assigned specific heat capacities based on how much ice they could melt compared with an equal amount of water. Today it is customary to use joules rather than calories.

© Paul G. Hewitt and Phillip R. Wolf

It doesn't make sense to speak of a change in length for liquids. We instead speak of a fractional change in volume per degree, called the *coefficient of volume expansion*, which is given the symbol β, also with units 1/°C. Analogous to linear expansion, we have the equation

$$\beta = \frac{\Delta V/V}{\Delta T}, \quad \text{or} \quad \Delta V = \beta V \Delta T.$$

Here is some information that will be useful in solving problems in this chapter:

 1 milliliter of water has a mass of 1 gram.
 1 liter of water has a mass of 1 kg.
 1 calorie = the amount of energy needed to raise the temperature of 1 gram of water by 1°C = 4.18 J.
 1 Calorie = 1 kcal = 1000 calories.

SPECIFIC HEAT CAPACITIES	
Material	c, Specific heat capacity (J/g·°C or kJ/kg·°C)
Water	4.18
Aluminum	0.900
Clay	1.4
Copper	0.386
Lead	0.128
Olive Oil	1.97
Silver	0.23
Steel (iron)	0.448

COEFFICIENTS OF LINEAR EXPANSION	
Material	α, Coefficient of Linear Expansion (1/°C)
Steel	11×10^{-6}
Brass	19×10^{-6}
Aluminum	23×10^{-6}
Glass	10×10^{-6}
Gold	14.3×10^{-6}

Sample Problem 1

Joan wishes to take a hot bath. She fills the tub with x liters of water at temperature T_{hot}.
(a) Write an equation for the amount of energy needed to bring the water from an initial temperature of T_{cold} up to T_{hot}. (As stated above, 1 liter of water has a mass of 1 kg.)

Focus: $Q = ?$

The energy required to change the water's temperature depends on the specific heat capacity of the water, c_w, the mass of water, and the temperature change $T_{hot} - T_{cold}$. Since 1 liter of water has a mass of 1 kg, m_{water} (in kg) = x (in liters). So
$Q = cm\Delta T = c_w x(T_{hot} - T_{cold})$.

(b) Calculate the amount of energy required to raise the temperature of 320 L of water from 18°C to 52°C.

Solution: $Q = c_w x(T_{hot} - T_{cold}) = \left(4.18 \frac{\text{kJ}}{\text{kg·C°}}\right)\left(320\,\text{L} \times \frac{1\,\text{kg}}{1\,\text{L of water}}\right)(52°\text{C} - 18°\text{C}) = \mathbf{45{,}000\ kJ}$.

This is equivalent to the energy content of 1.3 liters of gasoline, or enough to drive a reasonably fuel-efficient car about 10 miles.

Sample Problem 2

Harmon has taken the 2nd place silver medal at the County Fair for his collection of belly-button lint, but he doubts that his medal is really silver. He takes the medal, mass m_{medal}, and dunks it into a pot of boiling water in order to heat it to 100°C. He puts water of mass m_w into a Styrofoam cup and measures its initial temperature, T_0. He removes the medal from the boiling water and puts it into the Styrofoam cup. After stirring, the water in the cup ends up at T_f.

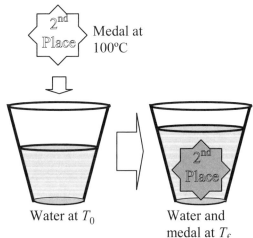

(a) What is the specific heat capacity of his medal?

Focus: $c_{medal} = ?$

Harmon's method depends on the law of conservation of energy as applied to heat transfer. Whatever energy the medal gives up, Q_{medal}, is gained by the water, Q_{water}.* That is,

$$|Q_{medal}| = |Q_{water}|.\ ^\blacklozenge$$

From $|Q_{medal}| = |Q_{water}| \Rightarrow c_{medal} m_{medal} |\Delta T_{medal}| = c_w m_w |\Delta T_w|$.

The water ends up warmer than before so $|\Delta T_w| = T_f - T_0$. The medal's final temperature is *less than* its initial temperature so $|\Delta T_{medal}| = 100°C - T_f$.

$$\Rightarrow c_{medal} = \frac{c_w m_w |\Delta T_w|}{m_{medal} |\Delta T_{medal}|} = \frac{c_w m_w (T_f - T_0)}{m_{medal} (100°C - T_f)}.$$

Harmon merely has to compare the value for his calculated specific heat capacity with the specific heat capacity of silver.

(b) Harmon's medal has a mass of 68 grams, the cup initially holds 155 grams of water at 21.0°C, and the final temperature of the water and medal is 24.5°C. Determine whether or not his medal is silver.

Solution: $c_{medal} = \dfrac{c_w m_w (T_f - T_0)}{m_{medal}(100°C - T_f)} = \dfrac{(155\text{g})(24.5°C - 21.0°C)\left(4.18 \frac{\text{J}}{\text{g}\cdot\text{C°}}\right)}{(68\text{g})(100°C - 24.5°C)} = 0.44 \dfrac{\text{J}}{\text{g}\cdot\text{C°}}.$

This result is far from the specific heat capacity of silver (0.23 J/g·°C), so the medal is definitely not silver. What might that metal be?

---

* In this problem and in all the others in this chapter, we assume that the system is *closed*—that is, no energy leaves or enters the system. (We approximate this condition by doing the experiments in a Styrofoam cup.) We also assume that the system has reached *equilibrium*—that is, that the water and the medal both have exactly the same temperature throughout. We're also assuming that no hot water "takes a ride" on the medal when we transfer it to the cup. Careful technique in the lab can help one *approach* these ideal, assumed conditions.

♦ Officially, Q_{medal} is negative because heat flows *from* it, while Q_{water} is positive because heat flows *into* it. From conservation of energy, $-Q_{medal} = Q_{water}$. But conceptually it is easier to think of a certain (positive) quantity of heat flowing from the hot medal to the cool water, and to keep all of these quantities positive. Operationally, this means we will write $Q = cm|\Delta T|$ and we will need to make sure that ΔT is always calculated as a positive quantity.

© Paul G. Hewitt and Phillip R. Wolf

Sample Problem 3

Suppose that you are laying sections of steel railroad track L meters long on a cool day where the temperature is T_c.

(a) Write an equation for the gap size you should leave between sections of track to allow expansion on the hottest day (temperature T_h) without buckling.
(There is a nice photograph of this in Figure 15.14 on page 276 of your *Conceptual Physics* textbook.)

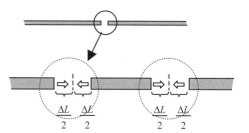

Gap = ?

As each steel rail gets hotter, it expands into the gap between the rails. Each rail expands by an amount $\Delta L/2$ into the gap on each end of the rail. Thus the total gap size needs to equal ΔL for each rail.

$$\text{Gap} = \Delta L = L_{rail}\alpha_{steel}\Delta T_{rail} = L\alpha_{steel}(T_h - T_h).$$

(b) Calculate the size of the gap for 18-meter long steel rails laid down on a day when it's 9°C outside, where the hottest day on record is 42°C.

$$\text{Gap} = \Delta L = L\alpha_{steel}(T_h - T_c) = (18\,\text{m})\left(11\times10^{-6}\,\frac{1}{°C}\right)(42°C - 9°C) = 0.0065\,\text{m} = \mathbf{6.5\ mm}.$$

Specific Heat Capacity Problems

15-1. People in the pioneering days placed hot potatoes in their pockets on cold winter days to keep their hands warm.
  (a) Write an equation for the quantity of heat released by a hot potato of mass m and at a temperature T_0 in cooling to T_f. (Assume that a potato is mostly water.)
  (b) Calculate the heat released by a 350-g potato that cools from 85°C to 15°C.

15-2. Your thin plastic water bottle holds m_w grams of water at temperature T_0. You put it into the refrigerator to cool it to temperature T_f.
  (a) Write an equation for the quantity of heat the refrigerator removes from the water.
  (b) Calculate how much heat the refrigerator removes from 500 mL of water in cooling it from 28°C to 4°C.

15-3. An amount of energy Q raises the temperature of some water from T_0 to T_f.
  (a) Derive an equation for the mass of water being heated.
  (b) Calculate the mass of water if 21 kJ increases the water's temperature from 0°C to 100°C.

15-4. You decide to try the "Ice-Water Diet" and drink water at 0°C. The energy stored in your body fat goes to warming the ice water up to your body temperature, 37°C.
  (a) How much ice water must you drink and warm up to 37°C to "burn" 3500 Calories (the approximate energy content of 1 pound of fat)? (Each Calorie is 4180 J.)
  (b) Would you recommend this diet to overweight friends? Explain.

15-5. An amount of heat Q raises the temperature of m grams of metal from T_0 to T_f.
  (a) Derive an equation for the specific heat capacity of the metal.
  (b) Metal having a mass of 48.4 grams experiences an increase in temperature from 34.0°C to 47.5°C when 252 J of heat is added to it. Calculate the specific heat capacity of the metal.

15-6. An amount of energy Q raises the temperature of a mass m of liquid with specific heat capacity c to T_f.
   (a) Derive an equation for the initial temperature of the liquid.
   (b) Calculate the initial temperature of 68.0 grams of kerosene ($c = 2.01$ J/g·°C) if its temperature increases to 32.2°C when 1530 J of energy is added.

15-7. The initial temperature of m grams of olive oil is T_0. Energy Q is added to the oil.
   (a) Derive an equation for the final temperature of the oil.
   (b) 980 grams of olive oil, initially at 18.0°C, is warmed by the addition of 43.3 kJ of heat. Calculate the final temperature of the olive oil.

15-8. You connect your bicycle to an electrical generator connected to a heating element. You place the heating element into an insulated cup containing m grams of water so that by pedaling your bicycle you can raise the temperature of the water from T_{cold} to boiling (100°C). Your power output is P, and we'll assume that all of the energy you put into turning the generator goes into heating up the water. (Recall that Power = Energy/time.)
   (a) Derive an equation for the amount of energy you add to the water in a time t.
   (b) Derive an equation for the pedaling time needed to heat your water to boiling for tea.
   (c) Assume that you can produce 110 Watts (that is, 110 J/s) and that you want to heat 300 g of water from 21°C to 100°C. Calculate how long you must pedal.

15-9. Equal amounts of energy are added to equal masses of two different metals. The first metal experiences a temperature change of ΔT_1.
   (a) Derive an equation for the temperature increase of the second metal compared with the temperature increase of the first metal.
   (b) A certain amount of heat added to 65 grams of lead causes a temperature change of 15.0°C. Calculate the temperature change when the same amount of energy is added to 65 grams of copper.

15-10. You want to find the specific heat capacity of some vegetable oil. You place a mass of oil m_{oil} at temperature T_0 into a Styrofoam cup. Then you add a mass m_{Al} of hot aluminum at temperature T_h to the oil. The final temperature of the two is T_f.
   (a) Derive an equation for the specific heat capacity of the oil.
   (b) A 57-g piece of aluminum at 100°C is added to a Styrofoam cup containing 65 g of oil at 23°C. The combination is stirred, and the final temperature is 46°C. Calculate the specific heat capacity of the oil.

15-11. A copper block of mass M and at a temperature T_{cop} is placed into an insulated cup containing a mass m_w of water at a temperature T_0.
   (a) To bring the final temperature of the water to T_f, what should be the initial temperature of the copper?
   (b) Water with a mass of 250 g at 24.0°C is in a Styrofoam cup. Calculate the initial temperature of a 300-g copper block that will produce a final temperature of 21.0°C.

15-12. Your (perfectly insulated) bathtub has x liters of water in it, but it has cooled down to a temperature T_{cw}. You'd like to add just the right amount of hot water at temperature T_{hw} to make your bath water the perfect temperature T_{perf}.
   (a) How much hot water must you add? (Recall that 1 liter of water has a mass of 1 kg.)
   (b) You like your bath water to be at 42°C, but the 82 liters of water presently in the tub are at 35°C. Calculate the amount of hot water at 50°C you must add to the tub.

15-13. You pour x mL of hot coffee at T_h into a steel coffee cup of mass m at an initial temperature of T_c. Heat from the coffee goes into warming the steel of the cup.
  (a) Derive an equation for the final temperature to which the coffee immediately cools.
  (b) Calculate the final temperature of 250 mL of coffee at 87°C after it is poured into an 85-g steel coffee cup originally at 22°C.

15-14. Your half-finished cup of coffee has cooled down to a temperature T_c. You like your coffee to be at the perfect temperature T_p. You put your cup, containing x mL of coffee, into a microwave oven. (1 mL of coffee has a mass of 1 gram.)
  (a) Write an equation for the amount of energy it takes to restore your coffee to its "perfect" temperature. (Assume that the coffee has the same thermal properties as water and that the cup itself gains negligible heat from the microwave oven.)
  (b) The oven delivers energy to the coffee at a rate of P watts (J/s). Derive an equation for the time required for the microwave oven to bring the coffee to its perfect temperature.
  (c) If the microwave oven delivers 750 W to the coffee, how long will it take to reheat 140 mL of coffee from 22°C to 83°C?

15-15. A bomb calorimeter consists of a sealed stainless steel container (the "bomb") into which is placed a sample to be burned. The container is filled with oxygen under pressure, and the whole thing is surrounded by water in an insulated container. When a small fuse ignites the sample, the temperature of the water (and the bomb) increases, from which you can determine how much energy was released. You decide you want to know how many Calories there are in peanuts. You place x grams of peanuts into a steel bomb of mass m_b and surround the bomb by a mass m_w of water. Then you ignite the fuse.
  (a) Suppose that the temperature of the water and steel bomb increases by ΔT. Write an equation for the energy released by the peanuts. (Assume all of the released energy goes into the bomb and water. The small part of the energy that goes into heating the products of the combustion can be ignored, as can the energy released in burning the fuse itself.)
  (b) Suppose that the energy to warm the water and the steel bomb came from x grams of peanuts. Write an equation for the energy provided by each gram of the peanuts.
  (c) Suppose that you use 0.95 g of mashed peanuts, the mass of the bomb is 750 g, the mass of the water is 450 g, the initial temperature of the water is 19.1°C, and the final temperature is 29.4°C. What is the calorific value of peanuts in Cal/g? (Recall that 1 Calorie = 1000 calories = 4180 J.)

15-16. Some homes have "on-demand" water heaters. Rather than storing hot water in a large tank, these heaters activate when you turn the hot water on, and provide hot water only as long as it is needed. In taking a shower, suppose that you use x liters of hot water each minute. (Recall that each liter of water has a mass of 1 kg.)
  (a) Write an equation for the quantity of heat required to change the temperature of x liters of water from T_{cold} to T_{hot}.
  (b) Write an equation for the energy required per second (that is, the power) to heat x liters of water from T_{cold} to T_{hot} in a time interval of 1 minute.
  (c) Calculate the power rating for a perfectly efficient electric heater designed to heat 10.0 liters of water from 15°C to 50°C each minute.

15-17. A light bulb of wattage P is placed in an insulated cup filled with a mass m of oil with a specific heat capacity c_{oil}. You turn on the light bulb. The oil begins to get hotter.
 (a) How much energy does the bulb transfer to the oil in a time t?
 (b) How much warmer will the oil be after the bulb has been submerged for a time t?
 (c) Calculate the temperature change of 150 g of olive oil when a 60-watt bulb is submerged in the oil for 45 seconds.

15-18. Two identical blobs of clay of mass m moving toward each other at speed v have a head-on collision. They stick together.
 (a) Use conservation of momentum to calculate the final speed of the combined blobs of clay.
 (b) What was the initial KE of the two-clay-blobs system?
 (c) What is the final KE of the two-clay-blobs system?
 (d) By how much did the KE of the two-clay-blobs system change?
 (e) Assuming that all of the "lost" KE is converted to heat within the clay itself, by how much does the temperature of the clay blobs increase?
 (f) Calculate the increase in temperature for two colliding 12-g clay blobs, each initially moving at 5 m/s. (Note: Pay attention to units when you do this calculation.)

15-19. • A nuclear power plant typically produces 2 joules of heat for every joule of electric energy that it produces. The heat is often removed by water pumped through a secondary cooling loop. Suppose that your power plant produces x MW (megawatts, or million watts) of electric power and is next to the ocean. The water your reactor takes in and discharges back into the ocean is limited to a temperature rise of no more than ΔT_{max}.
 (a) Write an equation showing how much heat the reactor releases every second.
 (b) Write an equation showing how much heat a mass of water m_w absorbs when undergoing a temperature increase of ΔT_{max}.♦
 (c) Write an equation for the mass of cooling water you must take into your reactor each second if you want the water to change temperature by ΔT_{max}.
 (d) How much cooling water per second is required to cool a 1000 MW reactor if we stipulate that the cooling water can only have a maximum temperature rise of 4.0°C?

15-20. • A solar water heater has a collector plate of area A (in units m^2). It receives solar energy at a rate I_0 (watts/m^2) and transfers a fraction ε of that energy to water passing through the collector plate at a rate R (mL/second). (The fraction ε is called the *efficiency* of the collector.)
 (a) Come up with an equation that states how much solar energy the collector receives per second.
 (b) Derive an equation for the energy transferred to the water each second if the collector delivers the collected solar energy to the water with an efficiency of ε.
 (c) Suppose that the flow rate of water through the collector is R mL/s (R grams/s). Derive an equation for ΔT for water passing once through the collector.
 (d) Calculate the change in temperature for water that flows through a 1.2 m^2 solar collector at a rate of 15 mL per second at a time of day when the sunlight strikes the collector with an intensity of 850 W/m^2 and transfers that energy to the water with an efficiency of 48% ($\varepsilon = 0.48$).

♦ If the water returned to the ocean is too hot, it can significantly impact the sea life near the return pipe.

14-23. Air flows through a horizontal pipe, flowing first through a section of area A and then through a narrower section of area $0.40\,A$.
   (a) Compare the relative speeds of air in the wide and narrow sections of the pipe, assuming that the air's density doesn't change appreciably.
   (b) In which section of the pipe is air pressure greater?

14-24. A volume V of gasoline flows from a gas pump into a car's gas tank in a time t through a pipe of diameter D.
   (a) Derive an equation for the speed of the gasoline as it flows through the pipe.
   (b) Calculate the speed of the gasoline if 1 liter flows through a 2.0-cm diameter pipe into the car's gas tank every 8 seconds.

14-25. • A carburetor employs a *venturi tube*, which is a tube with a narrow constriction. Air flows at speed v through the air intake of diameter D and then speeds up as it travels through the narrow venturi neck of diameter d. Since the air is moving faster through the neck, the pressure in the neck is reduced. The difference between this lower pressure and the outside atmospheric pressure forces the fuel into the carburetor.

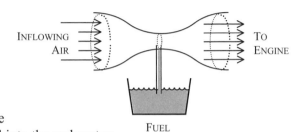

   (a) Starting with Bernoulli's principle, derive an equation that gives the pressure difference between the neck and the air-intake.
   (b) Calculate the pressure drop as 3.0 m/s air with density 1.20 kg/m³ moves from a 7.5-cm diameter intake tube to a 1.5-cm diameter venturi neck.

14-26. • As part of a lecture demonstration of Bernoulli's principle, Paul blows compressed air from a tank across the opening of a hollow straw of area A to lift a metal cylinder of mass m that sits inside the straw.

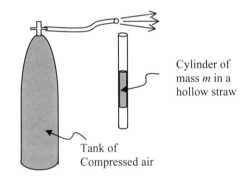

Cylinder of mass m in a hollow straw

   (a) Derive an equation for the minimum air speed that will raise the cylinder in the straw.
   (b) Determine the minimum air speed that will lift a 0.17-gram cylinder in a straw with $A = 0.30$ cm².

Show-That Problems for Gases

14-27. We usually don't notice atmospheric pressure, even though it's quite enormous.
   Show that the total weight of air pressing down on a sheet of newspaper of area 0.200 m² is 20,200 N.

14-28. A party balloon is squeezed to 2/3 of its initial volume.
   Show that the pressure in the balloon increases 1.5 times.

15-28. Water in a cylindrical reservoir of radius R and average depth D has an average temperature T.
  (a) Derive an equation for ΔD, how much water level rises due to thermal expansion, if the temperature of the reservoir increases by ΔT. (For simplicity, assume the reservoir itself doesn't expand.)
  (b) Calculate the approximate rise in level for a cylindrical reservoir of diameter 40 m and a depth of 20 m when water temperature increases by 10°C. (Assume that $\beta_{water} \approx 210 \times 10^{-6}/°C$.)

15-29. A popular demonstration of linear expansion involves trying to place a brass ball of diameter $d_{0(ball)}$ through a brass ring of a slightly smaller diameter, $d_{0(ring)}$. The initial temperature of the ball and ring is T_0. The ring is placed in a flame and heated until it has expanded enough for the ball to pass through.
  (a) Derive an equation for the minimum temperature T_f to which the ring must be heated before the ball will fit through it.
  (b) Calculate the value for T_f for a 2.540-cm diameter brass ball and a brass ring with an inner diameter of 2.530 cm, both with an initial temperature of 21°C.

15-30. • Suppose that NASA tool designers want to design a steel wrench that will be used by astronauts to tighten or loosen a d-mm brass nut on the outside of their lunar lander. The challenge is to design and manufacture a wrench here on Earth at 20°C where it will be a little bit too big, but will fit just right at a higher temperature when both the nut and the wrench have expanded. The temperatures on the lunar surface can reach a maximum temperature T_{max}.

  (a) Write an equation for the width of the brass nut at $T = T_{max}$.
  (b) Let $d_{0(wrench)}$ be the width of the wrench at 20°C. Write an equation for the width of the wrench at $T = T_{max}$.
  (c) To have the wrench and the nut the same size at T_{max}, how big does the gap in the wrench have to be at 20°C?
  (d) The average temperature on the Moon's surface during the lunar day is 107°C. The width of the brass nut at 20°C is 12.00 mm. Calculate the width of the "gap" in the wrench at 20°C if it is going to fit onto the brass nut when both of them are at 107°C.

Show-That Problems for Specific Heat Capacity

15-31. A 62.3-gram sample of aluminum is heated.
  Show that adding 516 J will raise its temperature by 9.2°C.

15-32. Aluminum has a specific heat capacity of 0.900 J/g·C°.
  Show that when a 10.0-gram piece of aluminum at 20°C gains 225 J of heat, its final temperature will be 45°C.

15-33. A heat input of 35.9 J raises the temperature of 47.0 grams of a material by 2.5 C°. Show that the specific heat capacity of the material is 0.306 J/g·°C.

15-34. Show that mixing 1 liter of 20°C water with 2 liters of 40°C water yields a final temperature of 33°C.

15-35. Show that mixing 101 g of 26°C water with 75 g of 40°C water yields a final temperature of 32°C.

15-36. A cook adds 100 grams of ice water at 0°C to 375 grams of water in a pot and finds that the final temperature is 60°C. Show that the initial temperature of the water in the pot was 76°C.

15-37. A 50.0-g piece of metal is removed from 100°C water and placed into 400 grams of 20.0°C water. The final temperature of the water is 22.0°C. Show that the specific heat capacity of the metal is 0.86 J/g·°C.

15-38. An 88.9-g piece of silver at 93.0°C is placed in 175 g of olive oil at 18.0°C in an insulated container. Show that the final temperature will be 22.2°C.

15-39. A certain amount of heat raises the temperature of a sample of iron by 10°C.
Show that the same amount of heat will raise the temperature of an equal mass of lead by 35°C.

15-40. • A spherical 300-µm diameter iron ball coming off a 2000°C sparkler lands on your 30°C hand. Show that the amount of energy transferred from the sparkler to your hand is about 0.1 Joules. Use ρ_{iron} = 7.87 g/cm³ = 7870 kg/m³.

15-41. • Show that the energy required to heat a quantity of water from 20°C to 100°C would lift that same quantity of water by more than 34 km. (That's over 20 miles!)

Show-That Problems for Thermal Expansion

15-42. A brass rod is 2.400 m long at 21°C. Show that it will be 3.6 mm longer at 100°C.

15-43. A metal rod, initially 0.910 m long, expands by 960 µm when heated from 19°C to 100°C. Show that the coefficient of thermal expansion of this metal is 13×10^{-6}/°C.

15-44. Suppose that you are laying 10.0 km of steel railroad track on a day when the temperature is 3°C. Show that the total amount of gap you need to leave in the track to accommodate a 45°C day is 4.6 m.

15-45. Suppose that you want to place an aluminum rod 2.000 cm in diameter into a round hole 1.998 cm across.
Show that you have to lower the temperature of the aluminum rod by 44°C to get it to fit.

15-46. A 30.0-m length of steel rod expands 8.00 mm when heated.
Show that the temperature change is 24°C.

15-47. Suppose that at 10°C you have a steel rod that is 1.0065 m long and an aluminum rod that is 1.0049 m long.
Show that both rods will have the same length at 143°C.

15-48. An aluminum rod grows 0.0033 m in length when its temperature is raised from 10°C to 90°C. Show that its initial length before being heated was 1.8 m.

16 Heat Transfer

The three primary methods of heat transfer are *conduction*, *convection*, and *radiation*. Transfer of heat by conduction occurs when there is direct contact between two objects at different temperatures. Transfer of heat by convection occurs by movement of warmed fluids, such as air rising off hot asphalt. Heat transfer by radiation is the direct emission of energy in the form of electromagnetic waves, such as from our Sun.

Conduction

Consider a copper rod wrapped in insulation and which is positioned between a container of boiling water and a container of ice water.

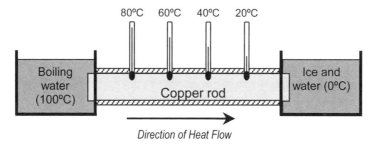

Heat will flow through the copper rod from the hot side to the cold side. At equilibrium, the rate of heat flow through the rod will depend on several factors:

1. What the rod is made out of—Some materials conduct heat better than others. How well a material conducts heat is given by its *coefficient of thermal conductivity*, κ, which has units of W/m·°C. (The units will make sense below.)
2. The cross-sectional area of the rod, A (in unit m^2)—Doubling the cross-sectional area is like two adjacent rods conducting heat instead of one rod.
3. The temperature difference across the ends of the rod, ΔT (in °C)—A higher temperature difference across the rod causes a proportionally larger flow of heat through the rod.
4. The length of the rod, L (in unit m)—The rate of heat flow lessens with increased length of the rod. As the example below indicates, the rate of heat flow is inversely proportional to the length of the rod.

> Left: The temperature difference over a rod of length L is 100°C. Right: But for a rod of length 2L, temperature difference is only 50°C over each length L, so the rate of heat flow is half as much. The rate of heat flow is inversely proportional to the rod's length.

In summary, the quantity of heat Q conducted during a time t through a rod or a bar of length L and cross-sectional area A is $Q = \dfrac{(\kappa A \Delta T)}{L} t$, where ΔT is the temperature difference between the ends of the rod and κ is the thermal conductivity of the material. If we solve for κ, we get $\kappa = \dfrac{QL}{tA\Delta T}$, so the SI unit of thermal conductivity is J/(s·m·°C), or equivalently, W/m·°C.

Sample Problem 1

A glass window in a house has a height h, width w, and a thickness L. The temperature inside the house is T_{inside}, while the lower temperature outside is $T_{outside}$.

(a) What quantity of energy is conducted through the window in time t?

Focus: $Q = ?$ Heat will flow through the window from the warmer to the cooler side.

$$Q = \frac{(\kappa A \Delta T)t}{L} = \frac{\kappa_{glass}\, wh(T_{inside} - T_{outside})t}{L}.$$

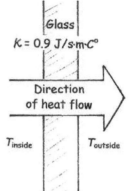

(b) Consider a glass window 4.0 mm thick, 63 cm wide, and 84 cm high. The temperature inside the house is 22°C, while temperature outside is 4°C. Calculate the quantity of heat that flows due to conduction through the window in 1 hour. Thermal conductivity κ for the glass is 0.90 J/s·m·C°.

Answer:

$$Q = \frac{\kappa_{glass}\, hw(T_{inside} - T_{outside})t}{L}$$

$$= \frac{\left(0.90 \frac{J}{s \cdot m \cdot °C}\right)\left[\left(84\,cm \times \frac{1m}{100\,cm}\right) \times \left(63\,cm \times \frac{1m}{100\,cm}\right)\right](22°C - 4°C)(3600\,s)}{\left(4.0\,mm \times \frac{1m}{1000\,mm}\right)} = 7.7 \times 10^6\,J.$$

(c) How would the amount of heat loss differ if the glass were 5.0 mm thick instead of 4.0 mm thick?

Answer: For a larger L, the quantity of heat is smaller. Doing the same calculation as in (b), with $L = 5.0$ mm, gives 4/5 the previous answer—**0.80 times as much**.*

(d) The quantity of heat that transfers through the window is given by $Q = \frac{(\kappa A \Delta T)t}{L}$. What is the *rate* at which this quantity of heat energy flows through the glass?

Answer: The rate of energy flow per unit time would be $\Delta Q/\Delta t$, or simply the above expression divided by time t. Then we have

$$\frac{Q}{t} = \frac{(\kappa A \Delta T)}{L}.$$

---

* Actual heat transfer is more complicated than this. As heat flows through the window, a thin layer of air on the cold side of the window warms and a thin layer of air on the warmer side of the window cools. Since total ΔT will be across the window *plus* a layer of air on either side of the window (and air is a poor conductor of heat), the rate of heat flow will drop. Meanwhile, convection will also kick in as the warmer air tends to rise and be replaced by colder air. In addition, the warmer parts of the glass will be losing energy by radiation to the cooler outdoor environment, and the room is radiating some of its heat out the window! What we are working on here is a first, *simple* model of heat transfer. Actual cases that heat engineers tackle are more complicated, but comprehensible. For now, we keep it simple.

(e) Calculate the rate at which heat flows through the glass window.

Solution:

$$\frac{Q}{t} = \frac{7.7 \times 10^6 \text{ J}}{3600 \text{ s}} = 2100 \text{ W}.$$

Radiation

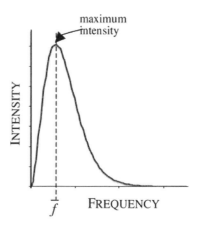

Radiant energy is in the form of electromagnetic waves (light, for example). Although every body radiates electromagnetic waves at a range of frequencies, the *peak* frequency $\overline{f}$ (at which this radiation intensity is maximum) is proportional to the body's surface temperature. In the textbook we express this as $\overline{f} \sim T$, where $\overline{f}$ is the peak frequency of radiation, and T is the temperature of the emitter in kelvins.* We go further here and express this relationship as an equation, called *Wien's Law*:

$$\overline{f} = \left(5.88 \times 10^{10} \frac{\text{Hz}}{\text{K}}\right) T.$$

Even at relatively modest Kelvin temperatures, the frequency is going to be a huge number. The frequency of visible light ranges from about 4.3×10^{14} Hz (an incredible 4.3×10^{14} waves per second) for red light to 7.5×10^{14} Hz for violet light.

Sample Problem 2

The red supergiant Antares has a peak radiation frequency of 2.0×10^{14} Hz.
(a) What is the approximate surface temperature of Antares?

Solution: From $\overline{f} = \left(5.88 \times 10^{10} \frac{\text{Hz}}{\text{K}}\right) T \Rightarrow T = \frac{\overline{f}}{5.88 \times 10^{10} \frac{\text{Hz}}{\text{K}}} = \frac{2.0 \times 10^{14} \text{ Hz}}{5.88 \times 10^{10} \frac{\text{Hz}}{\text{K}}} = \textbf{3400 K}.$

(b) What would be the peak frequency of radiation from Antares if it had twice the temperature?

Solution: From our relation $\overline{f} \sim T$, twice the temperature would be twice the frequency.
Twice 2.0×10^{14} Hz = **4.0 × 10¹⁴ Hz**.

$$\overline{f} = \left(5.88 \times 10^{10} \frac{\text{Hz}}{\text{K}}\right) T = \left(5.88 \times 10^{10} \frac{\text{Hz}}{\text{K}}\right) \times 2(3400 \text{ K}) = \textbf{4.0} \times \textbf{10}^{14} \textbf{ Hz}.$$

---

*The relationship between Kelvin and Celsius temperatures is T (Kelvin) = T (Celsius) + 273. The Kelvin scale is discussed further in Chapter 18.

Problems for Heat Transfer

16-1. Q joules of heat flow each second through a copper rod of cross-sectional area A and length L.
  (a) Derive an equation for the temperature difference between the two ends of the rod.
  (b) Calculate the temperature difference between the two ends of the rod. You need to know that $\kappa_{copper} = 401$ W/m·°C, 125 joules flow each second, the length of the rod is 12.0 cm, and its diameter is 1.26 cm.

16-2. Suppose that the air temperature directly below the plaster ceiling of your house is a warm T_0, but your (noninsulated) attic is at a chillier T_a.
  (a) Write an equation for the rate of heat loss through your ceiling if the area of the ceiling is A and the plaster has a thickness L.
  (b) Suppose that the air temperature just below the ceiling is 23°C and the temperature of the air in the attic is 5°C. Calculate the rate of heat loss for a 1.6-cm thick plaster ceiling having an area of 110 m². Assume $\kappa_{plaster} = 0.21$ W/m·°C.

16-3. A rod of length L and diameter d transfers Q joules of energy in time t when the hot end is kept at T_{hot} and the cold end is kept at T_{cold}.
  (a) Write an equation for the coefficient of thermal conductivity of the material that makes up the rod.
  (b) Calculate κ_{rod} if the length of the rod is 25.0 cm, the diameter is 2.54 cm, the hot end is kept at 78°C, the cold end is kept at 22°C, and 254 J of heat flows through the rod in 8.0 s.

16-4. An architect plans to place a glass window of thickness x in the wall of a room where room temperature is T_{room} and the temperature outside the room is $T_{outside}$. Assume that the architect knows κ_{glass}.
  (a) Derive an equation for the area A of a window that allows Q joules of heat to flow through the window each second.
  (b) Calculate the area of a 5.0-mm thick glass window with $\kappa_{glass} = 0.90$ W/m·°C that will allow 257 J/s to flow through the window when $T_{outside} = 32$°C and $T_{room} = 21$°C.

16-5. An old laboratory oven is shaped like a cube of length s on each side. The walls and oven door are made from asbestos ($\kappa_{asbestos} = 0.20$ W/m·°C) of thickness a with thin sheets of metal on either side. (For all practical purposes, the contribution of the metal sheets to resisting heat flow through the walls can be ignored.) The oven has an electric heating element whose maximum power output is P. When the oven is first turned on, temperature inside the oven rises. As the oven gets hotter, more heat inadvertently flows out through its walls. At some point, it is hot enough inside so that the oven loses heat at the same rate that heat is being added, at which point the temperature of the oven stops increasing.
  (a) Derive an equation for the maximum temperature that the oven can maintain in a room at T_{room}. Ignore losses to radiation and convection.
  (b) Suppose that the laboratory oven is 30 cm on each side, the asbestos walls are 1.2 cm thick, and the heating element puts out a maximum 2000 watts. The oven sits in a 20°C room. Calculate the maximum possible temperature inside the oven.

16-6. A pond in winter has a layer of ice of thickness x. The temperature of the air above the pond is T_{air} (below 0°C), while the temperature of the water directly below the ice is 0°C.
   (a) Derive an equation for the quantity of heat passing through an area A of the ice layer each second.
   (b) If the temperature outside is –10°C, how much heat passes through each square meter of ice each second? Assume the ice is 5.0 cm thick and that $\kappa_{ice} = 2.2$ W/m·°C.
   (c) A 1.00-m² layer of ice 1 mm thick has a mass of 0.916 kg. Removal of 335,000 joules of energy from the water at 0°C forms 1 kilogram of ice. If the temperature remains –10°C outside, how long will it take for the ice to get 1 mm thicker?
   (d) What happens to the rate of ice formation as the ice grows thicker?

16-7. A typical lab experiment for measuring the coefficient of thermal conductivity of a metal bar involves taking an insulated bar of length L and diameter d and placing it between two insulated chambers, similar in setup to the first diagram in this chapter (page 179). In one chamber, steam at 100°C is blown onto the hot end of the bar. In the other chamber, cold water with an initial temperature T_0 flows past the other end of the bar and leaves at a slightly higher temperature T_f. The average temperature the cold end of the bar "sees" is $(T_0 + T_f)/2$. A beaker is used to catch a mass m of cold water that flows past the cold end of the bar in a time t.
   (a) Write an expression for the energy gained by the water in time t.
   (b) Write an expression for the average ΔT between the hot and cold ends of the bar.
   (c) Write an expression for the quantity of heat conducted through the bar in time t.
   (d) Equate your expressions from (a) and (c) for the heat flow through the bar. What is the coefficient of thermal conductivity of the bar?
   (e) A brass bar has a diameter of 2.54 cm and a length of 25.0 cm. One end is in 100° steam and the other end encounters incoming water temperature of 22.3°C, with the outgoing temperature being 24.7°C. 102 grams of water flow by the cold side of the bar in 55 seconds. Calculate κ of the bar.

16-8. A block of ice keeps the inside of an ice chest at 0°C. The ice chest is made from Styrofoam of thickness L, a total surface area A, and hangs by thin ropes from a tree.
   (a) Write an expression for the rate at which the ice chest absorbs energy when the outside temperature is T.
   (b) Suppose that you start out with a block of ice of mass m in the ice chest. Also suppose it takes x joules of energy to melt a kilogram of ice. Derive an expression showing how long it takes for all of the ice in the ice chest to melt.
   (c) Calculate answers to (a) and (b) for a Styrofoam ice chest with total surface area 1.28 m² made with walls 2.5 cm thick and which holds 7.5 kg of ice on a 25°C day. Use $\kappa_{Styrofoam} = 0.026$ W/m·°C. (Note: 335,000 J are required to melt 1 kg of ice.)

16-9. In a lab experiment a heated plywood box l meters long, w meters wide, and h meters deep provides warmth to some biological samples to be kept at a constant temperature T_{hot}. The walls of the box have a thickness x.
   (a) Derive an equation for the power provided to the box's interior if the box is kept in a room whose temperature is T_{room}. Ignore radiation and convection losses from the box.
   (b) Suppose that the box is 2.0 m long, 0.60 m wide, and 0.50 deep, and the plywood is 12.7 mm thick. Room temperature is 22°C. How much power must be provided to the interior of the box to maintain its contents at 55°C? Use $\kappa_{plywood} = 0.11$ W/m·°C.

© Paul G. Hewitt and Phillip R. Wolf

16-10. Referring to the previous problem, suppose that the box's plywood walls were replaced by the same thickness of Styrofoam ($\kappa = 0.026$ W/m·°C).
   (a) How much power would have to be supplied to the box to maintain its interior temperature at 55°C?
   (b) What would be the rate of heat flow if the Styrofoam were twice as thick?

16-11. • Suppose that the box of Problem 16-9 were made with a layer of plywood of thickness L_p lined with a Styrofoam layer of thickness L_f. The interior of the box is still to be kept at T_{hot} and the room is still at temperature T_{room}.
   (a) Derive an expression for the rate at which power must be provided to the interior of the box. (Assume that the surface area of the plywood and Styrofoam are the same).
   (b) Assume that the conditions of the problem are the same as in 16-9(b). How much power must be provided to the interior of the box?

16-12. Earth has an average surface temperature of about 18°C.
   (a) Write an expression for the peak frequency at which Earth radiates.
   (b) In which part of the electromagnetic spectrum is this terrestrial radiation? (Note the chart of the electromagnetic spectrum on page 458 of your *Conceptual Physics* textbook.)

16-13. The tungsten lamp filament in a 60-watt incandescent light bulb typically operates at a temperature of 2800 K.
   (a) What is the peak frequency at which the filament radiates?
   (b) In comparison to red light, where in the electromagnetic spectrum does this frequency occur?

Show-That Problems for Heat Transfer

16-14. Each second 29.1 joules flow through a 22-cm long nickel rod ($\kappa_{nickel} = 91$ W/m·°C) when the temperature difference between the two ends of the rod is 56°C.
Show that the diameter of the rod is 4.0 cm.

16-15. A 10-cm thick concrete wall ($\kappa_{concrete} = 0.8$ W/m·°C) separates two sections of a water treatment plant. On one side the water is 18°C, and on the other side it is 23°C.
Show that the rate of heat flow through each square meter of wall is 40 J/s.

16-16. Each second 120 joules of heat flow through a 6.0-mm-thick piece of acrylic whose area is 0.86 m². $\kappa_{acrylic} = 0.19$ W/m·°C.
Show that the temperature difference across the acrylic sheet is 4.4°C.

16-17. A 2.5-cm diameter rod made from aluminum ($\kappa \approx 240$ W/m·°C) runs between a 56°C temperature bath and a 42°C temperature bath. Heat flow through the rod is 33 W.
Show that the length of the rod is 5.0 cm.

16-18. The dim red dwarf star *Wolf 359* has a surface temperature of 2430 K and is 7.8 light-years from Earth.
Show that the peak frequency at which *Wolf 359* radiates is 1.43×10^{14} Hz.

16-19. The peak frequency of the light radiated by the star is 3.4×10^{14} Hz.
Show that the star has a surface temperature of approximately 5800 K.

17 Change of Phase

When a material is heated, the particles in that material gain energy and move faster. At some point in the heating process, the particles may move fast enough to overcome the forces holding them to one another, which will cause the material to undergo a physical change called a *phase change*. For example, a solid can melt to a liquid or a liquid can vaporize to a gas. At the phase-change temperature, any energy input to the material goes into separating the molecules from one another rather than into raising the material's temperature. The heat added to cause the phase change is called the *latent heat* and is given the symbol L. In this view, temperature is akin to kinetic energy (and actually *is* kinetic energy per molecule in the system), and latent heat is akin to potential energy.

The quantity of heat involved in a phase change from solid to liquid (or from liquid to solid) is called the *latent heat of fusion*, L_{fus}. For the similar process of changing phase from liquid to vapor (or vice versa), the heat involved is called the *latent heat of vaporization*, L_{vap}. The heat involved in the process of changing phase from solid directly to vapor is called the *latent heat of sublimation*, L_{sub}.

Conservation of energy is central when doing heat problems. We note the initial state of a system, the final state of that system, and calculate the quantity of energy involved in going from one to the other. Or if the system is *closed* (no energy entering or leaving the system), then the total energy in the system simply changes form, without loss or gain.

Suppose that we place a very cold ice cube into a warm room and want to know how much energy is involved in the ice cube melting and the resulting water warming up. We know, physically, that the ice on the surface of the cube will begin melting before the center of the ice cube has warmed to 0°C. But we can think of this melting and warming process as occurring in discrete steps.

First step: The very cold ice warms up to 0°C. The quantity of heat involved in this step is
$Q_1 = c_{ice} m_{ice} \Delta T_{ice}$.
Second step: The ice at 0°C melts to water at 0°C. $\Rightarrow Q_2 = m_{ice} L_{fus}$.
Third step: The water at 0°C warms up to water at T_f. $\Rightarrow$
$Q_3 = c_{water} m_{ice} \Delta T_{ice\ water} = c_{water} m_{ice} (T_f - 0°C) = c_{water} m_{ice} T_f$.

The total heat involved in the process is the sum of these three quantities = $Q_1 + Q_2 + Q_3$.

Here are some useful numbers for solving problems in this chapter:

L_{fus} (water) = 335 J/g = 335,000 J/kg = 335 kJ/kg

L_{vap} (water) = 2260 J/g = 2,260,000 J/kg = 2260 kJ/kg

c_{ice} = 2.09 J/g·°C = 2.09 kJ/kg·°C

c_{water} = 4.18 J/g·°C = 4.18 kJ/kg·°C

c_{steam} = 2.03 J/g·°C = 2.03 kJ/kg·°C

Other specific heat capacities are listed on page 170 of this book.

Sample Problem 1

(a) Write an equation that shows the quantity of heat required to melt a mass m_{ice} of ice initially at T_0 (below 0°C) to liquid water at 0°C.

Focus: $Q = ?$ We think of this process occurring in two steps:

```
[Ice at T_0]  ==>  [Ice at 0°C]  ==>  (Liquid water at 0°C)
 Q_1 = c_ice m_ice ΔT_ice         Q_2 = m_ice L_fus
```

$Q_1 = c_{ice} m_{ice} \Delta T_{ice}$ $\qquad$ $Q_2 = m_{ice} L_{fus}$

We need to add enough heat to first warm the ice from T_0 to 0°C ($c_{ice} m_{ice} \Delta T_{ice}$) plus enough heat to melt the ice to liquid water at 0°C ($m_{ice} L_{fus}$). So $Q = c_{ice} m_{ice} |\Delta T_{ice}| + m_{ice} L_{fus}$.

(b) Calculate the quantity of heat required to melt 450 g of ice initially at –3.0°C to liquid water at 0°C.

Solution:

$$Q = c_{ice} m_{ice} |\Delta T_{ice}| + m_{ice} L_{fus} = \left(2.09 \frac{J}{g \cdot °C}\right)(450 \text{ g})(3.0°C) + 450 \text{ g}\left(335 \frac{J}{g}\right) = 1.54 \times 10^5 \text{ J} = 154 \text{ kJ}.$$

Sample Problem 2

- **An 18-g ice cube at –4.0°C is placed into 75 g of water at 10°C in an insulated container.**

(a) What will be the final temperature of the system?

Focus: $T_f = ?$

Analysis: When energy flows from the water to warm and maybe melt an ice cube, the system can fit into one of three possible scenarios:

1. *A sliver of ice goes into a cup of warm water,*

   ($Q_{\text{to warm the ice to 0°C and melt the ice}} < Q_{\text{to cool the liquid water to 0°C}}$): All of the ice will melt and the ice water that forms will warm up. The end result is that the warm water becomes a little cooler than it was initially.

2. *You take several ice cubes and put them in a glass of warm water,*

   ($Q_{\text{to warm up ice to 0°C}} < Q_{\text{to cool the liquid water to 0°C}} < Q_{\text{to warm up ice to 0°C and melt all of the ice}}$): The ice will warm to 0°C and start to melt, but there won't be enough heat available from the warm water to melt all of the ice. The final system is a mixture of ice and water at 0°C.

3. *You put a large, very cold ice cube into a small amount of warm water,*

   ($Q_{\text{to warm up ice to 0°C}} > Q_{\text{to cool the liquid water to 0°C}}$): The warm water will cool to 0°C before the ice has warmed to 0°C and begun to melt. The liquid water will continue to lose heat to the sub-zero ice and some will freeze onto the ice cube even as the ice is still warming.

   We need to determine where among these possible outcomes our particular system lies.

Solution:

Quantity of heat to warm the ice cube to 0°C

$$= c_{ice}m_{ice}|\Delta T_{ice}| = c_{ice}m_{ice}(T_f - T_0) = \left(2.09\frac{J}{g\cdot °C}\right)(18.0 \text{ g})(4.0°C) = 150 \text{ J}.$$

Quantity of heat needed to warm the ice cube to 0°C and melt *all* of the ice

$$= c_{ice}m_{ice}|\Delta T_{ice}| + m_{ice}L_{fus} = 150 \text{ J} + (18.0\text{g})\left(335\frac{J}{g}\right) = 6180 \text{ J}.$$

Quantity of heat the liquid water can release as it cools to 0°C

$$= c_w m_w |\Delta T_w| = c_w m_w (T_0 - T_f) = \left(4.18\frac{J}{g\cdot °C}\right)(75\text{g})(10.0°C - 0°C) = 3135 \text{ J}.$$

The liquid water cooling to 0°C can give up more than enough energy to warm the ice to 0°C but not enough energy to completely melt the ice. The final temperature of the water is therefore **0°C**.

(b) How much of the ice melts?

Focus: $m_{melted} = ?$

We know that the entire ice cube gains heat to warm to 0°C. Then some of the ice cube (say, x grams) will melt. The heat to warm up the ice cube and melt some of the ice comes from the 75 g of water cooling from 10°C to 0°C.

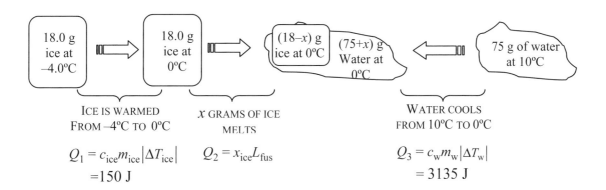

From conservation of energy, $Q_1 + Q_2 = Q_3$

$\Rightarrow Q_2 = Q_3 - Q_1 = 3135 \text{ J} - 150 \text{ J} = 2985 \text{ J}$ of energy left to melt the ice.

Since $Q_2 = x_{ice}L_{fus} \Rightarrow x_{ice} = \dfrac{Q_2}{L_{fus}} = \dfrac{2985 \text{ J}}{335\frac{J}{g}} = \textbf{8.9 g}$. So almost half of the ice melts.

Sample Problem 3

You place a 300-W travel immersion heater into a cup that holds 245 grams of water at 24°C. The heat entering the cup is negligible compared with the heat entering the water.

(a) What is the temperature of the water in the cup after 180 seconds?

Focus: $T_f =$? Energy flows from the immersion heater into the water at a rate of 300 J/s for 180 seconds, which raises the temperature of the water. But if water temperature reaches 100°C before the end of the 180-second interval, any energy added after reaching that point will change the phase of at least some of the water into steam.

Recall that energy = power × time, so $Pt = Q = c_w m_w \Delta T_w$ (+ possibly $m_{steam} L_{vap}$?).

First let's assume that no water boils, and find the final temperature of the water.

$$Q = Pt = (c_w m_w)(T_f - T_0) \Rightarrow T_f = T_0 + \frac{Pt}{c_w m_w}$$

$$= 24.0°C + \frac{\left(300 \frac{J}{s}\right) 180 s}{4.18 \frac{J}{g \cdot °C}(245 \text{ g})} = 24.0°C + 52.7°C = \mathbf{76.7°C}.$$

The resulting temperature of the water and cup is 76.7°C. No energy goes to making steam.

(b) What is the temperature of the water in the cup after 360 seconds?

Analysis: Careful here! Another 180 seconds would increase the temperature of the water by an additional 52.7°C, which would produce a final temperature 76.7°C + 52.7°C = 129.4°C, an impossibility at normal atmospheric pressure. When the water reaches its boiling point, 100°C, continued heat input changes the phase of the water instead of increasing its temperature. So the final temperature of the water in the cup is **100°C**.

(c) How much water remains in the cup after 360 seconds?

Focus: $m_f =$? The mass of water left will be the difference between the initial mass and the mass that boiled away. The energy from the heater transferred to two places—first heating all of the water to 100°C, and then turning some of that water to steam.

From $Q_{\text{from heater}} = Q_{\text{to warm the water from 24°C to 100°C}} + Q_{\text{boil some of the water}}$

$$\Rightarrow Q = Pt = (c_w m_w)(T_f - T_0) + m_{\text{boiled}} L_{\text{vap}} \Rightarrow m_{\text{boiled}} = \frac{Pt - (c_w m_w)(T_f - T_0)}{L_{\text{vap}}}$$

$$= \frac{\left(300 \frac{J}{s}\right) 360 \text{ s} - \left[\left(4.18 \frac{J}{g \cdot C°}\right)(245 \text{ g})\right](100°C - 24.0°C)}{2260 \frac{J}{g}} = 13.3 \text{ g} \approx 13 \text{ g}.$$

The mass of water remaining in the cup is about $245 \text{ g} - 13 \text{ g} = \mathbf{232 \text{ g}}$.

Problems for Change of Phase

17-1. Energy must be removed from water at 0°C to make ice.
   (a) Write an equation for the quantity of heat that must be removed from m grams of liquid water at 0°C to turn it into ice at 0°C.
   (b) Calculate the quantity of heat that must be removed from 45 grams of liquid water at 0°C to turn it into ice at 0°C.

17-2. Consider removing m grams of ice from a freezer at a temperature T_0 ($T_0 < 0°C$). Later the ice is a puddle of water at T_f.
   (a) Write an equation for the quantity of heat that the ice absorbs.
   (b) Calculate the quantity of heat that a 22-g chunk of ice absorbs when removed from the freezer at –5°C, and which later becomes a puddle of water at 17°C.

17-3. Consider starting with m grams of liquid water at T_0 ($T_0 < 100°C$) and turning it into steam at 100°C.
   (a) Write an equation for the quantity of heat that must be added to the water.
   (b) Calculate the quantity of heat that must be added to 133 grams of water at 32°C to turn all of it into steam at 100°.

17-4. Freezing soup is like freezing water. Assume the same thermodynamic parameters for soup as for water.
   (a) How much energy must be removed from m grams of soup at temperature T_0 to freeze it into frozen soup at T_f ($T_f < 0°C$)?
   (b) Calculate the quantity of heat that has to be removed from 565 grams of soup at 78°C to freeze it into frozen soup at –5°C.

17-5. You decide to try the "South-Pole Diet," which consists of eating ice and burning calories to melt the ice. You eat shaved ice at 0°C, and your body melts the ice and warms it to body temperature, 37°C.
   (a) How much ice do you have to consume to "burn" 3500 Calories (roughly the energy stored in a pound of fat)? (1 Cal = 1 kcal = 1000 cal = 4.18 kJ.)
   (b) Would you recommend the "South-Pole Diet" to overweight friends? Explain.

17-6. An M-gram chunk of hot iron is placed on a large piece of 0°C ice, whereupon m grams of ice melt.
   (a) What will be the final temperature of the system?
   (b) What must have been the initial temperature of the hot iron?
   (c) A 250-gram chunk of hot iron is placed on the top of a large piece of 0°C ice, whereupon 48 grams of ice melt. Calculate the initial temperature of the iron.
   (The specific heat of iron is 0.448 J/g·°C.)

17-7. M grams of ice at a temperature T_c ($T_c < 0°C$) are placed in an insulated container holding a large quantity of liquid water at 0°C.
   (a) Derive an equation for the mass of the liquid water that freezes onto the ice.
   (b) Calculate how much additional ice is formed on a 25-g piece of ice initially at –12°C.

17-8. A mass m_0 of water sits boiling in a pot. After time t, the mass of water left in the pot is m_f.
  (a) At what rate is energy being transferred into the pot?
  (b) Calculate the rate of energy transfer in watts if 763 grams of water were initially in the pot and, 12.0 minutes later, 696 grams of water remain in the pot.

17-9. One way to measure the heat of fusion of ice is to place a known mass of ice at 0°C into a Styrofoam cup with some water whose temperature and mass have already been measured, let the ice melt, and then measure the final temperature.

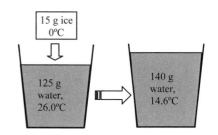

The following data were collected in the lab:

Initial mass of water	125 g
Initial temperature of water	26.0°C
Mass of ice added to the cup	15.0 g
Final temperature of the water in the cup	14.6°C

Calculate the latent heat of fusion of ice from the data in the table, assuming that no significant amount of energy goes into the Styrofoam cup holding the water. (Remember to include the warming of the melted ice water as well as the cooling of the original water.)

17-10. One way to measure the heat of vaporization of water in the lab is to boil water in a container that is sealed except for a hose that allows steam to escape. The steam is bubbled into some water in an insulated cup and condenses to form additional liquid water in the cup. The following data were collected in the lab:

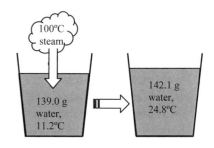

Original mass of water	139.0 g
Original temperature of water	11.2°C
Mass of water after the steam bubbles in	142.1 g
Final temperature of the water in the cup	24.8°C

Calculate the latent heat of vaporization of water from the data in the table, assuming that no significant amount of energy goes into the Styrofoam cup holding the water. (*Hint*: Don't forget the cooling of the 100°C water resulting from the condensing steam as well as the heating of the original water.)

17-11. • One way to cool a swimming pool is to pump some water out of the pool and shoot this water as a spray of water back onto the surface of the pool. As a mass of water m_w travels through the air, some of the water evaporates. Assume that the energy to evaporate the water is $m_{evap}L_{vap}$ and that only the water left behind ($m_w - m_{evap}$) cools to T_f. (Note that water can evaporate at ordinary temperature. It doesn't have to boil.)
  (a) Derive an equation for the fraction of the water that evaporates while the spray travels through the air, m_{evap}/m_w. Assume that the temperature of the water leaving the pool is T_0 and the temperature of the water returning to the pool is T_f.
  (b) The water returning into the pool is 4°C cooler than the water leaving the pool. What percentage of the water evaporated while in the air?

17-12. Air conditioners are sometimes rated in "tons" of cooling capacity. A ton here is defined as the amount of cooling provided when a ton (1 ton = 2000 lb = 907 kg) of ice is melted over a period of 24 hours.
(a) How many joules does it take to melt a ton of 0°C ice?
(b) How many watts are equivalent to a "2-ton" air conditioner?

17-13. A six-pack of canned soda contains m grams of liquid. Suppose that you wanted to place ice into a perfectly insulating ice chest to cool the soda down from T_s to a final temperature T_f. The ice is initially at a temperature T_{ice}. Assume that soda has the same thermal characteristics as water. Neglect the energy needed to cool the cans themselves.
(a) Derive an equation for the mass of ice you'd need. (Think: The final state of the ice will be liquid water at T_f.)
(b) Calculate the mass of ice, initially at –5°C, needed to cool 2.1 kg of canned soda from 23°C to a final temperature of 4°C.

17-14. You are lying on the beach on a hot day, neither gaining nor losing energy through conduction with the air. The energy your body gains from sunshine is counteracted by energy given off by evaporation of perspiration.
(a) Suppose sunlight with intensity I (in watts/m^2) shines directly upon an area A of exposed skin and that you can cool your body only through evaporation (sweating). The energy to evaporate the water (sweat) is mL_{vap}. Derive an equation for the mass of water you must sweat each hour to maintain a constant body temperature.
(b) Calculate the mass of water you would have to sweat each hour to balance out the heat gained from a solar intensity of 700 W/m^2 on 0.50 m^2 of exposed skin.

17-15. • "Swamp coolers" are sometimes used in dry climates as a form of air conditioning. Hot dry air is drawn through damp cloth or damp wood shavings before entering a building. Some of the heat in the air is used to vaporize the water, thus cooling and humidifying the air at the same time. The density of air is approximately 1.2 kg/m^3. Assume that the specific heat capacity of air is about 1000 J/kg·°C.
(a) Derive an equation for the mass m_w of water that must be evaporated to cool a volume V of air (in cubic meters) by an amount ΔT.
(b) Calculate the mass of water that evaporates each hour to cool 8.5 m^3 of air per minute by 7°C.

17-16. A mass m of hot water at temperature T_h is poured into a cavity in a very large block of 0°C ice.
(a) Derive an equation that shows what mass of ice melts.
(b) Calculate how much ice melts when 45 grams of 71°C water is poured into the ice.

17-17. You're about to head out the door with some coffee, mass m_c, which you've poured from the pot into a Styrofoam cup. But the coffee's temperature T_h is hotter than you'd like it to be.
(a) Derive an equation for the mass of ice at 0°C that must be added to lower the coffee's temperature to T_c. (Remember that you have to warm up the resulting "ice water" to T_c.)
(b) Calculate the mass of ice at 0°C that should be added to 275 g of coffee at 83°C to lower the coffee's temperature to 71°C.

17-18. You, mass M, get onto your sled of mass m (in kg) at the top of a slope. You slide down the hill of height h and come to rest a short distance past the bottom of the hill on the flat area below.
  (a) Derive an equation for the mass of snow that melts due to your slide down the hill. Assume that all of your change in gravitational potential energy goes into melting snow and that the temperature of the snow, both before and after melting, is 0°C.
  (b) Calculate the mass of snow melted if your mass is 65 kg and the mass of your sled is 4 kg, for a 140-m high hill.

17-19. Steam at 100° C is blown onto ice at 0°C.
  (a) Derive an equation for the mass of steam required to combine with m grams of 0°C ice to end up with only liquid water at 0°C.
  (b) Calculate the mass of steam combined with 80 grams of ice to produce 0°C water.

17-20. Tetrafluoroethane (also known as R134a) is a common refrigerant. It has a heat of vaporization of about 216 J/g and boils at –10°C in a typical application.
  (a) Derive an equation for the mass of R134a that must be vaporized in order to cool m grams of water from T_{hot} to T_{cold}. (Assume that all of the heat removed from the water is used to vaporize the R134a.)
  (b) Calculate the amount of R134a that must be evaporated to cool a 2-L bottle of soda from 25°C to 4°C. Assume that specific heat capacity and density of soda and of water are the same, and recall that 1 L of water has a mass of 1 kg.

17-21. • A 1.00-gram sample of iron, initially at 20°C, was given a constant energy input of 5.00 J/s. The following temperature vs. time graph was obtained for the iron sample:

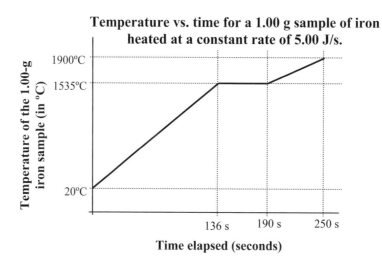

  (a) From the graph determine the specific heat capacity of solid iron.
  (b) From the graph determine the latent heat of fusion of iron.
  (c) From the graph determine the specific heat capacity of liquid iron.

Show-That Problems for Change of Phase

17-22. Energy is needed to change water to steam.
Show that 220 kJ of energy are required to turn 87 grams of water at 36°C to steam.

17-23. After steam from a steam kettle enters a room and condenses, the resulting water then cools to 22°C (room temperature).
Show that each gram of steam gives up about 2600 J to warm the room.

17-24. Show that it takes 3013 J to change the phase of 1 g of ice at 0°C to steam at 100°C.

17-25. The specific heat capacity of ethanol is 2.44 J/g·°C while its heat of vaporization is 841 J/g. Ethanol boils at 78°C.
Show that the energy required to heat up and boil 100 g of 21°C ethanol is 98,000 J.

17-26. One kilogram of ice at –10°C is transformed to steam at 120°C.
Show that 3.1×10^6 J of energy are required.

17-27. Show that placing 12.0 grams of ice at 0°C into 82.0 g of water at 23.9°C will produce a final temperature of the system of 10.6°C.

17-28. Show that if 57.3 grams of metal at 100°C is placed onto a large chunk of ice at 0°C, and 7.5 grams of the ice melts, the specific heat capacity of the metal is 0.44 J/g·C°.

17-29. An ice cube at 0°C is placed in a cup of 100°C water.
Show that if the mass of the ice cube is one-eighth the mass of the 100°C water, then the final temperature of the water will be 80°C (assuming negligible energy transfer to the cup).

17-30. Equal masses of ice at 0°C and steam at 100°C are added together.
Show that the temperature of the resulting mixture is 100°C.

17-31. When certain amounts of 0°C ice and 100°C steam are added together, the result can be 100°C water.
Show that this will occur when the mass of the ice is three times the mass of the steam.

17-32. When certain amounts of 0°C ice and 100°C steam are added together, the result can be water at 50°C.
Show that this will occur when the mass of the ice is about 4.5 times the mass of the steam.

17-33. Energy is added to 2.00 kg of ice at 0°C to melt it, then bring it up to the boiling temperature, 100°C, and then turn it to 100°C steam.
Show that the total quantity of heat required is 6.0×10^6 J.

17-34. Fresh-cut "green" timber is moist. Suppose that each kilogram of construction-grade lumber starts out as 1.6-kg of "green" wood.

Show that 1.4×10^6 J of energy must be supplied to dry the wood. Assume that all of the added energy goes to evaporating water. (Wood drying uses as much as 80% of the total energy consumed in a lumber mill.)

17-35. • You bubble 45.9 grams of steam at 100°C into a mixture of 325 g of water and 141 g of ice in an insulated 113-g copper bucket (c_{copper} = 0.386 J/g·C°).
Show that the final temperature of the system will be close to 34.7°C.

17-36. • In 2000, 39% of all U.S. freshwater withdrawals went to cooling thermoelectric power plants. Typically, hot water from the plant is cooled by evaporation in a cooling tower and the remaining water is recirculated to cool the plant again.
Suppose that you are operating a 500 MW coal-fired plant that is 35% efficient.
(a) Show that heat is input to the power plant at a rate of 1430 MW. (Recall that efficiency = Energy$_{out}$/Energy$_{in}$= Power$_{out}$/Power$_{in}$.)
(b) Show that the plant produces heat at a rate of 930 MW. (Any power input that doesn't go to electricity is output as heat.)
(c) The heat produced has to be removed from the power plant. Show that to remove this heat just by evaporating water requires evaporating 6500 gallons of water each minute.
(1 L of water = 1 kg = 0.264 gallons.)
(d) Show that this power plant will evaporate 3.4 billion gallons of water each year.

> For beginners, wisdom is knowing not to throw a rock straight up. Advanced wisdom is knowing what to overlook.

18. Thermodynamics

Thermodynamics involves the movement of energy from one system to another, and also the relationship between heat and work.

Internal Energy

We define U to be the *internal energy* of a system. We think of it as having two parts:

1. The sum total of the kinetic energies of the individual particles or molecules within the system, in all of its different forms—the linear (*translational*) motion of the particles (related to the temperature of the system) as well as the internal vibrations and rotations of the particles themselves

2. The sum total of all of the potential energy stored in the attractions and repulsions among all of the particles in relation to one another

The internal energy of the system can be increased by adding heat to the system—the particles within it will move faster and the temperature will increase, or perhaps the added heat will cause a phase change and increase the potential energy of the system. Likewise, the internal energy of the system can increase if you do work W on the system (say, on a gas by compressing it—applying a force for some distance on the gas particles).

In sum, $\Delta U = Q + W$.

If heat flows out of the system, Q is negative. Likewise, if you let the system do work on the environment (say, an expanding gas that supplies a force for some distance on the environment), the sign of W is negative.

Sign of Q	Sign of W
Heat flow into the system $\Rightarrow Q$ is $+$	Work is done on the system $\Rightarrow W$ is $+$
Heat flow out of the system $\Rightarrow Q$ is $-$	Work is done by the system $\Rightarrow W$ is $-$

A physicist's favorite system when thinking about thermodynamics is a gas-filled cylinder fitted with a movable piston. Heat flowing into or out of the gas through the cylinder walls raises or lowers the temperature and internal energy of the gas. Work done compressing the gas accelerates the gas particles as they collide with the approaching piston. If instead the gas expands to do work on the environment, particles of the expanding gas slow down and lose energy due to collisions with the receding piston. In both cases, work changes the internal energy of the gas.

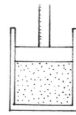

Sample Problem 1

Q joules of heat flow into a container of gas while W joules of work are done in compressing it.
(a) Write an equation for the change in the internal energy of the gas.

Focus: $\Delta U = ?$

The change in internal energy generally depends on both the amount of heat flowing into or out of the system, and the amount of work done on or by the system. In this case, heat flows into the gas while work is done on it, so $\Delta U = Q + W$.

(b) Calculate the change in the internal energy of the gas when 100 J of heat are put into it while at the same time 60 J of work are done in compressing the gas.

Solution: Heat flows *into* the gas so Q is positive. Work is done *on* the gas so W is likewise positive.
So $\Delta U = Q + W = (+100 \text{ J}) + (+60 \text{ J}) = \mathbf{+160 \text{ J}}$.

Sample Problem 2

A sample of gas absorbs heat Q while it expands and does an amount of work W on its surroundings.
(a) Write an equation for the change in the internal energy of the gas.

Solution: In the equation $\Delta U = Q + W$, Q refers to heat absorbed *by* the gas and W is work done *on* the gas. But here we see that heat is added to the gas (so Q in the equation is positive), and work is done *by* the gas (W in the equation is negative). So, $\Delta U = Q + (-W) = \mathbf{Q - W}$.

(b) Calculate the change in internal energy if the sample absorbs 120 J of heat while expanding to do 80 J of work on its surroundings.

Solution: $\Delta U = Q - W = 120 \text{ J} - 80 \text{ J} = \mathbf{+40 \text{ J}}$.

Temperature Scales and Gas Laws

For a gas in a container, *volume* measures the total space available to the gas, *pressure* measures the sum total force of impact per unit area due to collisions of the gas particles on the inner walls of the container, and *temperature* measures the average translational kinetic energy of the particles in a sample—when the sample is hot, the particles move faster than when the sample is cold.

We can seal a sample of gas in a container and measure the pressure of the gas at different temperatures. As we increase the temperature, the particles of gas hit the walls of their container harder and more often, and the pressure increases. If we decrease the temperature, the particles hit the walls less hard and less often, and the pressure decreases. If we plot our measurements of pressure vs. temperature, it turns out that we get a straight line, and if we extend that line, we find that it intersects a pressure of zero at a temperature of –273°C.

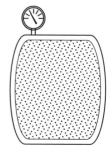

Here is a sample of gas trapped in a sealed container. As the temperature of the gas rises, the pressure increases.

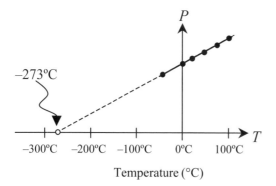

PRESSURE OF AN IDEAL GAS (AT CONSTANT VOLUME) VS. CELSIUS TEMPERATURE

Temperature (°C)

Since temperature measures the average energy of motion of the particles in the gas, the lowest possible temperature would be one at which the particles stop moving. If the particles stop moving, then they stop hitting the walls of the container, and the pressure in the container would be zero. Our graph tells us that this would occur at a temperature of –273°C. We call this temperature *absolute zero*.* This process defines

---

* In reality, every gas will change to a liquid before it reaches absolute zero, and even at absolute zero, the particles will retain some small amount of motion.

"nature's temperature scale," the *Kelvin* scale, which starts at absolute zero but keeps the same-sized temperature division as the Celsius scale. The relationship between the two scales is:

$$T(\text{K}) = T(°\text{C}) + 273.$$

Measurements of temperature on the Kelvin scale are proportional to measurements of pressure. This is seen by a plot of measured pressures versus absolute (Kelvin) temperatures:

$$\frac{P_1}{T_1} = \frac{P_2}{T_2}.$$

This relationship is known as Gay-Lussac's Law, after the French chemist Joseph Luis Gay-Lussac.

Consider an experiment consisting of a movable piston in a cylinder with a weight on top of the piston that maintains a constant external pressure on an enclosed gas. The volume of the gas is measured at increasing temperatures. A plot of volume vs. temperature measurements produces a straight line, which when extended to a volume of zero, again indicates a temperature of –273°C.♦

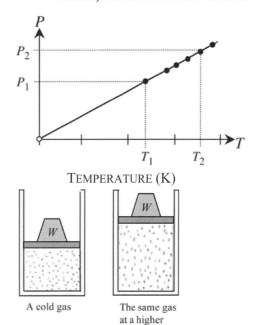

PRESSURE OF AN IDEAL GAS (AT CONSTANT VOLUME) VS. KELVIN TEMPERATURE

TEMPERATURE (K)

A cold gas

The same gas at a higher temperature

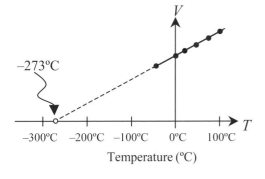

VOLUME OF AN IDEAL GAS (AT CONSTANT PRESSURE) VS. CELSIUS TEMPERATURE

Temperature (°C)

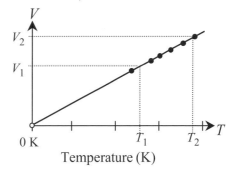

VOLUME OF AN IDEAL GAS (AT CONSTANT PRESSURE) VS. KELVIN TEMPERATURE

Temperature (K)

When we measure temperatures on the Kelvin scale, volume and temperature are proportional. This relationship is seen by a plot of measured volumes vs. absolute (Kelvin) temperature, as above.

$$\frac{V_1}{T_1} = \frac{V_2}{T_2}.$$

This is known as Charles' Law, after the French chemist Jacques Charles.

♦ This doesn't mean that the volume of the gas becomes zero! Gases consist of particles that have a finite size, and the gas will condense to a liquid before reaching absolute zero. But while the gas *is* a gas, its volume is proportional to its Kelvin temperature.

Recall from Chapter 14 that the pressure and volume of a gas held at constant temperature are inversely proportional—that is $P_1V_1 = P_2V_2$ (Boyle's Law).

Combining these relationships between pressure, volume, and temperature for a gas sample results in the combined gas law

$$\frac{P_1V_1}{T_1} = \frac{P_2V_2}{T_2},$$

with temperatures measured in kelvins.

Sample Problem 3

A sample of gas in a cylinder is initially at pressure P_1 and Kelvin temperature T_1, and it occupies a volume V_1.

(a) Derive an equation for the volume the gas will occupy if it is compressed to a new pressure P_2 and cooled to a new Kelvin temperature T_2.

Focus: $V_2 = ?$ From the combined gas law, $\frac{P_1V_1}{T_1} = \frac{P_2V_2}{T_2} \Rightarrow V_2 = \left(\frac{P_1}{P_2}\right)\left(\frac{T_2}{T_1}\right)V_1.$

Notice that the new volume depends on the original volume, a ratio of pressures, and a ratio of temperatures. The equation tells us that if the pressure doubles, the volume will be halved, and if the absolute temperature doubles, the volume will double, just as we would expect.

(b) A cylinder holds 0.500 L of air at a pressure of 101 kPa and a temperature of 20°C. Calculate the volume of air at a pressure of 125 kPa and a temperature of 5°C.

Solution: Before we can apply the combined gas law, we need to convert our temperatures to kelvins. We have $V_1 = 0.500$ L; $P_1 = 101$ kPa; $T_1 = 20 + 273 = 293$ K; $P_2 = 125$ kPa; and $T_1 = 5 + 273 = 278$ K. Now

$$V_2 = \left(\frac{P_1}{P_2}\right)\left(\frac{T_2}{T_1}\right)V_1 = \left(\frac{101 \text{ kPa}}{125 \text{ kPa}}\right)\left(\frac{278 \text{ K}}{293 \text{ K}}\right)(0.500 \text{ L}) = \mathbf{0.383 \text{ L}}.$$

Note that the ratio of the pressures is less than one, consistent with our idea that increasing the pressure should lower the final volume. Note too that the temperature ratio is less than one, consistent with our idea that a decrease in temperature should lower the volume.

Heat Engines and Efficiency

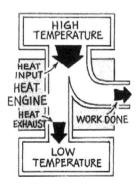

A *heat engine* is any device that converts heat to work in a cyclic fashion. For example, an automobile engine is a heat engine. Burning fuel adds heat to the cylinders. The expanding combustion products do work to turn a shaft that moves the car. As the gas in each cylinder is compressed and readied for the next burst of input heat, a smaller amount of heat is exhausted from the engine. The diagram at the left shows an idealized view of a heat engine. At some high temperature, heat Q_{in} is added to the engine, which then does work W on its surroundings and ejects heat Q_{out} at a lower temperature. By energy conservation, $Q_{in} = Q_{out} + W \Rightarrow W = Q_{in} - Q_{out}.$

The *efficiency* of a heat engine $\varepsilon = \dfrac{\text{Work out of the system}}{\text{Heat into the system}} = \dfrac{W}{Q_{in}} = \dfrac{Q_{in} - Q_{out}}{Q_{in}}$.

The French engineer Sadi Carnot determined that there was a limit to how efficient a thermodynamic cycle could be—you can never convert all the heat input to a system into work. When energy flows into a heat engine at a high temperature T_{hot}, some energy has to flow out to a cooler environment at T_{cold}. The best possible efficiency for a heat engine operated between a high-temperature reservoir and a low temperature reservoir is given by:

Ideal efficiency $\varepsilon = \dfrac{T_{hot} - T_{cold}}{T_{hot}} = 1 - \dfrac{T_{cold}}{T_{hot}}$ (with temperatures in *kelvins*).

Carnot's calculations assume that the cycle happens in a series of tiny incremental steps, as though the piston was being moved slowly, one micron at a time, with plenty of time for the system to reach equilibrium before the next one-micron movement of the piston. Actual (real) heat engines do work in a finite amount of time, and their efficiency is always lower than that of a Carnot heat engine.

Sample Problem 4

In the 1990s a 210-kW OTEC (Ocean Thermal Energy Conversion) power plant in Hawaii was designed to run between the warm ocean surface waters at T_{hot} and the colder deep ocean waters at T_{cold}.

(a) Write the expression for the theoretical maximum efficiency of this power plant.

Answer: $\varepsilon = \dfrac{T_{hot} - T_{cold}}{T_{hot}}$ with temperatures expressed in kelvins.

(b) Theoretically, how much heat must be extracted from the warmer seawater each second to produce 210 kW?

Focus: $Q_{in} = ?$

210 kW = 210 kJ of energy *out* each second. This useful work out of the power plant is only a fraction ε of the total heat taken in by the plant.

From $\varepsilon = \dfrac{W_{out}}{Q_{in}} \Rightarrow Q_{in} = \dfrac{W_{out}}{\varepsilon} = \dfrac{210 \text{ kJ}}{\varepsilon}$.

(c) Calculate answers for (a) and (b) when $T_{hot} = 26°C$ and $T_{cold} = 6°C$.

Solution: First we must express the given temperatures in kelvins.

We have $T_{hot} = 273 + 26 = 299$ K and $T_{cold} = 273 + 6 = 279$ K. Then

$\varepsilon = \dfrac{T_{hot} - T_{cold}}{T_{hot}} = \dfrac{299 \text{ K} - 279 \text{ K}}{299 \text{ K}} = \mathbf{0.067}$.

Ideally, 6.7 joules out of every 100 joules of heat taken in by the plant would be converted to electrical energy.

$Q_{in} = \dfrac{210 \text{ kJ}}{\varepsilon} = \dfrac{210 \text{ kJ}}{0.067} = \mathbf{3100 \text{ kJ}}$ each second.

Since real power plants do not operate at ideal efficiencies, the plant would have to take in more than 3100 kJ of energy per second to produce 210 kW of power.

Adiabatic Processes in the Atmosphere

Compressing or expanding a gas while no heat enters or leaves the system is said to be an *adiabatic process*. As a parcel of warm air rises through the atmosphere, the pressure around it decreases and the parcel of air expands. As the air within the parcel expands, it does work against the external pressure of the surrounding air. This work comes at the expense of the internal energy of the air parcel, so its temperature decreases—about 10°C for each kilometer the air rises in the atmosphere. Likewise, as a parcel of cool air descends, the increasing pressure of the atmosphere compresses the parcel and makes it hotter (also 10°C per km).[*]

Sample Problem 5

Glider pilots usually don't have heaters in their gliders. As the gliders gain altitude, the air temperature drops according to the 10-degrees-per-kilometer rule (also described on page 320 in the *Conceptual Physics 11th edition* textbook). So glider pilots should dress warmly even though the temperature at the glider port is comfortably warm.

(a) Suppose that the temperature at a glider port is a comfortable T_0. Write an equation that indicates what temperature a pilot can expect when soaring a distance Δh above the ground.

Focus: $T = ?$

As the glider rises, we can assume that the temperature in the cabin is much the same as that of the air around it, decreasing approximately 10°C for each 1 km (1000 m) increase in elevation. So

$\Delta T = -\dfrac{10°C}{1000 \text{ m}} \Delta h$ (negative because the temperature decreases as we ascend)

and $T = T_0 + \Delta T = T_0 - \dfrac{10°C}{1000 \text{ m}} \Delta h$.

(b) The temperature at the glider port is 22°C. Calculate the temperature at an altitude 3000 m above the glider port.

Solution: $T = T_0 - \dfrac{10°C}{1000 \text{ m}} \Delta h = 22°C - \left(\dfrac{10°C}{1000 \text{ m}}\right) 3000 \text{ m} = \mathbf{-8°C}.$

Again, the pilot should dress warmly!

(c) Suppose that on a different day the temperature at an altitude of 3000 m happens to be a somewhat warmer 3°C. Further suppose that the glider suddenly descends from this altitude to an altitude of 1600 m. What would be the temperature at this new altitude?

Solution: In this case the altitude decreases. The temperature of the surrounding air increases as the glider descends. This time

$\Delta T = -\dfrac{10°C}{1000 \text{ m}} \Delta h$ (Δh is negative when we descend, so ΔT will be positive.)

so $T = T_0 - \Delta T = T_0 - \dfrac{10°C}{1000 \text{ m}} \Delta h = 3°C - \left(\dfrac{10°C}{1000 \text{ m}}\right)(-1400 \text{ m}) = \mathbf{17°C}.$

---

[*] Meteorologists refer to this 10°C change in temperature per km of altitude as the *dry adiabatic lapse rate*.

Problems Involving Heat, Work, and Internal Energy

18-1. In each case below calculate the missing quantity in the table:

	Q	W	ΔU
a.	100 J of heat flows in	System does 40 J of work	?
b.	200 J of heat is added	No work is done	?
c.	500 J of heat is added	System does 350 J of work	?
d.	70 J of heat is removed from a gas	70 J of external work is done to compress the gas	?
e.	?	Expanding gas does 500 J of work	Internal energy decreases by 340 J
f.	You lose 9.0×10^5 J of heat playing basketball	?	Your internal energy decreases by 1.20×10^6 J
g.	No heat is added	40 J of external work done to compress a gas	?

18-2. A gas in an insulated cylinder expands and does work W on the environment. (Since the cylinder is insulated, no heat can flow into or out of the gas.)
(a) Write an expression for the change in the internal energy of the gas.
(b) Calculate the internal energy change of an insulated gas that does 60 J of work on the environment.
(c) Does the gas in the cylinder warm up or cool down? Defend your answer.

18-3. Work W is done to compress a gas in an insulated cylinder.
(a) What is the change in the internal energy of the gas?
(b) Write an expression for the internal energy change of an insulated gas that has 530 J of work done on it.
(c) Does the gas in the cylinder warm up or cool down? Defend your answer.

18-4. An external force does work W on a gas to compress it. As the gas is being compressed, an amount of heat Q flows out of the gas to the environment.
(a) Write an expression for the change in the internal energy of the gas.
(b) Calculate the internal energy change of a gas that has 530 J of work done to compress it and that loses 420 J of heat to its environment.
(c) Does the gas in the cylinder warm up or cool down? Defend your answer.

Gas Laws Problems

18-5. A half-filled Mylar party balloon has volume V_0 on a cool morning when the ambient temperature is T_0 (°C), but it expands a little as the temperature warms up to T_f (°C).
(a) Write an expression for the new volume of the balloon. (Note that the pressure stays constant.)
(b) Calculate the new volume at 27°C if the gas in the balloon occupied 1.50 L at 10°C.

18-6. A SCUBA tank's pressure gauge reads P_0 while sitting on the deck of a boat on a day when the temperature is T_0.
 (a) Write an expression for the pressure gauge reading if the tank falls off the boat and sinks to the bottom of a lake, where the lake water is at T_f (°C). (Note that the volume of gas remains constant.)
 (b) Calculate the new pressure at 6°C if the tank pressure was 197 atmospheres at 15°C.

18-7. A spray can containing only gases is heated from T_1 to T_2.
 (a) Write an expression for the new pressure inside the can.
 (b) Calculate the new pressure at 49°C (120°F) if the initial pressure was 485 kPa at 20°C.
 (c) Often hair spray cans have warning labels that say, "Do not store above temperatures of 120°F." Why this cautionary note?

18-8. A weather balloon is partially filled with helium at sea level, where the pressure is P_0 and the temperature is T_0.
 (a) By what factor will the volume have changed when the balloon ascends to high altitude, where the pressure is P and the absolute temperature is T?
 (b) Calculate the change in volume for a balloon that is filled at 101 kPa and 10°C and ascends to an altitude of 11,000 m, where the pressure is 22.6 kPa and the temperature is 216 K.

18-9. A particular automobile engine has a compression ratio of 9.5—that is, for the gas in the engine's cylinders, $\dfrac{\text{initial volume of the gas}}{\text{final volume of the gas}} = 9.5$. The intake air has initial pressure P_1 and temperature T_1. After compression, the gas temperature is T_2.
 (a) Write an expression for the pressure P_2 of the compressed gas in the engine cylinder.
 (b) Calculate the final pressure if the intake air was at 35°C at 0.95 atm and the temperature of the compressed air was 1330 K.

18-10. You open your plastic bag while in the mountains, where the temperature is T_1 and the atmospheric pressure is P_1. Then you seal the bag shut, trapping a volume V_1 of air within the bag.
 (a) You return to lower altitude, where the pressure is P_2 and the temperature is T_2. Write an expression for the volume occupied by the air in the bag.
 (b) Calculate this final volume if the volume of the mountain air was 0.50 L at 10°C, and 71 kPa and the temperature and pressure at lower altitude are 24°C and 98 kPa, respectively.

18-11. A car tire is filled to a gauge pressure of P_1 on a cold morning when the ambient temperature is T_1. (Gauge pressure is the difference between the actual pressure in the tire and the outside atmospheric pressure—that is $P_{gauge} = P_{actual} - P_{atm}$.)
 (a) Assume that the volume of gas in the tire doesn't change and the atmospheric pressure is P_{atm}. Derive an equation for the gauge pressure inside the tire after it has warmed up to a temperature T_2.
 (b) Calculate the final pressure if the gauge pressure was 32.0 psi (pounds per square inch) at 5°C when the atmospheric pressure is 14.7 psi, the temperature of the tire increases to 32°C.
 (c) If the tire actually expands a little bit, what effect will this have on the pressure inside the tire?

Efficiency Problems

18-12. An automobile engine runs at T_h (°C) on a day when the temperature outside is T_c (°C).
   (a) Write an expression for the maximum possible efficiency of the automobile engine.
   (b) Calculate the maximum possible efficiency if the engine runs at 630°C on a 25°C day.

18-13. An ideal heat engine takes in heat Q_{in} at a temperature T_h. It exhausts heat Q_{out}.
   (a) Write an expression for the amount of work done by the engine.
   (b) Write an expression for the efficiency of the engine.
   (c) Write an expression for the exhaust temperature of the engine.
   (d) Calculate answers to the above for a heat input of 465 J at a temperature of 620 K, and a heat output of 285 J.

18-14. A heat engine exhausts heat Q_{out} while performing useful work W.
   (a) Derive an expression for the efficiency of the engine.
   (b) Calculate the efficiency if the engine does 1200 J of work while exhausting 3800 J of waste heat to the environment.

18-15. A heat engine takes in heat Q_{in} and exhausts heat Q_{out}.
   (a) Write an expression for the efficiency of the engine.
   (b) Calculate the efficiency if the engine takes in 3600 J of heat and exhausts 1400 J of heat.

18-16. A particular heat engine has efficiency ε.
   (a) Write an expression for the quantity of heat taken in to do work W.
   (b) Write an expression for the quantity of heat exhausted to its surroundings.
   (c) Calculate answers for (a) and (b) for a 27% efficient heat engine that does 1400 J of work.

18-17. A geothermal power plant operates between hot groundwater at T_h and a river at T_c.
   (a) Write an expression for the theoretical maximum efficiency of this power plant.
   (b) Calculate the maximum efficiency for $T_h = 185°C$ and $T_c = 15°C$.

18-18. A geothermal power plant runs between hot groundwater at T_h and a river at T_c.
   (a) If the plant produces x kW of power, derive the quantity of heat absorbed from the hot groundwater every second.
   (b) Calculate the minimum amount of heat that must be absorbed each second by a 35-kW geothermal power plant that operates between $T_h = 185°C$ and $T_c = 15°C$.

18-19. • A Stirling engine is a type of heat engine with the same theoretical efficiency as a Carnot engine. It typically uses two joined piston-cylinder combinations to obtain work from an external heat source. Consider a Stirling engine that heats a confined gas to T_h and rejects heat to its surrounding at T_c. Its heat source is a solar collector of area A (m²), which is illuminated by sunlight of intensity I (W/m²).
   (a) Write an expression for the theoretical maximum efficiency of this Stirling engine.
   (b) Write an expression for the amount of sunlight energy hitting the collector each second. (Answer will be in units J/s = W.)
   (c) In theory, how much useful work can we get out of this Stirling engine each second?
   (d) In practice, the engine operates at 75% of its theoretical efficiency. How much useful work per second do we actually get out of the engine?

(e) This Stirling engine has a solar collector area of 3.0 m², it operates with a high temperature of 300°C, and its surroundings are at 30°C. Calculate the actual power produced by the engine if the sunlight intensity is 800 W/m².

(f) Another Stirling engine using the same solar collector operates with a high temperature of 100°C in 30°C surroundings. It runs at 70% of its ideal efficiency. Calculate the actual power produced by this lower-temperature engine if the sunlight intensity is 800 W/m².

18-20. • A small research geothermal electrical generation site uses hot water from the ground as the high temperature reservoir, and the outside air as the low temperature reservoir. Hot groundwater enters the plant at a temperature T_0 and leaves the plant at a slightly lower temperature T_f. Waste heat from the plant is rejected to the environment at a temperature T_c. A mass m of hot groundwater is pumped through the power plant each second.

(a) Derive an expression for the heat supplied to the power plant every second from the hot groundwater. (The hot groundwater is cooled from T_0 to T_f.)

(b) Derive an expression for the theoretical maximum efficiency of the geothermal plant if the temperature of the hot "reservoir" is taken to be the average of T_0 and T_f. What fraction of the heat input can this heat engine ideally turn into work?

(c) Derive an expression for the maximum power output of the geothermal plant.

(d) At this research plant, 5.8 kg per second of hot groundwater enters the plant at $T_0 = 79.6°C$ and leaves the plant at 79.4°C. In the winter, the temperature of the outside air is $T_c = -0.4°C$. Calculate the theoretical efficiency and maximum theoretical power output of this geothermal plant.

18-21. • Turbine generators are like jet engines—they draw in large quantities of air, compress it, and mix it with fuel. The combustion of the fuel turns the turbine, which draws in and compresses more air, and also turns a generator. Typical combustion temperatures are around 1100°C and typical exhaust temperatures are around 500°C.

(a) If this turbine operates like an ideal Carnot heat engine, what would be the efficiency of the turbine?

(b) For every 100.0 J of heat the engine takes in, how much work is done?

(c) For every 100.0 J of heat the engine takes in, how much heat is rejected into the exhaust?

(d) *Combined-cycle* turbines use the hot exhaust as a source of heat to produce steam to run a steam turbine. If the exhaust from the gas turbine (now our heat *input*) is at 500°C, and the cooling water for the steam turbine is at 20°C, what is the maximum efficiency of this second part of the cycle?

(e) In part (c) you calculated the amount of waste heat from the first part of the cycle, which is also the heat input for the second part of the cycle. Use the efficiency that you calculated in part (d) to calculate how much work you get out of the steam turbine.

(f) How much work is done overall in the *combined*-cycle turbine for every 100.0 J of heat input? [This should be the work you calculated in (b) *plus* the work you calculated in (e).]

(g) What is the theoretical efficiency of the combined-cycle turbine?[*]

(h) Why does using a combined-cycle turbine make more sense than using a single-cycle turbine?

---

[*] Real combined-cycle turbines have efficiencies of 55–60%.

Problem Involving Adiabatic Processes in the Atmosphere

18-22. A mass of air descends from a height h_0 above sea level to a height h_f above sea level.
   (a) Write an expression for the change in air temperature.
   (b) Calculate the change in temperature for dry air that descends from 6200 m above sea level to 2500 m above sea level.

Show-That Problems for Thermodynamics

18-23. In a certain operation, 160 J of work are done on a gas while 50 J of heat are removed from the gas.
   Show that its change in internal energy is +110 J.

18-24. A gas does 250 J of work while absorbing 130 J of heat.
   Show that its change in internal energy is –120 J.

18-25. A certain system experiences an internal energy change of +670 J while absorbing 1350 J of heat.
   Show that the system does 680 J of work on its surroundings.

18-26. 25.0 mL of gas at 105 kPa is compressed at constant temperature to 340 kPa.
   Show that the new volume is 7.7 mL.

18-27. An oxygen tank is at 136 atm of pressure at 23°C.
   Show that the pressure in the tank will be 129 atm at 8°C.

18-28. A plastic bag that holds 0.72 L of air is moved from a 25°C kitchen to the inside of a – 4°C freezer.
   Show that the new volume of the bag will be 0.65 L.

18-29. 5.0 L of gas, initially at 293 K, is at a pressure of 98 kPa. The pressure on the gas is increased to 212 kPa and the final volume of the gas is 6.2 L.
   Show that the final temperature of the gas is 786 K.

18-30. A gas bubble ascends from a deep lake bottom where the temperature is 4°C and the pressure is 650 kPa to the surface of the lake, where the temperature is 15°C and the pressure is 101 kPa.
   Show that the volume of the bubble increases to 6.7 times its original volume.

18-31. • A helium tank has internal volume 6.88 L (1 L = 10^{-3} m³) and a pressure of 130 atm when filled at 20°C. The tank is going to be used to fill spherical balloons that are 23 cm in diameter on a 30°C day. The pressure inside each filled balloon will be 1.03 atm. Assume that after the balloons are filled, 6.88 L of gas at 1.03 atm are left in the tank. Show that you can fill approximately 140 balloons from this tank.

18-32. A certain heat engine takes in 25 kJ of heat and exhausts 17 kJ.
   Show that its efficiency is 0.32.

18-33. A heat engine operates between $T_h = 750°C$ and $T_c = 35°C$.
   Show that the theoretical maximum efficiency is about 0.70, or 70%.

18-34. A power plant has a theoretical maximum efficiency of 0.46 and an exhaust temperature of 35°C.
   Show that the high-temperature reservoir must be at 297°C.

18-35. A 420-kW power plant operates between $T_h = 540°C$ and $T_h = 30°C$.
   Show that the plant requires a heat input of at least 670 kJ per second.

18-36. Dry air rises from an elevation of 800 m above sea level, where the temperature is 23°C, to an elevation of 4500 m above sea level.
   Show that when this occurs, the air's temperature likely drops to –14°C.

18-37. The 86th floor observation deck of the Empire State Building is 320 m above the ground.
   Show that the temperature there is normally 3.2°C cooler than the temperature at ground level.

19 Vibrations and Waves

Many things vibrate or oscillate—a pendulum, an object on the end of a spring, a tuning fork, the strings of musical instruments, or the air columns in wind instruments. Oscillations can be described by:

Amplitude, A—the maximum disturbance from the object's equilibrium position, typically measured in a unit of length (meters or centimeters) or angle (degrees or radians)

Period, T—the time taken to make one complete oscillation, usually measured in seconds

Frequency, f—the number of complete oscillations per second, usually measured in Hertz (Hz) = oscillations/second

Period and frequency are related by

$$T = \frac{1}{f} \quad \text{or} \quad f = \frac{1}{T}.$$

Sample Problem 1

Maria displaces a simple pendulum bob a short sideways distance x and then releases it. She finds that the pendulum makes n complete oscillations in t seconds.

(a) What is the period of the oscillation?

Answer: The period is the time for one oscillation = $\dfrac{t \text{ seconds}}{n \text{ oscillations}} = \dfrac{t}{n}$.

(b) What is the frequency of the oscillation?

Answer: The number of oscillations per unit time = $\dfrac{n \text{ oscillations}}{t \text{ seconds}} = \dfrac{n}{t}$.

Notice that this is the reciprocal of the period calculated in part (a).

(c) Derive an equation for the length of the pendulum.

Answer: Begin with the period of a pendulum, which is related to its length and the strength of the local gravitational field: $T = 2\pi\sqrt{\dfrac{L}{g}}$, so $L = g\left(\dfrac{T}{2\pi}\right)^2 = g\left(\dfrac{\left(\frac{t}{n}\right)}{2\pi}\right)^2 = g\left(\dfrac{t}{2\pi n}\right)^2$.

(d) Calculate the period, frequency, and length for a pendulum that completes 15.0 oscillations in 42.0 s.

Solution: $T = \dfrac{t}{n} = \dfrac{42.0 \text{ s}}{15.0 \text{ oscillations}} = 2.80 \text{ s}; \quad f = \dfrac{n}{t} = \dfrac{15.0 \text{ oscillations}}{42.0 \text{ s}} = 0.357 \text{ Hz}.$

$$L = g\left(\frac{T}{2\pi}\right)^2 = 9.8 \frac{\text{m}}{\text{s}^2}\left(\frac{2.80 \text{ s}}{2\pi}\right)^2 = 1.95 \text{ m}.$$

Oscillations can produce waves, and waves can be either *standing* (as in the vibration of a violin string) or *traveling* (as in the sound wave that carries energy through the air from the violin to your ear). Waves are characterized by an amplitude A, a wavelength λ, a frequency f, and a speed v.

Sample Problem 2

While sitting at the dock of the bay, Otis notices incoming waves with distance d between crests. The incoming crests lap against the pier pilings at a rate of one every 2 seconds.

(a) What is the frequency of the waves?

Answer: The frequency of the waves is given, one per 2 s, or $f = $ **0.5 Hz**.

(b) What is the speed of the waves?

Focus: Each crest of the wave travels a distance d in 2 seconds.

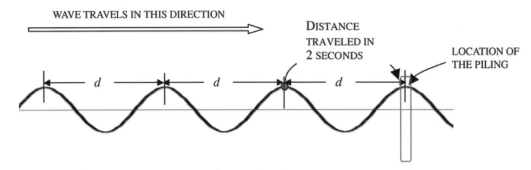

From $\text{speed} = \dfrac{\text{distance}}{\text{time}} \Rightarrow v = \dfrac{d}{2\text{ s}} = \left(\dfrac{1}{2}\cdot\dfrac{1}{\text{s}}\right)(d) = \left(\dfrac{1}{2}\text{Hz}\right) \times d$. This example illustrates a central relationship between wave speed, frequency, and wavelength: $v = f\lambda$.

(c) Calculate the speed of the waves if the distance d between crests is 1.8 m.

Solution: $v = fd = 0.5\text{ Hz}(1.8\text{ m}) = 0.5\left(\dfrac{1}{\text{s}}\right)(1.8\text{ m}) = 0.9\dfrac{\text{m}}{\text{s}}$.

The Doppler Effect

The pitch of a sound depends on its frequency. When a car honking its horn approaches you, the compressions of the sound waves arrive at your ear closer together and with a higher frequency than they do from a car at rest. You hear a higher pitch than the horn is producing. When a car recedes from you, the compressions of the sound waves arrive at your ear farther apart and with a lower frequency. Now the pitch you hear is lower. This shift in pitch is due to the Doppler effect.

The frequency heard by an observer, f_o, is related to the source frequency f_s, source speed v_s, and speed of sound v in the medium by these formulas: *

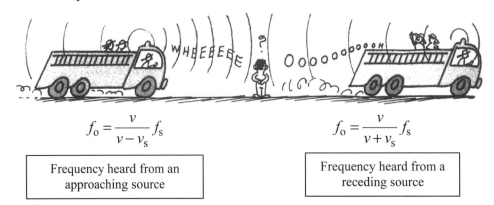

$$f_o = \dfrac{v}{v - v_s} f_s$$

Frequency heard from an approaching source

$$f_o = \dfrac{v}{v + v_s} f_s$$

Frequency heard from a receding source

---

* Derivations of these formulas, and formulas for cases where the observer is in motion as well, can be found in many algebra-based physics textbooks. We won't treat these extensions of the Doppler effect here.

Sample Problem 3

A fire engine, moving at speed v_s with its siren blasting, approaches you and soon thereafter recedes. The frequency of the siren is f_s.

(a) Calculate the frequency you hear when the vehicle approaches at a speed of 31 m/s with its siren blasting at 500 Hz. (Assume here that the speed of sound in air is 340 m/s.)

Focus: $f_o = ?$ Which formula do you use? Since the siren is approaching you, you expect to hear a higher frequency. Both expressions for the observed frequency are of the form $f_o = \text{ratio} \times f_s$. You need the ratio that is greater than 1, where you divide v by a number less than v. In this case you choose the formula with $v - v_s$ in the denominator.

Solution: $f_o = \dfrac{v}{v - v_s} f_s = \dfrac{340 \frac{m}{s}}{340 \frac{m}{s} - 31 \frac{m}{s}} 500 \text{ Hz} = \textbf{550 Hz}.$

(b) Calculate the frequency you hear when the vehicle recedes at the same speed.

Focus: $f_o = ?$ Since the siren is receding, you will hear a lower frequency. You need the ratio in the formula to be less than 1, so you choose the expression with $v + v_s$ in the denominator.

Solution: $f_o = \dfrac{v}{v + v_s} f_s = \dfrac{340 \frac{m}{s}}{340 \frac{m}{s} + 31 \frac{m}{s}} 500 \text{ Hz} = \textbf{460 Hz}.$

(c) Why do firemen on the moving vehicle hear no Doppler shift in the siren?

Answer: They hear no Doppler shift because there is no velocity difference between them and the siren. The Doppler effect results from *differences* in velocities between the source and the receiver.

Shock Wave

A supersonic (faster-than-sound) aircraft produces a conical shock wave composed of overlapping spheres of sound crests, similar to the bow wave of a boat composed of overlapping circular waves. We see in the figure that during the time the aircraft moves a distance $v_s t$, sound waves from the aircraft move a distance vt (where v_s is the speed of the source and v is the speed of sound). Note that $\sin\alpha = \dfrac{vt}{v_s t} = \dfrac{v}{v_s}$. The faster the aircraft, the smaller $\sin\alpha$ is and the narrower the cone. Interestingly, the inverse of this ratio, v_s/v, is called the Mach number. For example, at twice the speed of sound, $v_s/v = 2$ and we say the craft is moving at Mach 2. The angle α shown in the sketch is $\sin^{-1}\left(\dfrac{1}{2}\right) = 30°$. This is half the angle of the V shape produced by the aircraft, which is 60°.

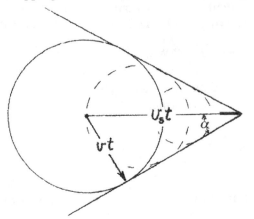

Sample Problem 4

When the Concorde used to fly at its cruising speed of Mach 1.75, it created a sonic boom.

(a) How fast in km/h was its cruising speed? (Use 340 m/s for the speed of sound in air.)

Solution: Cruising speed was $1.75\left(340\tfrac{m}{s}\right) = 595\tfrac{m}{s}$. Converted to km/h,

$$595\tfrac{m}{s} \times \tfrac{1\,km}{1000\,m} \times \tfrac{3600\,s}{1\,h} = 2140\tfrac{km}{h}.$$

(b) From the moment the Concorde is 8.0 km directly overhead, how many seconds would elapse before you hear its sonic boom?

Focus: $t = ?$ You can imagine that the nose of the sonic boom cone starts forming when the Concorde is directly above you. We're looking for the time for the sound waves to travel to you, 8 km below.

From $v = \dfrac{d}{t} \;\Rightarrow\; t = \dfrac{d}{v} = \dfrac{8000\,m}{340\,\tfrac{m}{s}} = 24\,s.$

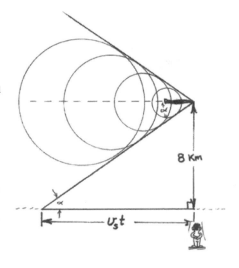

(c) How far past the overhead point would the Concorde be when the boom reached you?

Solution: During the 24 seconds it would take for the shock to reach you, the Concorde would have traveled a horizontal distance $v_{Concorde}\, t = 595\tfrac{m}{s}(24\,s) = 14,000\,m = 14\,km.$

At high altitudes, where it is cold, the speed of sound is somewhat less than 340 m/s, so in fact the Concorde would have been even farther ahead.

Problems for Vibrations and Waves

19-1. A simple pendulum swings to and fro once in time t.
    (a) What is the frequency of the pendulum?
    (b) If the pendulum were shorter, would the frequency be more or less? Defend your answer.

19-2. Jacqueline ties a small lead ball onto a string of length L and gives the ball a gentle tap.
    (a) Write an equation for the period of oscillation for this pendulum.
    (b) Calculate the period of a 35-g lead ball tied onto the end of a string 82 cm long.

19-3. When a radio station broadcasts a radio signal, electrons in the broadcast tower oscillate at the broadcast frequency f.
    (a) Write an equation for the time it takes for the electrons to complete one oscillation.
    (b) Calculate the time for electrons oscillating at 103 MHz. (MHz = megahertz = 10^6 Hz.)

19-4. A mass on a vertical spring vibrates over a distance d from its lowest to its highest point and completes n full up-and-down vibrations in time t.
    (a) What is the amplitude of vibration?
    (b) What is its frequency?
    (c) What is its period?
    (d) Calculate the amplitude, frequency, and period if d is 0.30 m and the mass completes 10 oscillations in 20.0 s.

19-5. • The piston in an engine makes one complete oscillation for each rotation of the engine. Suppose that the engine is turning at n revolutions per minute and that the piston stroke (the distance between the top and bottom of the piston's path) is d.
    (a) Derive an equation for the average speed of the piston.
    (b) Calculate the piston's average speed when the engine turns at 2800 rpm and the piston stroke is 95 mm.

19-6. A wave travels along a stretched rubber tube. The vertical distance from crest to trough for the wave is Y and the horizontal distance from crest to nearest trough is X.
    (a) What is the amplitude of the wave?
    (b) What is the wavelength of the wave?

19-7. Light travels at speed $c = 3.00 \times 10^8$ m/s through a vacuum, and for most practical purposes, through air as well.
    (a) Write an equation for the wavelength of a light wave given its frequency f.
    (b) Write an equation for the frequency of a light wave given its wavelength λ.
    (c) Determine the wavelength of light of frequency 5.09×10^{14} Hz emitted by a sodium lamp.
    (d) Determine the frequency of light whose wavelength is 5.5×10^{-7} m.

19-8. Radio waves travel at the speed of light, $c = 3.00 \times 10^8$ m/s. The wavelength of radio waves is much longer than the wavelength of visible light waves.
    (a) Write an equation for the wavelength of a light wave in terms of its frequency f.
    (b) Determine the wavelength of radio waves having a frequency of 500 kHz.
    (c) If the frequency were higher, would the wavelength be longer or shorter? Defend your answer.

19-9. High frequency sound waves (ultrasound) are reflected off the internal structures of bodies to provide interior images of the body. The level of detail that can be seen is approximately equal to the wavelength of sound waves being used.
   (a) Derive an equation for size of the smallest internal structural detail that ultrasound waves can see if the speed of sound in tissue is v and the frequency is f.
   (b) Calculate the level of structural detail that can be imaged with 3.0-MHz ultrasound waves if the speed of sound in tissue is 1500 m/s.

19-10. A mosquito flaps its wings at frequency f, which produces the annoying buzz that travels at the speed of sound, v.
   (a) What is the frequency of the buzz?
   (b) How far does the sound travel between wing beats?
   (c) What is the wavelength of the annoying sound?
   (d) Calculate the wavelength of the buzzing sound if the frequency of vibration is 600 Hz and the speed of sound is 340 m/s.

19-11. While sitting on a pier, Lillian notices water waves incident upon the pier pilings at regular intervals. The crests are separated by a distance d, and hit the pilings n times per minute.
   (a) Derive an equation for the speed of the water waves.
   (b) Calculate the speed of the waves if the distance between their crests is 3 m and they hit the pilings 20 times per minute.

19-12. Two buoys float a distance d apart in a channel. The buoys rise and fall once every second.
   (a) Calculate the wavelength of the waves if the speed of the waves is 5.0 m/s.
   (b) Calculate the number of waves between the buoys if they are 200 m apart.

19-13. Steve produces waves in a small pond by dipping his fingers into the water at regular intervals. He observes the waves traveling between a pair of reeds that stick out of the water a distance d apart.
   (a) It takes t seconds for the first wave to travel past the first reed to the second reed. What is the speed of the waves?
   (b) Derive an equation for the wavelength of the waves produced when Steve dips his fingers into the water at frequency f.
   (c) At any moment, how many waves are there between the reeds?
   (d) Calculate the speed and wavelength of the waves, and how many waves there are between the two reeds, given this information: When Steve dips his finger into the water 2 times each second, the waves take 3.0 second to pass between the first reed and the second reed, which are 1.5 m apart.

19-14. Two canoes tied to a dock are distance d apart. One is at a crest at the same time the other is at an adjacent trough (one-half wave between them). They each bob up and down at frequency f.
   (a) What is the speed of the waves?
   (b) Calculate the speed if the canoes are 6.0 m apart and they bob up and down once every 2.0 s.

19-15. • You take two photographs in quick succession of a wave rolling by two bottles floating in a lake, a distance d apart from each other. The photographs are separated by a time interval t.
(a) What is the slowest possible speed of the water wave if it is traveling to the right?
(b) What is the slowest possible speed of the water wave if it is traveling to the left?
(c) Calculate the slowest possible wave speed in each case for bottles 5.0 m apart and photographs taken 0.67 s apart.

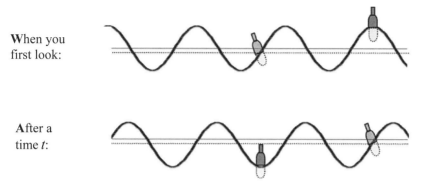

When you first look:

After a time t:

19-16. The diagram shows a standing wave set up in a rubber tube of length L secured to a wall when the other end is shaken at a frequency f.
(a) What is the wavelength of the wave shown?
(b) This standing wave can be treated as the superposition of two waves traveling in opposite directions. What is the speed of these waves?
(c) Calculate the wavelength and wave speed for a 3.3-m rubber tube shaken at 1.3 Hz.

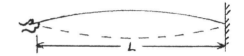

19-17. Suppose that the frequency of vibration is increased 3-fold for the tube of the previous problem.
(a) Make a sketch of the new wave. In terms of L, what will be its new wavelength?
(b) Calculate the frequency and wavelength for the new wave.
(c) What do you notice about the wave *speed* when you change the frequency?

19-18. A fire engine horn blasts at frequency f_s when it approaches you at speed v_s.
(a) Write a formula for the frequency that you hear.
(b) Write a formula for the frequency that you hear after the engine passes you and is moving away.
(c) Calculate the frequencies you hear from the 500-Hz horn of a fire engine that approaches you, and then passes you, at 15 m/s. (The speed of sound is 340 m/s.)

19-19. As a fire engine approaches you, its horn sounds at frequency f_s and your sound analyzer measures the frequency that you hear.
(a) The sound analyzer registers f_o. Derive an equation for the speed of the fire engine.
(b) Calculate the speed of the fire engine if the horn sounds at 500 Hz and you register a sound of 513 Hz. (The speed of sound is 340 m/s.)

19-20. While standing near a railroad crossing, you hear the horn of a receding train.
   (a) Why does the conductor on the train not hear a Doppler shift in the horn's frequency?
   (b) Derive an equation for the horn frequency that the train conductor hears when you hear a frequency of f_o, and the train is moving toward you at speed v_s.
   (c) Calculate the horn frequency that the train conductor hears when you hear a frequency of 431 Hz, and the train is moving toward you at 28 m/s (100 km/h).

19-21. You, at rest, hear the sound of an approaching source as having a higher pitch due to the Doppler effect.
   (a) Derive an equation for the speed of an approaching sound source that increases the sound frequency you perceive by x%.
   (b) Calculate the speed of an approaching car whose horn you perceive to be sounding at a frequency 10% higher than its actual frequency.

19-22. Stationary Stan is listening to the sound from an airplane flying along his line of sight. The frequency that Stan perceives is two-thirds the frequency emitted by the plane.
   (a) Would the airplane be approaching or receding?
   (b) Calculate the speed of the airplane if the speed of sound in air is 340 m/s.

19-23. A speedboat makes a bow wave as it knifes through the water.
   (a) What is the angle of the resulting V shape of the bow wave if the boat travels 5 times as fast as waves on the surface? (In the diagram of the shock wave shown on page 210, this angle is 2α.)
   (b) As the boat gains speed, does the angle of the bow wave increase or decrease?

19-24. A supersonic aircraft produces a conical shock wave in the air.
   (a) If the aircraft flies at 1.5 times the speed of sound, what is the angle (2α) of the conical wave?
   (b) If the aircraft flies faster, does the angle increase or decrease?

Show-That Problems for Vibrations and Waves

19-25. A cork floating in water bobs up and down four times each second as a small wave passes. The speed of the wave is 0.40 m/s.
Show that the wave peaks are 0.10 m apart.

19-26. A mechanical metronome can be set to vibrate at a desired frequency.
Show that if it makes 9 complete vibrations in 12 s, its frequency is 0.75 Hz.

19-27. A tall skyscraper swings to and fro in strong winds at a frequency of 0.15 Hz.
Show that the period of this vibration is 6.7 s.

19-28. Ultraviolet light has a higher frequency than visible light.
Show that the frequency of ultraviolet light of wavelength 360 nm is 8.33×10^{14} Hz. (1 nm = 10^{-9} m, and recall that the speed of light is 3×10^8 m/s.)

19-29. A nurse counts 84 heartbeats in 1 minute.
Show that the period and frequency of the heartbeats are 0.71 s and 1.4 Hz, respectively.

19-30. The crests of a long ocean wave are 24 m apart, and in 1 minute 10 crests pass by.
Show that the speed of the wave is 4.0 m/s.

19-31. A typical tidal wave (tsunami) can have a speed of 750 km/h and a wavelength of 320 m. Show that the period of the wave is about 1.5 s.

19-32. A seismic P wave traveling at 8000 m/s has a wavelength of 2000 m.
Show that the frequency of the wave is 4 Hz.

19-33. Humans can hear sound frequencies between 20 and 20,000 Hz.
Show that the corresponding wavelengths range from 17 m to 1.7 cm for a speed of sound of 340 m/s.

19-34. Microwave ovens typically cook food by using microwaves with frequency 2.45 GHz (gigahertz, 10^9 Hz). (The speed of microwaves is the speed of light, 3×10^8 m/s.)
Show that the wavelength of these waves is 12.2 cm.

19-35. Planet Earth, like a spherical bell, vibrates in a discrete set of fundamental frequencies between 0.002 and 0.007 Hz.
Show that the periods of these vibrations range from 143 s to 500 s.

19-36. Seismologists attribute 0.2-Hz micro-seismic vibrations at Earth's surface to ocean wave interactions.
Show that the period of this small vibration is 5 s.

19-37. While watching water at the dock of the bay, Otis notices that ten waves pass him every 30 seconds, and that the crests of successive waves exactly coincide with posts that are 3.6 m apart.
Show that the speed of the waves is 1.2 m/s.

19-38. A standing wave between fixed supports consists of 3 half-wavelength segments.
Show with a sketch that the number of nodes in the standing wave is 4.

19-39. A standing wave between fixed supports consists of 3 half-wavelength segments.
Show with a sketch that the number of antinodes in the standing wave is 3.

19-40. The Foucault pendulum in the rotunda of the Griffith Observatory in Los Angeles has a 110-kg brass ball at the end of a 12.2-m long cable.
Show that the period of this pendulum is 7.0 seconds.

19-41. You want to make a simple pendulum that has a period of 2.00 seconds.
Show that the pendulum should be 99.3 cm long.

19-42. An astronaut on the moon finds that it takes 74.5 seconds for a 1.00-m long pendulum to complete 15 swings.
Show that the acceleration due to gravity on the surface of the Moon is 1.60 m/s^2.

19-43. • Our astronaut knows that Moon's radius is 1740 km.
Combine the result of the previous problem with Newton's law of universal gravitation, $F = GM_{moon}m/R_{moon}^2$, and $W = mg$, to show that the Moon has a mass of 7.3×10^{22} kg.

19-44. A hummingbird feeder is tied by a rope to a tree branch. You notice that in a gentle breeze, the feeder moves back and forth 15 times in 1 minute.
Show that the rope holding the feeder is about 4 m long.

19-45. You want to double the period of a certain pendulum.
Show that you have to make the pendulum four times as long.

19-46. While standing at a railroad crossing, you hear the whistle of an oncoming train.
Show that you'll hear the 800-Hz whistle as 863 Hz when the train's speed is 25 m/s and the speed of sound in air is 340 m/s.

19-47. A humpback whale produces a 99-Hz sound wave as she swims at 15 m/s toward her stationary mate. The speed of sound in seawater is 1530 m/s.
Show that her mate perceives the sound wave as 100 Hz.

19-48. The apparent frequency of sound from a source approaching you is higher than when the source is stationary.
Show that the apparent frequency increases by 5% when the source approaches you at 16 m/s. (Take the speed of sound in air to be 340 m/s.)

19-49. Your talent for perfect pitch comes in handy. While waiting at the bus stop, you note that the familiar 440-Hz ("concert A") train horn sounds like a B-flat (466 Hz).
Show that the bus is approaching you at a speed of approximately 19 m/s. (Take the speed of sound in air to be 340 m/s.)

19-50. A row of older male rabbits runs away from you, shrieking at 620 Hz. You hear the shriek as having a frequency of 613 Hz. (Take the speed of sound in air to be 340 m/s.)
Show that the speed of the receding hare line is 3.9 m/s.

19-51. In an ideal situation, a boat knifing through the water makes a bow wave very similar to the conical shock wave made by supersonic aircraft.
Show that the full angle of the bow wave is about 29° when the boat moves at 4 times the speed of water waves on the surface.

19-52. An airplane traveling at twice the speed of sound makes a conical shock wave.
Show that the angle of the cone from its axis (half angle for the shock wave) is about 30°.

20. Sound

This chapter logically extends the material of the previous chapter, focusing specifically on sound waves. The speed of sound is different in different materials. In air at 0°C and 1 atmosphere of pressure, sound travels at a speed of 331 m/s. The speed of sound in air increases approximately 0.60 m/s for each Celsius degree increase in temperature. In the problems below, we'll use an average speed of sound of 340 m/s in air (1224 km/h or 723 mi/h).

Sample Problem 1

Sound travels more than four times as fast in seawater as in air.
(a) If the frequency of a sound wave in air is 440 Hz, what is its frequency in seawater?

Answer: The frequency of a sound wave is the same in both air and water: **440 Hz**, in this case. The frequency of the sound wave mirrors the vibrational frequency of its source. If the source object vibrates 440 times per second, the medium will be compressed and expanded at exactly the same rate, 440 Hz, independent of what the medium is, even when a wave travels from one medium to another. Wave properties that *can* change as a wave passes from one medium to another are wavelength, amplitude, and speed.

(b) Calculate the wavelength of a 440-Hz note in air.

Focus: $\lambda = ?$

From $v = f\lambda \Rightarrow \lambda = \dfrac{v}{f} = \dfrac{340 \frac{m}{s}}{440\,Hz} = \dfrac{340 \frac{m}{s}}{440\left(\frac{1}{s}\right)} = 0.77\,m = \mathbf{77\,cm}.$

(c) Calculate the wavelength of a 440-Hz sound wave in seawater. The speed of sound in seawater is 1530 m/s.

Solution: $\lambda = \dfrac{v}{f} = \dfrac{1530 \frac{m}{s}}{440\,Hz} = \dfrac{1530 \frac{m}{s}}{440\left(\frac{1}{s}\right)} = \mathbf{3.48\,m}.$

Sample Problem 2

While hiking, you note a flash of lightning and hear the thunder after a time interval *t*.
(a) Write an equation for your distance from the bolt of lighting if the time interval between seeing the flash and hearing the thunder is *t*.

Focus: $d = ?$

The speed of light is so much greater than the speed of sound that we can assume that the light reaches you instantaneously. The sound wave arrives at your ears after a time *t*.

$d = v_{sound}\,t.$

© Paul G. Hewitt and Phillip R. Wolf

(b) How far are you from the bolt if you hear the thunder 3.0 seconds after you see the lightning flash?

Solution: $d = v_{sound} t = 340 \frac{m}{s}(3.0 \text{ s}) = 1000 \text{ m} = 1 \text{ km}$.

Sound travels about 1 kilometer in 3 seconds.

Sample Problem 3

A dolphin emits ultrasonic sound waves to locate prey. The closer the prey is, the shorter the time is for the echo of the ultrasound from the prey to return to the dolphin.
(a) Calculate the distance between a dolphin and its prey if the time between sending and receiving a pulse is 0.22 s. Assume the speed of sound in the water is 1530 m/s.

Focus: $d = ?$

Solution: $d = vt = 1530 \frac{m}{s}(0.11 \text{ s}) = \mathbf{170 \text{ m}}$.

Notice we use 0.11 s for the time, since the 0.22 s is the round-trip time.

(b) How does a Doppler shift in the echo let the dolphin know whether the prey is approaching or receding?

Answer: If the prey is approaching, the frequency of the echo will be higher; if the prey is receding, the frequency of the echo will be lower. Dolphins employ both the time of echo return and the Doppler effect to better assess their prey and their surroundings.

Problems for Sound

(For the following problems, assume that the speed of sound in air is 340 m/s.)

20-1. A factory whistle emits a loud blast of sound. The intensity of the sound diminishes with distance according to the inverse-square law. You measure a certain intensity of sound a distance d from the whistle.
  (a) Derive an equation for how far away from the whistle you would have to be to detect sound with half the intensity.
  (b) How far away would a listener detect sound one-tenth as intense as at distance d from the whistle?

20-2. A sound wave in air has speed v and frequency f.
  (a) Write an equation showing its wavelength.
  (b) Calculate the wavelength if the frequency is 880 Hz.

20-3. You watch a carpenter down the street hammering nails in a roof and note that the sound and sight are synchronized perfectly. Then, when he stops hammering, you hear one more hammer blow.
  (a) Suppose that the time delay between the time you see the last blow and the time you hear it is t. Write an equation for the distance between you and the carpenter.
  (b) Calculate your answer to the above for a time delay of 0.80 second.

20-4. At a musical concert, you sit far back and listen to the same concert broadcast on your local radio.
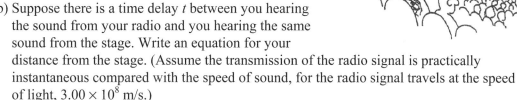
(a) Which do you hear first, the sounds from the radio or from the concert stage?
(b) Suppose there is a time delay t between you hearing the sound from your radio and you hearing the same sound from the stage. Write an equation for your distance from the stage. (Assume the transmission of the radio signal is practically instantaneous compared with the speed of sound, for the radio signal travels at the speed of light, 3.00×10^8 m/s.)
(c) Calculate your distance from the stage for a time delay of 0.50 s.
(d) Pretend that in some way, the radio signal traveled all the way around the world (a distance of 40,000 km) before reaching you. How far back from the stage would you have to sit so that the sound from your radio and the sound from the stage traveling through the air arrive at your ear at the same time?

20-5. Sound travels about 4.5 times faster in seawater than in air.
(a) Express the wavelength of a sound wave in seawater in terms of speed and frequency.
(b) Calculate the wavelength in seawater of a sound wave whose frequency is 1500 Hz.

20-6. Lizzie is listening to music underwater in a swimming pool that is equipped with underwater speakers.
(a) Write an equation for the wavelength of a sound wave in air in terms of speed and frequency f.
(b) Write an equation for the wavelength in water of the same frequency sound wave.
(c) Calculate answers for the above for a 264-Hz sound wave. (Assume that the speed of sound in fresh water is 1490 m/s.)
(d) Would a 264-Hz note have a different pitch underwater than in air? Explain.

20-7. The speed of sound through a railroad track is $v_{\text{in track}}$. Tom, some distance away, taps the track with a hammer. Pat, with her ear on the track, hears the sound a time t later.
(a) Write an equation for the distance that separates Tom and Pat.
(b) Write an equation for the time taken for the sound of the tap to reach Pat through air.
(c) Write an equation for the time it takes between the tap and Pat hearing it through the air.
(d) Calculate answers for the above for a sound travel time through the track of 0.40 s and a speed of sound in the track of 4500 m/s.

20-8. At a baseball game you hear the crack of the bat hitting a ball a time interval t after observing it.
(a) Write an equation showing your distance from home plate.
(b) Calculate your distance from home plate for a time interval of 0.35 s.
(c) If the wind is blowing in your direction, will the time between seeing and hearing the hit change? If so, in what way? Defend your answer.

20-9. You're watching fireworks in the sky. You note a delay between seeing and hearing the explosions.
(a) How is the duration of the delay between seeing and hearing the fireworks affected if you are twice as far away?
(b) How does a three-times-as-great distance affect the duration of the time delay?

20-10. A whale emits a sound directed toward the bottom of the ocean, a distance y below.
  (a) If the whale is stationary in the water, how much time passes before it hears an echo?
  (b) If the whale is moving downward when it emits the sound, will the frequency of the echo be different from the original frequency emitted by the whale? If so, will the echo be of higher or lower frequency?

20-11. • Mala drops a stone into a well of depth h.
  (a) Derive an equation for the time that passes between her dropping the stone and hearing its splash.
  (b) Calculate the time if the depth of the well is 34 m.

20-12. • Allison is traveling on a lake at speed v in her canoe, in the same direction as water waves of speed of $2v$. She counts the waves that pass her as she knifes through the water.
  (a) Suppose she counts n waves in t seconds. Derive an equation for the wavelength of the waves.
  (b) Calculate the wavelength if her speed is 2.6 m/s and she counts four waves in 10 s.
  (c) What is the frequency of the waves as seen by a person on shore?

20-13. Erik visits a canyon. He shouts "hello" and hears his echo soon thereafter.
  (a) How far is Erik from a canyon wall when he hears the echo t seconds after shouting?
  (b) If it takes 5.0 s for him to hear the echo, calculate the distance to the canyon wall.
  (c) If instead Erik is moving toward the wall when he yells, will the echo have the same frequency as the original shout? Defend your answer.

20-14. Sonar equipment uses sound waves to map the ocean floor. Depth is determined by the time it takes for a pulse of sound to reach and then reflect from the ocean floor.
  (a) Write an equation for the ocean depth when a pulse is received in time t and the speed of the ultrasound in water is v.[*]
  (b) Calculate the distance to the ocean floor when a pulse is received 1.7 s after it was emitted. (Assume a 1530-m/s speed of ultrasound in water.)

20-15. Bats, like dolphins, use ultrasonic sound to locate prey.
  (a) If the time between the ultrasound pulse being sent out and the echo rebounding from a moth and returning is t, how far away is the moth? Assume that the bat is hovering.
  (b) Calculate the distance between the bat and the moth if the time between the original pulse and the echo's return is 0.15 seconds.

20-16. A bat emits a sound pulse when distance d away from an obstacle.
  (a) What is the maximum pulse duration that will allow the bat to hear the echo as distinct from the emitted pulse (assuming that the bat hovers in one place)?
  (b) Why are brief chirps used by bats in navigation?

---

[*] The speed of sound in seawater depends on the temperature, pressure, and salinity of the water, all of which can vary between the location of the submarine's initial sound pulse and the sea floor. Here we assume a very simple model of how sonar works.

20-17. Assume that no reflection occurs for objects smaller than a wavelength.
  (a) Write an equation for the minimum width detectable by an underwater sonar device that produces waves of speed v and of frequency f.
  (b) Calculate this minimum width for a sonar frequency of 5000 Hz when sonar waves travel at a speed of 1530 m/s in water.
  (c) Calculate the minimum width for a sonar frequency of 5 MHz.
  (d) To detect smaller objects, should the frequency of the sonar waves be higher or lower? Defend your answer.

20-18. • Sound waves reach your two ears simultaneously when the sound source is directly in front of you. But when a sound source is not in front of you, the direction of the source is perceived by a slight difference between the arrival times of the waves at each ear.

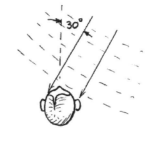

  (a) Write an equation for the difference in arrival times of sound waves from a source that is at an angle of 30° relative to your forward direction if the distance between your ears is d.
  (b) What is this time difference if $d = 0.18$ m? In this time, will a typical sound wave (of, say, 500 Hz) move more or less than one wavelength in air?
  (c) What happens to the duration of time difference when you rotate your head in a direction away from the sound source (effectively increasing the angle)?

20-19. • You and your friend Cheryl stand next to one another, facing the same direction. You both start walking forward at the same time. We are going to define a complete step as you moving your right foot, then your left foot, and then being just ready to move your right foot again:

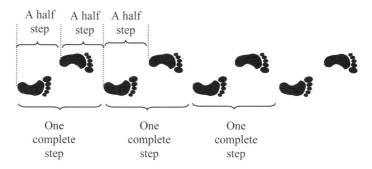

You take 30 complete steps per minute, starting with your right foot. Cheryl takes 28 complete steps per minute, also starting with her right foot.
  (a) How long will you have to wait until your steps again "line up"? (That is, until both of you are stepping with your right foot at the same time, or with your left foot at the same time.)
  (b) How many times per minute are you both "in step"?
  (c) How long will you have to wait for your steps to be exactly "out of step"? (That is, one of you is stepping with the right foot while the other is stepping with the left foot.)
  (d) How many times per minute are you exactly "out of step"?
  (e) How does this example relate the phenomenon of beats?
  (f) If you were to play two sound waves, one at 30 Hz and the other at 28 Hz, how many beats will you hear? Explain.

20-20. A piano tuner hears 2 beats per second when listening to the combined sound from his tuning fork and a note on the piano being tuned.
(a) If the fork has a frequency f, what are the possible frequencies of the piano string?
(b) Calculate the possible piano string frequencies if the tuner hears 2 beats when using a 440 Hz tuning fork.
(c) After a very slight tightening of the string, he hears 3 beats per second. Should the string be tightened further or loosened? Defend your answer.

Show-That Problems for Sound

20-21. The wavelength of a 340-Hz tone in air is 1.0 m. Sound travels about 1530 m/s in seawater.
Show that the wavelength of a 340-Hz tone in seawater is 4.5 m.

20-22. Six seconds after seeing a flash of lightning, the sound is heard.
Show that the observer is located about 2 km from the lightning flash.

20-23. Eight seconds after shouting "Hello" in a canyon, the echo returns to you.
Show that you are approximately 1400 m from the canyon wall.

20-24. The lowest frequency of sound that an average young person can hear is about 20 Hz. Show that the wavelength of this sound wave is about 17 m (when the speed of sound is 340 m/s).

20-25. The highest frequency of sound that an average young person can hear is about 20,000 Hz.
Show that the wavelength of this sound wave is about 1.7 cm.

20-26. After sitting too close to the speakers of too many concerts, the shortest sound waves that Paul can hear are 3.0 cm in wavelength.
Show that the highest frequency of sound he can hear is a little over 11,000 Hz.

20-27. Medical ultrasound of frequency 20 MHz is used to diagnose ailments in the body. The speed of ultrasound in body tissue is 1500 m/s.
Show that if penetration depth is about 400 wavelengths, the ultrasound can penetrate about 3 cm.

20-28. Some auto-focus cameras use ultrasonic pulses to sense range. For a time delay between emission and detection of 0.010 s, show that the object is 1.7 m in front of the camera.

20-29. A violin and a piano simultaneously sound notes with frequencies 434 Hz and 440 Hz. Show that the beat frequency heard is 6 Hz.

20-30. A violinist, tuning her instrument to a piano note of 440 Hz, detects 4 beats per second.
Show that the frequency of the violin can be either 436 Hz or 444 Hz.

20-31. A beat frequency is heard when notes of 240 Hz and 245 Hz are sounded together.
Show the same beat frequency occurs when notes of 200 Hz and 205 Hz are sounded together.

22 Electrostatics

Electrostatics is the study of electric charges at rest. The central rule of electrostatics and electricity in general is as follows: Like charges repel; unlike charges attract. Hence electrons (–) are attracted to protons (+) and are repelled by other electrons. The force of attraction or repulsion between two charges is given by Coulomb's law: $F = k\dfrac{q_1 q_2}{d^2}$, where k is the electrostatic constant $9.0 \times 10^9 \, \dfrac{\text{N·m}^2}{\text{C}^2}$, q_1 and q_2 are quantities of charge measured in coulombs, and d is the distance measured in meters between charges.[*] Note the similarity to Newton's law of gravitation.

Surrounding every charge is an electric field, which can exert a force on other charges. To determine the electric field strength, we measure the force F acting on a small test charge q_0 placed in the field.[*] The strength of the electric field is given by $E = \dfrac{F}{q_0}$. (Note the similarity of the electric field to the gravitational field $g = \dfrac{F}{m}$, discussed in Chapter 9.)

The electric field strength between the plates of a capacitor may be expressed as $E = \dfrac{V}{d}$, where V is the potential difference between the plates and d is the separation distance of the plates. Both the electric force and electrical field are vector quantities. Potential difference, on the other hand, is a scalar quantity.

Sample Problem 1

Two point charges, q_1 and q_2, are separated by a distance d.

(a) Write an equation for the force acting between them.

Solution: From Coulomb's law, $F = k\dfrac{q_1 q_2}{d^2}$.

(b) Calculate the electric force acting between a +2.0 μC charge ($\mu = 10^{-6}$) and a –3.0 μC charge separated by 35 cm.

Answer: Since k has units of $\dfrac{\text{N·m}^2}{\text{C}^2}$, charge should be expressed in coulombs and distance in meters.

$$F = k\dfrac{q_1 q_2}{d^2} = 9.0 \times 10^9 \, \dfrac{\text{N·m}^2}{\text{C}^2} \, \dfrac{(2.0 \times 10^{-6} \, \text{C})(-3.0 \times 10^{-6} \, \text{C})}{(0.35 \, \text{m})^2} = -0.44 \, \text{N}.$$

The negative sign indicates that the force is attractive.

---

[*] Strictly speaking, the equation as written applies to forces between point charges, or between uniformly charged spheres. If the charged objects are small compared with the distance between them, treating the objects as point particles gives an excellent approximation.

[*] By convention, q_0 is a positive point charge small enough in magnitude to not affect the field being measured, and the direction of the electric field is the direction of the force on a positive test charge.

Sample Problem 2

Two identically charged particles are separated by distance r.

(a) At what location between the particles would the force on a positive test charge q_0 be zero?

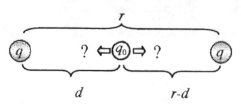

Focus: Find d such that $F_{net} = 0$.

The force on the test charge from the charge on the left is $F = k\dfrac{q\,q_0}{d^2}$ toward the right.

The force on the test charge from the charge on the right is $F = k\dfrac{q\,q_0}{(r-d)^2}$ toward the left. The net force on the test charge will equal zero at a location where it is equally attracted or repelled by the other two charges. That is, $F_{net} = 0$ when

$$k\frac{q\,q_0}{d^2} = k\frac{q\,q_0}{(r-d)^2} \Rightarrow \frac{1}{d^2} = \frac{1}{(r-d)^2}. \text{ Taking the square root of both sides gives}$$

$$\Rightarrow \frac{1}{d} = \frac{1}{r-d} \Rightarrow d = r - d \Rightarrow 2d = r \Rightarrow d = \frac{r}{2}.$$

This answer makes sense. Since the two charges are identical, the force from each on a third charge will be identical only if it lies the same distance away from each charge. So the test charge experiences zero force at distance $r/2$ from either charge—midway between the two.

(b) Does your answer depend on whether the identical charges are both positive or negative?

Answer: No. At the midpoint between the identical charges, the positive test charge would either be equally attracted to each if they were both negatively charged, or equally repelled by each if they were both positively charged. In either case $F_{net} = 0$.

Sample Problem 3

A particle of mass m and charge q experiences a force F when located at a point in an electric field.

(a) Write an equation for the electric field strength at this point.

Answer: $E = \dfrac{F}{q}$.

(b) The charge on an electron is given the symbol e. Write an equation for the force that an electron would experience at this point.

Solution: $F = ?$ From $E = \dfrac{F}{q} \Rightarrow F = qE = eE$.

(c) Write an equation for the acceleration that the electron experiences.

Solution: $a = \dfrac{F}{m} = \dfrac{eE}{m}$.

(d) Calculate the acceleration of an electron in an electric field of 540 N/C. The mass of an electron is 9.11×10^{-31} kg and the charge on an electron is 1.60×10^{-19} C.

Solution: $a = \dfrac{eE}{m} = \dfrac{(1.60 \times 10^{-19}\,\text{C})\left(540\,\frac{\text{N}}{\text{C}}\right)}{9.11 \times 10^{-31}\,\text{kg}} = 9.5 \times 10^{13}\,\dfrac{\text{m}}{\text{s}^2}$.

Sample Problem 4

Two parallel metal plates separated by distance d are given equal and opposite charges. The difference in potential between the plates is V.

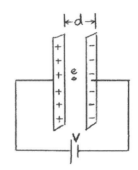

(a) How much work would be required to move an electron from the positively charged plate to the negatively charged plate?

Focus: $W = ?$ The potential difference V has units of volts $= \frac{\text{joules}}{\text{coulomb}}$, which tells us the work (in joules) required to move 1 coulomb of charge between the plates. So from $V = \frac{W}{q} \Rightarrow W = qV = eV$.

Note that the units check. Work is in joules, e is in coulombs, and V has units joules/coulomb, so eV has units joules. Checking units assures that you're either on, or not on, a correct track!

(b) If you were to now release the electron close to the negative plate, how much kinetic energy would it have when it reaches the positive plate?

Focus: KE = ?

Conservation of energy rules here. You did work eV to move the electron from the positive plate to the negative plate. As a result, the electron has electrical potential energy eV. Once the electron is released, that potential energy is transformed to kinetic energy. So the amount of KE that the electron has upon reaching the positive plate is eV.

(c) How much force is exerted on an electron placed between the plates?

Focus: $F = ?$ The force exerted on an electron is due to the electric field at its location between the plates. From $E = \frac{F}{q} \Rightarrow F = qE = eE$, where e is the charge of an electron.

The pair of plates comprises a capacitor, with electric field between the plates $E = \frac{V}{d}$.

So $F = e\left(\frac{V}{d}\right) = \frac{eV}{d}$. Another way to arrive at the same result is to recognize that the work W that was done to move the electron between the plates was eV.

From $W = Fd \Rightarrow eV = Fd \Rightarrow F = \frac{eV}{d}$. This force is the same anywhere between the plates since the electric field is constant between the plates of a parallel-plate capacitor.

Problems for Electrostatics

22-1. One penny is given a charge $-q$ while another penny is given a charge $+2q$. When the pennies are brought together and touch, the charges redistribute and equalize over their surfaces.
(a) What is the final charge on each penny?
(b) Calculate the final charge on each penny if q is 30 μC (30×10^{-6} C).

22-2. Two spherical inflated rubber balloons each have identical amounts of charge spread uniformly on their surfaces. They repel each other with a force F when their centers are distance d apart.
(a) Find the charge on each balloon.
(b) Calculate q for a 2.5-N repelling force when the centers of the balloons are 0.30 m apart.

22-3. In accord with Coulomb's law, a force of attraction or repulsion occurs between pairs of charged particles.
  (a) Calculate the force between a particle with charge $-5.0\ \mu C$ and a particle with charge $+3.0\ \mu C$ separated by a distance of 0.50 m.
  (b) Calculate the force between a particle with charge $+5.0\ \mu C$ and a particle with charge $-3.0\ \mu C$ separated by a distance of 0.50 m.
  (c) Does the magnitude of calculated forces depend on the signs of the charges? Explain.

22-4. Consider two small charged objects, one with charge q and the other of unknown charge. When they are separated by a distance d, each exerts a force F on the other.
  (a) Write an equation for the charge of the second object.
  (b) Calculate the charge of the second object if the attractive force between the two charges is 2.8 N, the charge on the first object is 15 μC, and the distance between them is 1.2 m.

22-5. Consider the pair of protons in the nucleus of a helium atom, each with charge e (same as the electron, but positive) and mass m_p. Assume their center-to-center distance to be d.
  (a) Derive an equation for their initial acceleration if they were free to accelerate away from each other. (We'll see in Chapter 33 that an attractive force, the strong nuclear force, prevents protons in the nucleus from accelerating away from one another.)
  (b) Calculate the hypothetical initial acceleration of each proton in the pair without the presence of the strong nuclear force. The mass of a proton is 1.67×10^{-27} kg, and assume that the protons in a helium nucleus are about 3×10^{-15} m apart.

22-6. A particle of charge q is at a distance d from the center of the charged dome of a Van de Graaff generator. Another particle of charge $9q$ experiences the same magnitude of electrical force at a different distance from the center of the dome.
  (a) Find the distance of the particle of charge $9q$ from the dome's center in terms of distance d.
  (b) Does your answer depend on the sign of the charges? Defend your answer.

22-7. After scuffing your shoes on a hotel rug and gaining a charge $-q$, you see a charming someone with an equal and opposite charge $+q$ across the hall a distance x away.
  (a) Write an equation for the approximate strength of your electrical attraction to the charming person.
  (b) What is the strength of the charming person's attraction to you?
  (c) Calculate the strength of the attraction if you each have identical magnitudes of charge 48 μC while you're 3.4 m apart.

22-8. A hydrogen atom is composed of a proton and an electron with an average separation distance of 5.3×10^{-11} m. The magnitude of charge on each is 1.6×10^{-19} C.
  (a) Calculate the magnitude of force between the proton and the orbiting electron at this average distance.
  (b) The proton pulls the electron into an orbit with the force you calculated above. How much force, in comparison, does the electron exert on the proton?

22-9. Assume the orbiting electron of the previous problem traces a circular orbit. Then the electric force would be a centripetal force.
  (a) Calculate the centripetal acceleration of the electron.
  (b) Calculate its speed.

22-10. A particle of mass m with charge q is held in the horizontal electric field to one side of the charged spherical dome of a Van de Graaff generator. When released, the particle's horizontal component of acceleration is equal to 8 times the acceleration due to gravity.
(a) Find the electric field at this location.
(b) What would be the particle's initial horizontal component of acceleration if it were instead released at a distance twice as far horizontally from the center of the charged sphere?

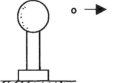

22-11. A particle of charge q is held near the charged dome of a Van de Graaff generator. When the particle is released, it undergoes acceleration a away from the generator. Ignore gravity.
(a) At what distance A from the center of the dome would the particle's acceleration be one-quarter as much as its initial acceleration a? (Answer in terms of the original distance from the generator's center.)
(b) At what distance B from the center of the dome would the acceleration of the particle be one-tenth a?
(c) Is the particle's acceleration when passing point A greater, the same as, or less than its acceleration when passing point B?
(d) Is the particle's *speed* when passing point A greater, the same as, or less than its speed when passing point B?
(e) Why are your answers to (c) and (d) different?

22-12. Two identical particles with charge q are separated by a distance d and repel each other with a force F.
(a) By how much does the force change if *each* charge is doubled, while the distance between them remains unchanged?
(b) By how much does the force change if *each* charge is doubled, while the distance between them also doubles?
(c) By how much does the force change if the charge of only *one* particle doubles, and the distance between them doubles?

22-13. An electric field occupies the space around every electric charge.
(a) Derive an equation for the strength of the electric field a distance d from an isolated charge q. (*Hint*: First write an expression for the force exerted on a test charge q_0 by the proton.)
(b) Find the electric field 0.010 m from an isolated proton. ($q_{proton} = e = 1.60 \times 10^{-19}$ C.)
(c) Find the electric field 0.010 m from an isolated helium nucleus (an alpha particle, with twice the charge of a single proton).
(d) What is the electric field 0.010 m from an isolated neutral helium atom?

22-14. A charge q_0 experiences a downward electrical force F at Earth's surface.
(a) What is the charge of Earth? Consider Earth to be a uniformly charged sphere with radius R_E.
(b) A 1.0-μC charge feels a downward electrical force of 1.0×10^{-4} N at Earth's surface. $R_E = 6.37 \times 10^6$ m. Find the charge of Earth.
(c) How many excess electrons does this represent?
(d) Find the strength of the electric field at Earth's surface.

© Paul G. Hewitt and Phillip R. Wolf

22-15. The electric field at distance d from a point charge q is 9.80 N/C. The field strength decreases with distance via the inverse-square law.
  (a) At what distance from the point charge is the electric field 0.098 N/C? (Express your answer as a multiple of the initial distance.)
  (b) At what distance from the point charge is the field 980 N/C?

22-16. A certain parallel-plate capacitor has a plate separation d and a potential difference between the plates of V. The electric field between the plates is uniform.
  (a) Write an equation for the strength of the electric field E.
  (b) Calculate the strength of the electric field if the distance apart is 0.050 m and a voltmeter shows a 100-V potential difference between the plates. (Note that the unit V/m is equivalent to N/C.)
  (c) If the acceleration of a charged particle between the plates is a when it is halfway between the plates, why will the particle experience the same acceleration if it is only one-quarter the distance between the plates? Defend your answer.

22-17. An electron is placed in the region between the parallel plates of a capacitor.
  (a) What is the magnitude of the force on the electron if the uniform field is 1000 N/C?
  (b) Which will be constant for the electron in this field, acceleration or speed? Defend your answer.

22-18. A simple electron gun consists of two oppositely charged plates, the negative one called the cathode and the positive one called the anode. Electrons are "boiled" off the cathode and accelerate toward the anode. A hole in the anode allows an electron beam to emerge from the device.
  (a) Write an expression for the KE of the electron when it reaches the positive anode.
  (b) If the potential difference between anode and cathode is 100 V, calculate the kinetic energy of an electron emerging from the gun.
  (c) Calculate the velocity of the electrons in the beam that emerges through the hole.

22-19. Ink-jet printers spray charged drops of ink onto paper. An electric field in the printer head produces a force on the ink drops.
  (a) Find the electric field strength that produces a force of 2.8×10^{-4} N on a drop having a charge of 1.6×10^{-10} C.
  (b) If the electric field were increased in strength by 10%, by what factor would the force on the charged drops increase?

22-20. Particles with charges q and $4q$ have the same mass and are moving in the same uniform electric field E.
  (a) How do the forces exerted by the field on these two particles compare?
  (b) How do the accelerations of these particles compare?
  (c) How would the accelerations compare if the particle with charge $4q$ had 4 times the mass of the other particle?

22-21. The potential difference between a particular storm cloud and the ground is 100 million volts.
  (a) If a charge of 2.5 C flashes in a bolt from the cloud to Earth, how big is the change of potential energy of the charge?
  (b) How big is the change in potential energy for a 5-C flash between the same cloud and ground?

22-22. In 1909 Robert Millikan was the first to find the charge of an electron in his now-famous oil drop experiment. He levitated charged drops of oil in an electric field so that an upward electrical force balanced the gravitational force on a drop.

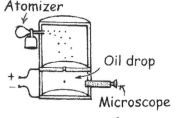

(a) Draw a force diagram for the floating oil drop.
(b) Given electric field strength is E and oil drop mass m, derive an equation for the charge on the oil drop.
(c) If a drop of mass 1.1×10^{-14} kg remains stationary in an electric field of 1.68×10^5 N/C, calculate the charge on the drop.
(d) How many extra electrons are on this drop?

22-23. An electric field does work to move charges placed in the field.
(a) How much work does an electric field E do when a charge q is moved a distance x in the direction of the field?
(b) Calculate the work done by a 7500 N/C electric field when a 55 μC charge moves 42 cm in the direction of the field.

22-24. When a small sphere of mass m and charge q_1 is suspended from a vertical spring, the spring stretches somewhat due to the gravitational force acting on m. (Recall Hooke's law: $F = kx$, where the stretch of a spring is proportional to the force on it.) The spring stretches an additional distance Δx when another sphere of charge q_2 is brought beneath it. The final distance between the centers of the spheres is d.
(a) With how much force are the spheres attracted? (*Hint*: Your answer can be given without using any property of the spring.)
(b) What is the spring constant k of the spring? (Don't confuse the spring constant k with the electrostatic constant k in Coulomb's law.)

22-25. • Suppose that on <u>Sample Problem 2,</u> the charge on the left was $+2q$ while the charge on the right was $-q$. A positive test charge q_0 is placed somewhere on the line passing through both charges.

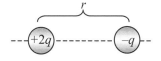

(a) Would the point where the net force on the test charge is zero be to the left of $+2q$, in between $+2q$ and $-q$, or to the right of $-q$? Defend your answer.
(b) Determine the distance from the $-q$ charge to the location where the net force on q_0 is zero.

22-26. • Three identically charged particles are placed on the corners of an equilateral triangle.
(a) Draw the forces acting on each particle due to the other two, indicating the directions and relative magnitudes.
(b) Determine the relative magnitude and angle of the resultant force on any particle.

Show-That Problems for Electrostatics

22-27. The charge on an electron is 1.60×10^{-19} C.
Show that 6.25×10^{18} electrons have a collective charge of 1 C.

22-28. In walking across a carpet, you might build up a net negative charge of 50 μC.
Show that you have acquired about 310 million excess electrons on your body.

22-29. The distance between two charged particles is doubled.
Show that the electrical force between them is reduced to one-fourth of its previous value.

22-30. The distance between two charged particles is halved.
Show that the electrical force between them is 4 times its previous value.

22-31. The distance between two charged particles is doubled, and the charge on one of the particles is halved.
Show that the electrical force between them is one-eighth of its previous value.

22-32. A test charge at a distance d from a charge q experiences a force F.
Show that the electric field in the location of the test charge is $E = kq/d^2$.

22-33. A particle of mass m and charge $2q$ is held in a uniform electric field E.
Show that the work needed to move the charge a distance x against the electric field is $2qEx$.

22-34. An isolated proton is acted on by an electrical force of 4.8×10^{-14} N.
Show that the magnitude of the electric field at the proton's location is 3.0×10^5 N/C.

22-35. A charge of mass m and charge q accelerates from rest from one plate of a parallel plate capacitor. The distance between the plates is d and the potential difference between them is V.
Show that the acceleration of the charge is given by $a = \dfrac{qV}{md}$.

22-36. A proton has weight and is pulled downward by gravity. An electric field could push it upward. ($m_p = 1.67 \times 10^{-27}$ kg.)
Show that the magnitude of a vertical electric field that would just support the weight of a proton near Earth's surface is 1.0×10^{-7} N/C upward.

22-37. Millikan discovered that a tiny charged drop would remain suspended in an electric field that counteracted the gravitational field.
Show that for a drop of mass m carrying charge q held motionless in a vertical electric field E, the following equation holds true: $qE/mg = 1$.

22-38. As the charge on a drop of given mass m gets smaller, the electric field required to hold the drop motionless gets larger.
Show that the maximum needed field is $E = mg/e$ when the drop carries a single quantum unit of charge, e.

22-39. The electron gun in the picture tube of yesterday's television sets accelerates electrons across a potential difference of 2000 V.
Show that the final speed of such an accelerated electron is about 2.7×10^7 m/s.

22-40. • A square with sides L has identical charges q on three of its corners.
Show that the magnitude of the electric field at the center of the square is $2kq/L^2$.

22-41. • A square with sides L has identical charges on three of its corners.
Show that the electric field at the remaining corner has a magnitude approximately equal to $1.9\, kq/L^2$.

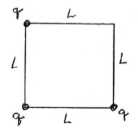

23 Electric Current

In this chapter we consider electric circuits—the relationships between voltage, current, resistance, energy, power, and the role that a battery plays in a circuit.

A *battery* in a circuit serves as a pump that provides the "electrical pressure" needed to push charges through the wires and electrical devices in the circuit. A battery does this by increasing the electric potential energy of charges and setting up an electric field in the wires. *Charges flowing through wires deliver their energy to a load (bulb, motor, resistor, etc.), and then return to the battery to repeat the cycle again.* Charge is measured in *coulombs* (equivalent to the amount of charge on 6.25×10^{18} electrons).

Energetic charges passing through the bulb filament deliver their stored PE to the filament, causing it to glow white-hot.

The amount of energy delivered to the load each second (that is, the *power*) equals the product of the *current* (amount of charge per second) passing through the load and the *voltage* (energy per charge) across that load.

$$\text{Power}\left(\frac{\text{Joules}}{\text{second}}\right) = \text{Current}\left(\frac{\text{Coulombs}}{\text{second}}\right) \times \text{Voltage}\left(\frac{\text{Joules}}{\text{Coulomb}}\right), \text{ or}$$

$$\frac{\text{Energy delivered to load}}{\text{second}} = \frac{\text{coulombs passing through load}}{\text{second}} \times \frac{\text{potential energy}}{\text{coulomb}}$$

For example, if 3 coulombs of charge pass through a bulb each second (current = 3 amps) and each coulomb delivers 2 joules of PE (voltage = 2 volts), the power delivered to the load is

$$3 \text{ Amps} \times 2 \text{ Volts} = 3\frac{\text{Coulombs}}{\text{second}} \times 2\frac{\text{Joules}}{\text{Coulomb}} = 6\frac{\text{J}}{\text{s}} = 6 \text{ watts}.$$

A larger voltage across the load causes a proportionally larger current in the load. This is

$$\textit{Ohm's Law}: \text{Current}\left(\frac{\text{Coulombs}}{\text{second}}\right) = \frac{\text{Voltage}\left(\frac{\text{Joules}}{\text{Coulomb}}\right)}{\text{Resistance(ohms)}} \text{ or } I = \frac{V}{R}.$$

If we double the voltage across a given resistance, twice as much current is produced.

Consider a very simple circuit that consists of a 6-volt battery connected to a single bulb. Suppose that we measure a current of 2 amps (= 2 coulombs per second) through the wire. We can say,

© Paul G. Hewitt and Phillip R. Wolf

- The 6-volt battery provides 6 joules of potential energy to each coulomb of charge that travels through the circuit.
- Each coulomb of charge flowing through the bulb filament is "delivering" 6 joules of energy to the filament.
- Since there are 2 coulombs of charge passing through the bulb each second, each delivering 6 joules of energy, there is a total of 12 joules per second (12 watts) being delivered to (and being radiated by) the bulb each second. That is: $P = IV = 2\text{ A} \times 6\text{ V} = 2\frac{C}{s} \times 6\frac{J}{C} = 12\frac{J}{s} = 12\text{ W}$.
- Since 2 amps (that is, 2 coulombs per second) are established in the bulb as a result of the 6-volt electric pressure difference across the bulb, the resistance of the bulb filament is
$$R = \frac{V}{I} = \frac{6 \text{ volts of electric pressure}}{2 \text{ amp current}} = 3 \text{ ohms, or } 3\text{ }\Omega.$$

Current is a flow of charge (I), pressured into motion by voltage (V), and hampered by resistance (R).

Here is a summary of all of the various quantities and units featured in this chapter:

Quantity	Symbol	Measured in ...	Unit symbol	Equivalent to ...
Charge	q	Coulombs	C	Charge on 6.25×10^{18} electrons
Voltage	V	Volts	V	$\frac{\text{Joules of energy}}{\text{Coulomb of charge}} = \frac{J}{C}$
Current	I	Amperes (or amps)	A	$\frac{\text{Coulombs of charge}}{\text{time in seconds}} = \frac{C}{s}$
Power	P	Watts	W	$\frac{\text{Coulombs of charge}}{\text{second}} \times \frac{\text{Joules of energy}}{\text{Coulomb of charge}} = I \times V = \frac{J}{s}$
Resistance	R	Ohms	Ω	$\frac{\text{electric pressure difference}}{\text{current}} = \frac{\text{Voltage}}{\text{current}} = \frac{V}{I}$

Sample Problem 1

An automobile headlight is lit by a current I.

(a) Write an equation showing how many coulombs of charge flow in its filament in time t.

Focus: $q = ?$

From the definition of current, $I = \frac{\text{charge}}{\text{time}} = \frac{q}{t} \Rightarrow q = It$. In SI units q is in coulombs, I is in amperes, and t is in seconds.

(b) Calculate the amount of charge in coulombs that flows through the lamp filament in 30 minutes when the current is 4.0 A.

Solution: $q = It$. Recognizing that 1 A = 1 C/s, and that 30 minutes is $30 \text{ min} \times \frac{60\text{ s}}{1\text{ min}} = 1800\text{ s}$

$\Rightarrow q = \left(4.0\frac{C}{s}\right)(1800\text{ s}) = \mathbf{7200\text{ C}}$.

(c) Current has direction in a circuit, conventionally from the positive to the negative terminal of the connected battery. Can we therefore say that current is a vector quantity?

Answer: **No.** The current in a wire is always along the length of the wire, whether the wire is straight or curved. No single vector could describe motion along a curved path. We arbitrarily call the positive direction the direction in which positive charge would flow (even though the actual flow is negative electrons moving in the opposite direction).

Sample Problem 2

A 12-volt automobile battery provides electrical pressure in an electric circuit.
(a) Find the amount of energy per coulomb of charge flowing in the circuit.

Focus: $\dfrac{\text{Energy}}{\text{coulomb}} = ?$

Voltage is defined as energy per charge. So by definition, 12 volts $= \dfrac{12 \text{ joules of energy}}{\text{coulomb of charge}}$.

(b) Write an equation showing the amount of energy provided to the circuit when q coulombs of charge pass through the battery.

Solution: From Voltage $= \dfrac{\text{Energy}}{\text{charge}} \Rightarrow$ Energy $=$ charge $\times$ Voltage $= qV$.

(c) Calculate the energy provided to the circuit when 8 coulombs of charge pass through a 12-V circuit.

Solution: Energy $= qV = $ (8 C)(12 J/C) $=$ **96 J**.

Sample Problem 3

In the simple circuit shown to the right we see an ideal battery (one that provides a constant voltage and has no internal resistance), a switch, and a single lamp. The battery voltage is V and the resistance of the lamp is R.
(a) Write an equation for the current in the lamp when the switch is closed.

Focus: $I = ?$ The current in the lamp depends on the potential difference V across the lamp, and the resistance R opposing the flow of charge in the circuit.*

The solution is the equation of Ohm's law, $I = \dfrac{V}{R}$.

(b) Calculate the current in the lamp if it has a resistance of 1 Ω and the battery voltage is 6 V.

Solution: $I = \dfrac{V}{R} = \dfrac{6 \text{ V}}{1 \text{ Ω}} =$ **6 A**.

(c) How much power is dissipated by the lamp?

Focus: $P = ?$ We can calculate power using $P = IV$, or combine this with Ohm's Law ($V = IR$) to come up with two other expressions for power:

$$P = IV = I(IR) = I^2R \quad \text{or} \quad P = IV = \left(\dfrac{V}{R}\right)V = \dfrac{V^2}{R}.$$

Answer: $P = IV = 6 \text{ A} \times 6 \text{ V} =$ **36 W** or $P = I^2R = (6 \text{ A})^2(1 \text{ Ω}) =$ **36 W** or $P = \dfrac{V^2}{R} = \dfrac{(6 \text{ V})^2}{1 \text{ Ω}} =$ **36 W**.

Of course, all three expressions for power give the same answer!

---

* Here the terms *potential difference* and *electric pressure difference* and *voltage* all mean the same thing.

© Paul G. Hewitt and Phillip R. Wolf

(d) Why does the filament of a lamp glow, but not the connecting wires?

Answer: Energy is dissipated in locations of circuit *resistance*. In a single-lamp circuit almost all of the resistance is in the filament of the lamp, where energy is converted to heat and light. Resistance generally increases with increases in temperature, so resistance is even greater once the lamp filament heats up. The resistance offered by the connecting wires and the battery interior are usually negligible compared with the lamp's resistance. That's why, for simplicity, we ignore all resistance in the circuit except that of the lamp.

Sample Problem 4

This is the same circuit as in Sample Problem 3, but with an additional lamp of resistance 2*R* added in series.

(a) Write an equation showing how much current is in the circuit when the switch is closed.

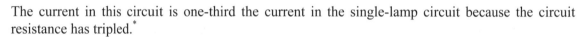

Focus: $I = ?$ These lamps are connected in series.
Resistances add in a series circuit, so the total *equivalent resistance* in the circuit is $R + 2R = 3R$.

The current is given by Ohm's law, $I = \dfrac{V}{R_{eq}} = \dfrac{V}{3R}$.

The current in this circuit is one-third the current in the single-lamp circuit because the circuit resistance has tripled.[*]

(b) Calculate the current in the lamps if *R* is 1 Ω and the battery voltage is 6 volts.

Solution: $I = \dfrac{V}{3R} = \dfrac{6V}{3(1\,\Omega)} = 2\,A$. That's 2 amps through each lamp and through the battery.

(c) Can we say that since the current in each lamp is 2 A that the total current in the circuit is 4 A?

Answer: **No!** For every electron that enters the lamp filament, another one leaves. Electrons don't "pile up" anywhere in the circuit. So if 2 coulombs of charge per second enter and leave the first lamp, 2 coulombs per second enter and leave the second lamp as well. The current through any part of this series circuit is 2 A.

(d) Through which of the two lamps does current first flow?

Answer: Although it may be useful to think of current occurring in one lamp at a time, current in fact is established in both lamps simultaneously. It doesn't matter which lamp is closer to which terminal of the battery. When the switch is closed, an electric field is established in all parts of the circuit at practically the same time. All free electrons throughout the circuit, including the battery interior, move in unison similar to the command "forward march" that makes each member of a marching band begin stepping at the same time.

---

[*] In this example, and in all others that we will consider in this chapter, we make the assumption that the resistance is independent of temperature effects.

(e) If we placed a voltmeter across each of the lamps, what voltage would we measure?

Focus: $V = ?$ Voltage measures the *difference* in PE per coulomb between two different points in the circuit. Each coulomb of charge leaving the battery possesses 6 J of electrical potential energy, which it must spend before returning to the battery. When there is only one lamp in the circuit, all 6 J/C is spent in the single lamp—the voltage across the lamp is 6 V. With two lamps in the series circuit, some of each coulomb's PE is spent in passing through one lamp while the rest is spent passing through the other lamp.

For the lamp with resistance R, $V_R = \text{Current} \times \text{Resistance} = IR = (2\ \text{A})(1\ \Omega) = \mathbf{2\ V}$. For the lamp with resistance $2R$, $V_{2R} = I \times (2R) = (2\ \text{A})(2\ \Omega) = \mathbf{4\ V}$. Note that the voltages add up to 6 V—all of the PE provided to the charges by the battery is simultaneously spent in the various resistance elements of the circuit.

(f) Which of the two lit lamps would be brighter?

Focus: Brightness has to do with *Power*—how much energy is being transformed into heat and light within each lamp each second. The power in the circuit comes from the energy provided by the battery and supplied to the charges passing through it.

$$\text{Power}\left(\text{in watts}, \frac{J}{s}\right) = \text{Current}\left(\text{in amps}, \frac{C}{s}\right) \times \text{Voltage}\left(\text{in volts}, \frac{J}{C}\right).$$

For the 1-Ω lamp $P = IV = 2\text{A} \times 2\text{V} = \left(2\frac{C}{s}\right)\left(2\frac{J}{C}\right) = 4\frac{J}{s} = \mathbf{4\ W}.$

For the 2-Ω lamp $P = IV = 2\text{A} \times 4\text{V} = \left(2\frac{C}{s}\right)\left(4\frac{J}{C}\right) = 8\frac{J}{s} = \mathbf{8\ W}.$ **(2-Ω lamp is brighter.)**

For comparison, the power from the battery is $P = IV = 2\text{A} \times 6\text{V} = \left(2\frac{C}{s}\right)\left(6\frac{J}{C}\right) = 12\frac{J}{s} = \mathbf{12\ W}.$

Note that conservation of energy nicely applies—the amount of energy transformed to heat and light in the lamps every second is exactly balanced by the amount of energy supplied by the battery each second.

Sample Problem 5

Here we have the same two lamps of the previous problem, but connected in *parallel* (not in series as in the previous problem).

(a) Calculate the current in each lamp when the switch is closed.

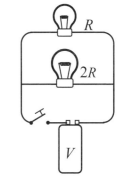

Focus: $I = ?$

In this arrangement voltage does not divide between the lamps as in a series circuit. Careful inspection will show that each lamp is connected across the terminals of the 6-V battery, so each lamp is "energized" with a full 6 volts. Thus the current through each lamp is

$$I_R = \frac{V}{R} = \frac{6\text{V}}{1\Omega} = \mathbf{6\ A}. \quad I_{2R} = \frac{V}{2R} = \frac{6\text{V}}{2\Omega} = \mathbf{3\ A}.$$

(b) What is the current in the battery?

Answer: Since the current in the 1-Ω lamp is 6 A (6 coulombs per second) and the current in the 2-Ω lamp is 3 A (3 coulombs per second), this means that current in the battery must be the sum of these, **9 amps** (9 coulombs per second).

(c) What single resistor connected to the battery would allow the same current in the circuit as the parallel connection of the 1-Ω resistor and 2-Ω resistor?

Focus: $R_{equivalent} = ?$

We want to find the resistance that will give a 9-A current in a 6-V circuit.

From $I = \dfrac{V}{R} \Rightarrow R = \dfrac{V}{I} = \dfrac{6\text{ V}}{9\text{ A}} = \dfrac{2}{3}\Omega$. So we can say that, resistance-wise, a 1-Ω resistor and 2-Ω resistor in parallel are equivalent to a single $\dfrac{2}{3}$-Ω resistor. To find the *equivalent resistance* of a pair of resistors in parallel, we use the "product-over-sum" rule—the product of the two resistors divided by their sum. That is, $R_{equivalent, parallel} = \dfrac{R_1 R_2}{R_1 + R_2}$.*

(d) Which circuit consumes more power, the series circuit or the parallel circuit?

Focus: $P_{parallel\ circuit} = ?\ P_{series\ circuit} = ?$

In each circuit, the total power is the sum of the power dissipated by each individual resistor. For the parallel circuit we have

$$P_{parallel} = P_{1\Omega} + P_{2\Omega} = (I^2 R)_{1\Omega} + (I^2 R)_{2\Omega} = (6\text{ A})^2 (1\ \Omega) + (3\text{ A})^2 (2\ \Omega) = 54\text{ W}.$$

Or, we could calculate $P_{parallel} = \dfrac{V^2}{R_{eq}} = \dfrac{(6\text{ V})^2}{\tfrac{2}{3}\ \Omega} = 54\text{ W}$. For the series circuit we have

$$P_{series} = \dfrac{V^2}{R_{eq}} = \dfrac{(6\text{ V})^2}{(1\ \Omega + 2\ \Omega)} = 12\text{ W}.$$

Clearly, 54 W for the parallel circuit is greater than the 12 W for the series circuit, so, **the parallel circuit consumes more power**. This makes sense. The equivalent resistance of the parallel circuit is less than the equivalent resistance for the series circuit. For the same battery voltage, there are more charges flowing through the parallel circuit than through the series circuit so the total energy delivered to the resistors by this greater current is greater.

The relationship between voltage and current for two-resistor circuits is summarized below:

	Voltage	Current
Series Circuit	The voltages across each individual resistor sum to the battery voltage.	The current through each resistor is the same and is equal to the current through the battery.
Parallel Circuit	The voltage across each resistor is the same and is equal to the battery voltage.	The currents through the individual resistors sum to the current from the battery.

---

* More generally, the equivalent resistance for *n* resistors in parallel is given by $\dfrac{1}{R_{eq}} = \dfrac{1}{R_1} + \dfrac{1}{R_2} + \ldots + \dfrac{1}{R_n}$.

Sample Problem 6

Three 2-Ω resistors and one 3-Ω resistor are connected in the circuit shown.
(a) What is the equivalent resistance of the circuit?

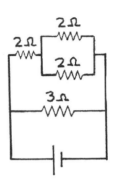

Focus: $R_{eq} = ?$

We can simplify the circuit in steps as shown below.
Note that there is a pair of 2-Ω resistors in parallel. This pair is in series with another 2-Ω resistor, and this set of resistors is in parallel with the 3-Ω resistor.

In Circuit (a) note the pair of 2-Ω resistors in parallel have an equivalent resistance of 1Ω (using the "product over sum" rule for a <u>pair</u> of resistors in parallel: As mentioned earlier, $R_{equivalent, parallel} = \dfrac{R_1 R_2}{R_1 + R_2}$).* Substituting a 1-Ω resistor for the pair of 2-Ω resistors gets us Circuit (b). In Circuit (b) note that the 2-Ω and equivalent 1-Ω resistors are in series and equivalent to a single 3-Ω resistor, which gives us Circuit (c). So the circuit becomes a pair of 3-Ω resistors in parallel. In Step (d) we see the equivalent resistance of a pair of 3-Ω resistors in parallel is 1.5 Ω. So the equivalent resistance of the circuit is **1.5 Ω**.

(b) If the battery is 12 V, calculate the current in the battery.

Answer: The current in the battery is the current in the equivalent resistance of the circuit,

which by Ohm's law is $\dfrac{V}{R_{eq}} = \dfrac{12 \text{ V}}{1.5 \text{ Ω}} = \textbf{8 A}$.

> Current in the line connected to the battery is the same as the current in the battery!

(c) If the battery is 12 V, what is the voltage across each resistor?

Working backwards from the diagrams, there will be 12 volts across each branch of the circuit in diagram (c), so the current in each branch is $I = \dfrac{V}{R} = \dfrac{12 \text{ V}}{3 \text{ Ω}} = 4 \text{ A}$.

This current in the 2-Ω resistor in diagram (b) gives a voltage of $V = IR = (4 \text{ A})(2 \text{ Ω}) = 8 \text{ V}$. This leaves 4V across the 1-Ω resistor in diagram (b), which means that there is 4 V across each of the 2-Ω resistors in parallel in diagram (a). So $V_{3Ω} = \textbf{12 V}$; $V_{single\ 2Ω} = \textbf{8 V}$; and $V_{parallel\ 2Ω} = \textbf{4V}$. You should confirm that a coulomb of charge following any path from the battery through the circuit would give up 12 J/C.

---

* For more than two resistors in parallel, the equation shown in the footnote of the previous page must be employed—something we'll skip here in the interests of avoiding "Information Overload"!

© Paul G. Hewitt and Phillip R. Wolf

Problems for Electric Current

23-1. A wire carries a current I. An amount of charge q moves past a point in the wire during some time interval.
  (a) Write an equation that shows the time it takes for q to move past a point in the wire.
  (b) Calculate the time it takes for 25 C to move past a point in a wire carrying 12.0 A.

23-2. An automobile engine's starter motor draws a current I.
  (a) Write an equation for the amount of charge q that flows through the motor operating for time t.
  (b) Calculate the quantity of charge that flows in a starter motor when it draws 45.0 A for 1.2 s.
  (c) How many electrons does this represent?

23-3. A silver wire transfers a charge q in time t.
  (a) Write an equation for the current in the wire.
  (b) Calculate the current in the wire when 420 C is transferred in 59 minutes.

23-4. When a clothes iron is connected to a voltage source, there is a current through the iron's heating element.
  (a) Write an equation for the current if the source voltage is V and the resistance of the heating element is R.
  (b) Calculate the amount of current the clothes iron draws when connected to 120 V if the resistance of the heating element is 18 Ω.
  (c) If the voltage were doubled, how would that affect the amount of current?

23-5. Some materials are better electrical conductors than others. The *resistivity* of a material tells us how well or how poorly a material conducts electricity. A material with a higher resistivity is a poorer conductor. The resistance of a wire is directly proportional to both the resistivity of the wire material and the length of the wire. Increasing either of these increases the wire's resistance. That is, $R \sim resistivity \times length$. (The resistance is also inversely proportional to the cross-sectional area of the wire, but we won't use that dependence in this problem.)
  (a) Consider two lengths of wire of the same thickness and the same resistance, one made of aluminum and the other made of copper. Aluminum has 1.6 times greater resistivity than copper for the same temperature. How much longer is one wire compared with the other?
  (b) If 20.0 m of aluminum wire has the same resistance and same thickness as the copper wire, how long is the copper wire?

23-6. Two lengths of wire have the same thickness. One is tungsten and the other is silver. Tungsten has 3.8 times greater resistivity than silver at the same temperature. (See information at the beginning of Problem 23-5.)
  (a) If the silver wire and the tungsten wire have the same thickness and the same resistance, how much longer is one compared with the other?
  (b) If 20.0 m of tungsten wire has the same resistance and the same thickness as the silver wire, how long is the silver wire?

23-7. Two lengths of wire have the same thickness. One is nichrome and the other is silver. Nichrome has 68 times greater resistivity than silver at the same temperature. (See information at the beginning of Problem 23-5.)
   (a) If the nichrome wire and the silver wire have the same thickness and the same resistance, how much longer is one compared with the other?
   (b) If 20.0 m of nichrome wire has the same resistance and the same thickness as the silver wire, how long is the silver wire?

23-8. Two resistors, R_1 and R_2, are connected in series with a battery of voltage V_0.
   (a) Write an equation for the current through each of the two resistors.
   (b) Write equations for the voltage across each of the two resistors.
   (c) Calculate the currents and voltages for a 20-Ω resistor and 60-Ω resistor connected in series to a 6.0-V battery.
   (d) What single resistor would give the same total battery current that you get from these two resistors in series?

23-9. Two resistors, R_1 and R_2, are connected in parallel with a battery of voltage V_0.
   (a) Write an equation for the current through each of the two resistors.
   (b) What is the voltage across each of the two resistors?
   (c) Calculate the currents and voltages for a 20-Ω resistor and 60-Ω resistor connected in parallel to a 6.0-V battery.
   (d) What single resistor would give the same total battery current that you get from these two resistors in parallel?

23-10. Four identical resistors, each of resistance R, are combined.
   (a) What is their equivalent resistance when all are connected in series?
   (b) What is their equivalent resistance when all are connected in parallel?
   (c) What is the equivalent resistance of two parallel branches, each with two resistors in series?

23-11. A loudspeaker with resistance R and another loudspeaker with resistance $2R$ are connected in series across the terminals of an amplifier.
   (a) What is the equivalent resistance of the two speakers?
   (b) What is their equivalent resistance if they are connected, more properly, in parallel?
   (c) Calculate the equivalent resistance in each case if the speakers have resistance of 8 Ω and 16 Ω.

23-12. Power is required to operate electric motors.
   (a) Write an equation for the power input to a motor that draws a current I when connected to a source voltage V.
   (b) Calculate the power input to a motor connected to a 120-V circuit when it draws 8.0 A.
   (c) Calculate the energy used when the motor operates for 8.0 hours.

23-13. The current drawn by a light bulb depends on its power and voltage rating.
   (a) Write an equation for the amount of current drawn by a light bulb rated at power P when connected to a voltage V.
   (b) Write an equation for the electrical resistance of this filament.
   (c) How much energy is "used" to light this bulb for a time t?
   (d) Calculate the above for a 120-V bulb rated for 60 W when it is left on for 8.0 hours.

23-14. A hair dryer requires power for operation.
- (a) Write an equation for the current drawn by a hair dryer of power P plugged into a circuit of voltage V.
- (b) Calculate the current drawn by a 1500-W hair dryer connected to a 115-V circuit.
- (c) What current will it draw if it is incorrectly connected to a 230-V circuit, and why might it burn out?

23-15. A hand dryer in a school restroom has a label indicating its power P and current I.
- (a) What circuit voltage is this dryer designed to be plugged into?
- (b) Determine the required voltage for a 2200-W dryer designed to draw 10.0 A.

23-16. Six identical lamps are connected in series across a line voltage V.
- (a) What is the voltage across each one?
- (b) If the current is I, calculate the resistance of each lamp.
- (c) What is the power dissipated by each lamp?
- (d) Calculate the voltage, resistance, and power for each lamp if the current is 0.50 A when the six lamps are connected in series across a 120-V line.

23-17. Six identical lamps are connected in parallel to a voltage V.
- (a) What is the voltage across each lamp?
- (b) If the total current in the circuit is I, calculate the resistance of each lamp.
- (c) What is the power dissipated by each lamp?
- (d) Calculate the voltage, resistance, and power for each lamp if the total current in the circuit is 5.0 A when the six lamps are connected in parallel across a 120 V line.

23-18. The circuit shown has three identical lamps, each of resistance R, in parallel across a battery of voltage V.
- (a) What is the voltage across each of the three branches?
- (b) What is the current through each of the branches?
- (c) If the battery voltage is 6 V and each lamp has a resistance of 2 Ω, calculate the current in each branch.
- (d) Find the equivalent resistance of the circuit.
- (e) Calculate the current produced by and in the battery.

23-19. Circuit A has battery voltage V_0 and three identical resistors R. (The resistors have been labeled R_1, R_2, and R_3 in the diagram so that you can keep track of which is which in your solutions.)
- (a) What is the voltage across each resistor?
- (b) What is the current in each resistor?
- (c) What power does the battery deliver to the circuit?
- (d) Calculate results for (a), (b), and (c) for a 12-V battery connected to three identical 2.0-Ω resistors.

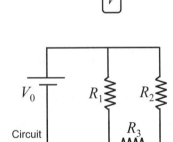
Circuit A

23-20. Circuit B has battery voltage V_0 and three identical resistors R.
(The resistors have been labeled R_1, R_2, and R_3 in the diagram
so that you can keep track of which is which in your solutions.)
Initially the switch is closed so that current runs through R_3.
(a) What is the current in each resistor?
(b) What is the voltage across each resistor?
(c) What power does the battery deliver to the circuit?
(d) What is the voltage across each resistor when the switch is open?
(e) What is the current in each resistor when the switch is open?
(f) Calculate results for all of the above for a 12-V battery connected to 3 identical 2.0-Ω resistors.

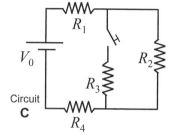

Circuit B

23-21. Circuit C has battery voltage V_0 and four identical resistors R.
(The resistors have been labeled R_1, R_2, R_3, and R_4 in the diagram
so that you can keep track of which is which in your solutions.)
Initially the switch is closed.
(a) What is the current in each resistor?
(b) What is the voltage across each resistor?
(c) What power does the battery deliver to the circuit?
(d) What is the current in each resistor when the switch is open?
(e) What is the voltage across each resistor when the switch is open?
(f) Calculate results for all of the above for a 12-V battery connected to four identical 2-Ω resistors.

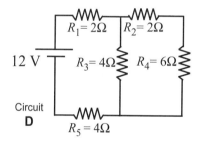

Circuit C

23-22. • Circuit D is powered with a 12-V battery, but has resistances of
different values as indicated.
(a) What is the voltage across each resistor?
(b) What is the current in each resistor?
(c) What power does the battery deliver to the circuit?

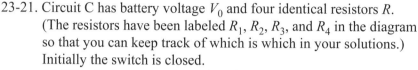

Circuit D

23-23. A strip of a certain wire is used as a fuse in a circuit whose current is to be limited to I amperes. The fuse melts after it has received E joules of energy in t seconds.
(a) What power is going to the fuse at the time immediately before melting?
(b) What is the resistance of the fuse wire?
(c) If the current in the circuit is to be limited to 10 A, and the fuse melts when it receives 9.0 J in 1.0 s, calculate the resistance of the fuse.

23-24. • A pair of resistors connected in series to a source voltage V uses one-fourth the power when connected in parallel to the same line.
(a) Why does it make sense that the resistors use more power when connected in parallel than when connected in series?
(b) Calculate the resistance of one of the resistors if the other is 1800 Ω.

23-25. • You have a 40-W bulb and a 100-W bulb, both rated for a 120-V circuit.
  (a) What is the resistance of each bulb?
  (b) What is the equivalent resistance when they are connected in series?
  (c) Calculate the power dissipation in each bulb when connected in series.
  (d) What is the equivalent resistance when they are connected in parallel?
  (e) In which type of circuit (parallel or series) will the 40-W bulb be brighter than the 100-W bulb?

Show-That Problems for Electric Current

23-26. A wire carries a current of 10 A.
  Show that the time it takes for a charge of 60 C to pass by a point in the wire is 6 seconds.

23-27. A current of 0.03 A through the human body is usually fatal.
  Show that when in contact with 120 V, a minimum body resistance of 4000 Ω is crucial.

23-28. When the human body is dry, it has a resistance of about 120,000 Ω. Suppose a human body is in contact with 120 V.
  Show that a current of 0.001 A in the body will be produced.

23-29. A 15-A fuse melts after it has received 12.0 J of energy in 1 s.
  Show that the resistance of the fuse is 0.053 Ω.

23-30. A 12-V battery energizes two bulbs in series, one with resistance 12 Ω and the other with resistance 18 Ω.
  Show that the power delivered by the battery is 4.8 W.

23-31. A 12-V battery energizes two bulbs in parallel, one with resistance 12 Ω and the other with resistance 18 Ω.
  Show that the power delivered by the battery is 20 W.

23-32. Three resistors of 6 Ω, 8 Ω, and 16 Ω are connected in series to an ideal 12-volt battery.
  Show that the current in the circuit is 0.40 A.

23-33. Three resistors of 6 Ω, 8 Ω, and 16 Ω are connected in series to an ideal 12-volt battery. Show that the power expended in the circuit is 4.8 W.

23-34. Five identical lamps are connected in series to a source that provides 550 volts, which produces a current of 1.1 A in the lamps.
  Show that the total resistance of all five lamps is 500 Ω.

23-35. The same five identical lamps of the previous problem are connected in parallel with a source that provides 550 volts.
  Show that the current in the source is 28 A.

23-36. A 60-Ω heater and a 144-Ω lamp are connected in parallel to a generator that supplies 120 V.
  Show that the current in the generator is 2.8 A.

23-37. Twelve 120-Ω lamps are connected in parallel to a 120-V house line.
  Show that the line current will not blow a 15-A fuse.

23-38. Twelve 120-Ω lamps are connected in parallel to a 120-V house line.
Show that the power that illuminates each lamp is 120 W.

23-39. Four 600-W heaters are connected in parallel to a 120-V line. Each heater can be turned on or off independently.
Show that the increase in line current as each heater is turned on is 5 A.

23-40. The total current in a parallel circuit of n branches is given by the sum of the currents in its branches. That is, $I_{total} = I_1 + I_2 + I_3 \ldots I_n$. Since the voltage V is the same across all branches, Ohm's law tells us
$$\frac{V}{R_{equivalent}} = \frac{V}{R_1} + \frac{V}{R_2} + \frac{V}{R_3} + \cdots \frac{V}{R_n}.$$
Show that dividing both sides of the equation by V gives an expression for the equivalent resistance of resistors in parallel.

23-41. The equivalent resistance for resistors of n branches in parallel can be found from *the reciprocal rule*:
$$\frac{1}{R_{equivalent}} = \frac{1}{R_1} + \frac{1}{R_2} + \frac{1}{R_3} + \cdots \frac{1}{R_n}.$$
Show that for two resistors in parallel, $R_{equivalent} = \dfrac{R_1 R_2}{R_1 + R_2}$. This is often called the *product-over-sum rule*, which applies only to *pairs* of resistors in parallel.

23-42. You have three 60-Ω resistors.
Show how to combine them to produce an equivalent resistance of 20 Ω.

23-43. You have three 60-Ω resistors.
Show how to combine them to produce an equivalent resistance of 90 Ω.

23-44. A 60-Ω resistor and a 20-Ω resistor are connected in parallel. This combination is then connected in series with a 33-Ω resistor. The circuit is connected to a 24-V battery. Show that the current in the 33-Ω resistor is 0.50 A.

23-45. A 100-W lamp is left on for 24 hours.
Show that the energy expended in keeping it lit is 8.64 MJ (1 MJ = 10^6 J).

23-46. A 100-W stereo system is connected to a 120-V outlet.
Show that the current drawn by the system is 0.83 A.

23-47. A string of 50 party lights connected in series operates at 120 V.
Show that the resistance of each bulb is 24 Ω if 0.10 A flows through the string of bulbs.

23-48. An electrical device (transformer) converts electrical power at one particular voltage and current to electrical power at another voltage and current. A typical laptop computer consumes electrical energy at 120 V, 1.5 A, and puts out a maximum 4.6 A at 18 V.
Show that this device is 46% efficient.

23-49. • Large power generators typically generate electric power at 12,000 V but transform this to 150,000 V before sending it along long-distance wires. Suppose that the electrical resistance in these wires is R, and that the transformer itself is 100% efficient—that is, $P_{output} = P_{input}$.
(a) Show that, for the same original transmitted power, the current through the wires at higher voltage is 0.080 times as much as the current at the lower voltage.
(b) Use $P = I^2R$ to show that the power loss to resistive heating in the wires will be more than 150 times smaller at the higher voltage than at the lower voltage.
(c) Explain why power companies use very high voltages on their transmission lines.

23-50. • One meter of a typical automotive copper wire has 1.1×10^{23} free electrons in it. Suppose that this wire is connected to a 12-volt, 36-W headlamp bulb and that the bulb is turned on.
(a) Show that 1.9×10^{19} electrons pass through this bulb's filament every second.
(b) Show that it would take about 1.6 hours for the electrons in this one meter of wire to pass through the bulb.
(c) Show that the average velocity of these electrons is therefore 1.7×10^{-4} m/s (0.17 mm/s).

23-51. The soda in a common can of soda contains 150 kilocalories of food energy (there are 4.18 kJ in each kcal). The can itself requires approximately 200 watt-hours of energy to manufacture from aluminum ore.
Show that more energy is required to manufacture the soda can itself than is in the beverage.

23-52. Batteries used for solar energy systems are commonly rated in amp-hours (the product of current × time). A common 12-V storage battery might be rated at 50 A·h.
(a) Show that 1.8×10^5 coulombs will circulate through a circuit if you run this battery from fully charged to dead.
(b) Show that this battery stores about 2.2 MJ of energy (1 MJ = 10^6 J).
(c) The energy content of gasoline is about 35 MJ/liter (132 MJ/gallon). Show that the energy stored in the battery is about the same as what is stored in 63 milliliters (about 2 ounces) of gasoline.

23-53. An iPod™ replacement battery is labeled "3.7 V, 500 mAh." (1 mAh means that the battery can produce a current of 1 milliamp for one hour.)
(a) Show that this battery stores a total of 6.7×10^3 J.
(b) Show that this is equivalent to the work a 68-kg (150-lb) person has to do to climb three flights of stairs (about 10 m).

Magnetism

We have seen that an electric charge is surrounded by an electric field that can affect other charges. Likewise, a mass is surrounded by a gravitational field that affects other masses. In this chapter we look at the effects of *magnetic fields* that surround magnets and moving charges. Just as the direction of the electric field points in the direction of the force acting on a positively charged particle, the magnetic field points in the direction of the force that would act upon a (theoretical) north magnetic pole.

When magnetic fields interact with moving charged particles, the force experienced by the particles depends upon three factors:

1. The charge q (its sign and magnitude).
2. The velocity v of the charged particle.
3. The magnetic field vector B (its strength and direction).

So our magnetic force equation looks something like

Magnetic force ≈ charge × velocity × magnetic field strength.

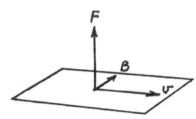

The magnetic force is peculiar in that it acts in a direction perpendicular to both the velocity vector and magnetic field vector. The magnitude of the force depends on the velocity-vector component $v_\perp$ that is perpendicular to the magnetic field vector B. In this book we will only consider particles moving perpendicular to the magnetic field.

The magnetic force equation is $\vec{F} = q v_\perp \vec{B}$.

The SI unit for B is the tesla (T), which is 1 newton per ampere-meter. So 1 T = 1 N/(A·m), or equivalently, 1 N/(C·m/s).

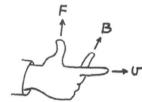

Notice in the above diagram that the force vector $\vec{F}$ is perpendicular to both $\vec{v}$ and $\vec{B}$. The correct direction for $\vec{F}$ can be determined using the *right-hand rule*. The diagram at right shows the relative directions of v, B, and F for a positive charge (forefinger pointing in the direction of v, middle finger in the direction of B, and thumb in the direction of the force F).[*]

When a current flows through a wire, the moving charges in it produce a magnetic field. Another version of the right-hand rule shows the direction of the magnetic field B around a current-carrying wire. Here you grab the (insulated!) wire with your right hand, with your thumb pointing in the direction of the (positive charge carrier) current. The direction of your fingers curling around the wire indicates the direction of the magnetic field that circles around the wire.

---

[*] For a negatively charged particle, the force would point in the opposite direction.

© Paul G. Hewitt and Phillip R. Wolf

Faraday's law is a highlight of Chapter 25 in *Conceptual Physics* and is treated qualitatively. We imagine magnetic lines of force passing through the area enclosed by a wire loop. The product of the magnetic field and the area of the loop is called the *magnetic flux*, with the symbol Φ. In accord with Faraday's law, any change in the "amount" of magnetic flux through the area of a loop (or coil of loops) induces a voltage in that loop (or coil). This flux can be changed by

1. Increasing or decreasing the strength of the magnetic field B;
2. Changing the area of the loop or coil of loops in the field; or
3. Changing the loop's angular orientation to the field.

Figure 25.1 in your textbook illustrates the first case. Moving the bar magnet into or out of the coil strengthens or weakens the field passing through the coil (that is, the flux) and induces a voltage. Figure 25.8 shows how the flux through a loop can change as the loop rotates in a uniform magnetic field. The relationship between induced voltage and changing flux is

$$\text{Voltage induced} \sim \text{number of loops} \times \frac{\Delta(\text{magnetic field} \times \text{loop area})}{\Delta \text{ time}} \Rightarrow \varepsilon = -N\frac{\Delta\Phi}{\Delta t},$$

where ε is the "electromotive force" (actually the induced voltage), N is the number of loops of wire in the coil, and $\frac{\Delta\Phi}{\Delta t}$ is the rate of change of the magnetic flux.*

The negative sign arises because experiments show that an induced voltage always gives rise to a current whose magnetic field opposes the change in flux. This is known as *Lenz's law*. If Lenz's law were not true, an induced current could produce a flux in the same direction as the change and grow in an ever-increasing crescendo—a conservation of energy no-no! Lenz's law is consistent with the conservation of energy.

Sample Problem 1

A particle of mass *m* with a positive charge *q* moves to the right at velocity *v* in a uniform magnetic field *B*. Velocity *v* and field *B* are at right angles to each other.

(a) Write an equation for the force exerted on the particle by the magnetic field.

Answer: $F = qvB$, a one-step solution.

(b) The magnetic field *B* points in a direction out of the page. In what direction will the particle be deflected by the magnetic field?

Direction of $F = ?$

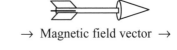
→ Magnetic field vector →

By convention, we represent the direct-ion of a magnetic field vector by an arrow with fletchings (rear feathers).

If the field points out of the page, we will just see the circular head of the arrow and the point of its cone. If the field points into the page, all we will see is the feather fletchings at the rear.

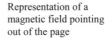
Representation of a magnetic field pointing out of the page

Representation of a magnetic field pointing into the page

---

* The magnetic flux Φ in a coil is the product of the field strength B and the area of the coil through which the magnetic field lines pass ($\Phi = BA$). Maximum Φ occurs when the field lines are perpendicular to the loop. If the coil steadily rotates, Φ changes in sinusoidal fashion from maximum to minimum and back again with each complete turn. (See Figure 25.8 in the textbook.) A quantitative treatment of changing Φ, as occurs in generators, requires familiarity with radian units that are not covered in the textbook.

We can determine the direction of the magnetic force using the right-hand rule. If you point your forefinger to the right (the direction of v) and your middle finger out of the page (the direction of B), your thumb points down the page (the direction of F). **The particle will be deflected downward.**

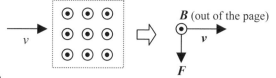

Had q been a negative charge moving in the same direction, the force would have been in the opposite direction, upward.

(c) Because the magnetic force on a charged particle is always perpendicular to the velocity, the trajectory of the charged particle in the magnetic field region will be a circular arc. Derive an equation for the radius of this arc. (Recall that any particle moving in a circular path is acted on by a centripetal force.)

Focus: $r = ?$

The magnetic force acts as the centripetal force on the particle.

From $qvB = \dfrac{mv^2}{r} \Rightarrow r = \dfrac{mv}{qB}$. This answer makes sense. A stronger magnetic field or a more highly charged particle will result in a large magnetic force, pulling the particle into a tighter circle. A more massive or faster particle will have a greater tendency to move in a straight line, so it would describe a circle with a larger radius.

(d) Calculate the radius of the circular arc if the mass of the particle is 1.5×10^{-9} kg, its charge is 5.0×10^{-5} C, its velocity is 1.2×10^3 m/s, and the magnetic field is 0.18 T.

Solution: $r = \dfrac{mv}{qB} = \dfrac{(1.5 \times 10^{-9}\text{ kg})(1.2 \times 10^3\text{ m/s})}{(5.0 \times 10^{-5}\text{ C})(0.18\text{ T})} = 0.20$ m.

Sample Problem 2

The force on a charged particle moving perpendicular to a magnetic field is given by $F = qvB$.
(a) Show that for an electric current, the equation for magnetic force becomes $F = ILB$, where I is the current, L is the length of wire in and perpendicular to the magnetic field, and B is the magnetic field strength.

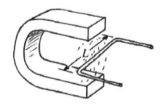

Focus: $F = ?$

Let's say that an amount of charge q takes a time t to travel the length of wire L. The speed v of the charge is therefore $v = \dfrac{L}{t}$. Since charge q will take time t to pass through a point in the wire, the current I in the wire is $I = \dfrac{q}{t}$. Putting this all together: $F = qvB = q\left(\dfrac{L}{t}\right)B = \left(\dfrac{q}{t}\right)LB = ILB$.

(b) Write an equation for the magnetic force on a length of wire L that carries a current I and is placed at right angles to a field of strength B.

Answer: The equation for force as found in part (a), $F = ILB$.

(c) Calculate the force on a wire of length 0.10 m carrying a 2.0-A current in a 0.0030-T magnetic field.

Solution: $F = ILB = (2.0\text{A})(0.10\text{ m})(0.0030\text{ T}) = \mathbf{0.00060\text{ N}}$.

(d) Calculate the force on 20 strands of the same wire, each with the same current in the same magnetic field.

Solution: The magnetic force acts on each of the 20 strands, so the total force on the stranded wire is 20 times greater than the force on each individual strand.
So $F = 20 \times ILB = 20 \times (2.0\text{ A})(0.10\text{ m})(0.0030\text{ T}) = \mathbf{0.012\text{ N}}$.

(e) If the current is heading "clockwise" through the wire in the diagram, and the magnetic field points up from the lower pole to the upper pole of the magnet, in what direction is the force on the wire?

Answer: The positive charge carriers head "into the page," between the poles of the magnet. B points up. The right-hand rule gives F pointing horizontally away from the magnet, **to the right**. So the wire tends to be ejected from the magnet.

(f) When the direction of the current is reversed, how does this affect the direction of the force?

Answer: Since the velocity vector for the charges reverses direction, the magnetic force acting on them also reverses direction. The wire tends to be pulled into the magnet, **to the left**. (A motor is a familiar example of a device in which the force can be alternated to and fro by changing the direction of the current.)

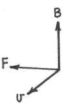

Sample Problem 3

A transformer (Figure 25.13 in the textbook) nicely employs Faraday's law. Consider an alternating voltage of 120 V that is applied to a step-up transformer having 140 turns on its primary and 1400 turns on its secondary.
(a) What is the voltage in the secondary?

Focus: Secondary voltage = ?

Page 448 in *Conceptual Physics* shows the relationship between primary and secondary voltages relative to the numbers of turns:

$$\frac{\text{Primary voltage}}{\text{Number of primary turns}} = \frac{\text{secondary voltage}}{\text{number of secondary turns}}$$

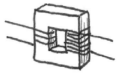

Solving for secondary voltage,

Secondary voltage = $\dfrac{\text{primary voltage} \times \text{number of secondary turns}}{\text{number of primary turns}} = \dfrac{(120\text{ V})1400}{140} = \mathbf{1200\text{ V}}$. This makes sense. The ratio of transformer turns is 10 to 1, so the voltage will be stepped up by a factor of ten.

(b) If the secondary coil is supplying a current of 0.1 A to a heater or other resistor, what is the current in the primary coil?

Answer: This transformer steps up the voltage from the primary coil to the secondary coil by a factor of ten. There is a corresponding step down of current in the secondary coil. The power into the transformer equals the power out, neglecting small losses. So $(IV)_{primary} = (IV)_{secondary}$ is consistent with the law of conservation of energy. Hence the current in the primary is ten times that in the secondary, or **1.0 A**.

(c) What is the power input to the transformer?

Focus: $P = ?$

From $P = IV$, we see that the power input is $(1.0 \text{ A})(120 \text{ V}) =$ **120 W**.

(d) By how much is the power stepped up in the secondary?

Answer: Whoa! Assuming no significant heat loss, the power in both coils is the *same*! Although voltages and currents can be stepped up or down in a transformer, energy, and therefore power, cannot. Ideally, the power in the secondary is also **120 W**. Conservation of energy rules!

Problems for Magnetism

It may be useful for several of the following problems to know that

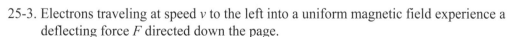

$q_{electron} = -1.60\times10^{-19}$ C; $q_{proton} = 1.60\times10^{-19}$ C; $m_{electron} = 9.11\times10^{-31}$ kg; $m_{proton} = 1.67\times10^{-27}$ kg.

25-1. A particle with charge $+q$ and velocity v is projected at right angles between the poles of a magnet of field strength B.
  (a) Write an equation that shows how large the magnetic force is on the particle while in the magnetic field.
  (b) In what direction is the magnetic force on the charged particle?
  (c) Calculate the magnitude and direction of the force on the particle if its charge is 0.0020 C, its velocity is 550 m/s, and the magnetic field between the pole pieces is 0.0020 T.
  (d) If the charge were instead $-q$, what would be the difference in the direction of the force?

25-2. A beam of electrons traveling to the right at speed v is directed perpendicularly into a uniform magnetic field of strength B pointing into the page.
  (a) Write an equation for the force on each of the electrons.
  (b) Calculate the deflecting force on each electron in the beam if they travel at 3000 m/s and the strength of the magnetic field is 0.15 T.
  (c) In what initial direction will the electrons be deflected?

25-3. Electrons traveling at speed v to the left into a uniform magnetic field experience a deflecting force F directed down the page.
  (a) Derive an equation for the strength of the magnetic field that produces the force F.
  (b) Calculate the strength of the magnetic field when the speed of the electrons is 2.0×10^{7} m/s and the deflecting force is 3.0×10^{-13} N.

25-4. Mona directs a beam of protons perpendicularly into a magnetic field B. Each proton experiences a deflecting force F.
   (a) Derive an equation for the speed of the protons in terms of charge, B, and F.
   (b) Calculate the proton speed in a magnetic field of strength 0.080 T, where it experiences a deflecting force of 1.2×10^{-13} N.

25-5. A proton in a magnetic field travels at speed v in a circular path of radius r.
   (a) Derive an equation for the magnetic field strength responsible for the circular path.
   (b) Calculate the magnetic field strength if the speed of the proton is 1.2×10^6 m/s and the radius of its curved path is 0.15 m.
   (c) Suppose that the magnetic field points into the page, and at one particular moment, the proton is heading to the right. Will the proton follow a clockwise or a counterclockwise path? Defend your answer.

25-6. A proton in a magnetic field B travels in a circular path of radius r.
   Derive an equation for the speed of the proton in terms of its mass, B, and r.
   (b) In terms of the same variables, derive an equation for the proton's kinetic energy.
   (c) Calculate the kinetic energy if its path has a radius of 0.40 m and the magnetic field strength is 0.85 T.

25-7. A length of current-carrying wire L is placed at right angles to a magnetic field B.
   (a) Write an equation for the force exerted on the wire when it carries a current I.
   (b) Calculate the force on the wire of length 0.14 m carrying a current 5.0 A in a field of 0.0050 T.
   (c) When the current heads counterclockwise between the poles of the magnet, the wire experiences a horizontal force into the magnet (toward the left). Which pole of the magnet is the north magnetic pole, the upper or the lower one?
   (d) If both the direction of the current and of the magnetic field are reversed, what effect, if any, will this have on the direction of the force?

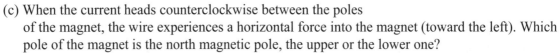

25-8. • A simple electron gun consists of two oppositely charged plates with a potential difference V between them. Electrons are "boiled" off the negative plate and accelerate toward the positive plate, many passing through a hole in its middle. This electron beam then passes through a uniform magnetic field B.
   (a) Derive an equation for the radius of the path traced out by the electron beam in the magnetic field.
   (b) In what direction does the electron beam bend when it encounters the magnetic field?
   (c) Calculate the radius of the beam's path if V = 1530 volts and B = 0.00120 T.

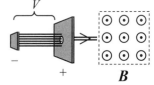

25-9. A horizontal rod of mass m, length L, and carrying a current I is aligned in a magnetic field for maximum magnetic force. The force is just right to keep the rod suspended in midair.
   (a) What is the direction of current through the rod—clockwise or counterclockwise? Defend your answer.
   (b) Find the magnetic field strength B.

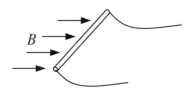

25-10. A horizontal rod of length L carrying a current I is suspended at right angles into a horizontal magnetic field (as shown in the sketch of the previous problem). The strength of the field is B. The magnetic force on the rod is measured by the balance and is found to be F.
   (a) What is the current in the rod in terms of the above variables?
   (b) Calculate the current if the force on the rod is 0.15 N, its length is 0.20 m, and the magnetic field is 0.054 T.
   (c) If the magnetic force on the rod is downward, in what direction is the current running in the circuit—clockwise or counterclockwise?

25-11. Recall that the force on a charged particle in an electric field is qE. The force on a charge $+q$ moving at right angles to a magnetic field is qvB.
   (a) In the diagram to the right, what is the direction of the electric force that will act on the particle?
   (b) What is the direction of the magnetic force that will act on the particle?
   (c) If adjustments are made so that the electric and magnetic forces on a moving charged particle are equal and opposite, the particle will be undeflected. Show that in this situation, the speed of the particle is given by $v = E/B$.
   (d) If the electric field is supplied by equal and oppositely charged parallel plates, where the electric field is $E = V/d$, show that a more practical version of the above equation is $v = V/(Bd)$.
   (e) In a mass spectrometer (Figure 34.14 on page 607 in *Conceptual Physics*) it is important that all ions have the same velocity before entering the magnetic field that will deflect them according to mass. Hence a "velocity selector" for ions is needed. How do the above equations provide the foundation for a device that selects only particular velocities of charged particles?

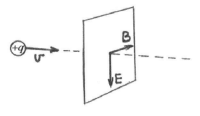

25-12. An electric field E exerts a force F on an ion of charge q. At right angles to the electric field is a magnetic field B.
   (a) Write an equation for the speed of the ion if the electric and magnetic forces are equal and opposite.
   (b) Calculate the speed if the charge of the ion is 1.6×10^{-19} C, the electric field is 6.0×10^{6} N/C, and the magnetic field is 0.83 T.
   (c) If the charge were twice as great, what strength of magnetic field would result in a straight-line path?
   (d) If the ion had instead double the mass, what strength of magnetic field would result in a straight-line path?

25-13. A charged particle q with initial velocity v enters a region of uniform electric and magnetic fields perpendicular to each other. The particle's path is perpendicular to the magnetic field of strength B.
   (a) Write an equation for the strength of the electric field if the particle is to pass through both fields undeflected.
   (b) Calculate the electric field strength if the charge of the particle is 1.6×10^{-18} C, its velocity is 5.5×10^{4} m/s, and the strength of the magnetic field is 1.35 T.

25-14. After passing through a velocity selector, a stream of ions moving at velocity v enters a uniform magnetic field B and is swept into a circular arc of radius r.
 (a) Derive an equation for the mass of each ion.
 (b) Calculate the mass of the ion if its charge is 1.6×10^{-19} C, its speed in the field is 2.05×10^5 m/s, the radius of its path is 0.308 m, and the magnetic field strength is 1.36 T.
 (c) Calculate the mass of the ion in atomic mass units (amu), given that there are 6.02×10^{26} amu in 1 kg. Can you guess what element this ion belongs to? (You may wish to refer to the periodic table on page 203 of your textbook.)

25-15. An alternating voltage V_p is applied to the primary coil of a step-up transformer having N turns on its primary and $20\,N$ turns on its secondary. The secondary current is I_s.
 (a) What is the voltage in the secondary?
 (b) What is the primary current?
 (c) What is the power input?
 (d) What is the power output, assuming no losses?
 (e) Calculate the answers to the above questions for a primary 120 V and a secondary current 0.1 A.

25-16. A transformer draws power P and delivers a current I_s in its secondary coil.
 (a) What is the secondary voltage?
 (b) Calculate the secondary voltage if the transformer draws 1000 watts of power, and delivers 25 A to its secondary.
 (c) What is the ratio of primary to secondary turns if the primary voltage is 120 V?

25-17. Roy uses a step-down transformer to power a toy electric train.
 (a) If the primary coil of the transformer has N turns, and the desired output is 6 volts from a household circuit of 120 V, how many turns should be on the secondary?
 (b) Calculate the number of turns on the secondary if the primary has 240 turns.
 (c) Why would it be a bad idea to reverse the input and output of the transformer (that is, to apply 120 volts to the secondary)?

25-18. The uniform magnetic field between the poles of an electromagnet increases at a steady rate and induces an emf ε in a single conducting loop of wire placed in the field.
 (a) Write an equation for the induced current in the loop if an emf ε is induced and the resistance of the loop is R.
 (b) Calculate the induced current for an induced emf of 4.0 V and a loop resistance of 0.8 ohms.
 (c) Does induced voltage depend on circuit resistance? Does induced *current* depend on circuit resistance? Defend your answers.

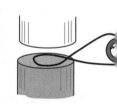

25-19. A coil of wire containing N circular loops, each of cross-sectional area A, is placed between the pole pieces of a magnet where the field is uniform and of field strength B. The plane of the loops of the coil is at right angles to the field. The field then decreases at a steady rate to zero in time t.
(a) Derive an equation for the magnitude of the induced voltage in the coil.
(b) Calculate the magnitude of the induced voltage if the field decreases from 0.36 T to zero in 2.0 s and the coil has 400 loops of cross-sectional area 0.050 m².

25-20. A coil of wire with N turns and an area A is placed in a magnetic field B and oriented such that the area is perpendicular to the field. The coil is then rotated a quarter-turn, 90°, in time t so it is aligned parallel to the field.
(a) Derive an equation for the average induced voltage in the coil.
(b) Calculate the average induced voltage if the magnetic field is 1.2 T, the number of turns in the coil is 20, the area of the loops is 0.045 m², and the coil is flipped a quarter-turn in 0.45 s.

Show-That Problems for Magnetism

25-21. An electron is projected at 3.0×10^6 m/s into a 0.13-T magnetic field. Show that the maximum deflecting force on the electron is 6.2×10^{-14} N.

25-22. An electron is projected into a magnetic field of strength 0.15 T and undergoes a deflecting force of 1.51×10^{-14} N.
Show that the speed of the electron is 6.3×10^5 m/s.

25-23. A particle with a velocity of 1.0×10^4 m/s and charge 0.00010 C encounters a perpendicular magnetic field and is deflected with a force of 3.5×10^{-2} N.
Show that the strength of the magnetic field is 3.5×10^{-2} T.

25-24. A vertical beam of particles with a mass 12 times that of a proton and a charge of 9.6×10^{-19} C enters a magnetic field of 0.30 T and is bent in a semicircle of radius 0.50 m.
Show that the speed of the particles is 7.2×10^6 m/s.

25-25. A particle of mass 19 mg (1 mg = 10^{-3} g) and charge 10.0 μC is moving at 3000 m/s. It encounters a perpendicular magnetic field of strength 0.44 T.
Show that it is swept into a curved path with a radius of about 13 km.

25-26. A particle moving at 600 m/s with charge of 20 μC enters a magnetic field of 1.6 T and is swept into a circular arc of radius 225 m.
Show that the mass of the particle is 12 milligrams.

25-27. A current of 10.0 A flows through a rod of length 0.20 m placed perpendicularly across a magnetic field of 0.050 T.
Show that the magnetic force on the rod is 0.10 N.

25-28. A force of 0.60 N acts on an 8.00-A current-carrying rod of length 0.300 m placed at right angles to a magnetic field.
Show that the strength of the magnetic field is 0.25 T.

25-29. A step-up transformer boosts 12 V to 120 V.
Show that the current in the secondary is one-tenth as much as the current in the primary.

25-30. The secondary coil of an ideal transformer has 600 turns and the primary has 30 turns. Show that when the primary is connected to a 120-V line, the output of the secondary will be 2400 V.

25-31. The secondary coil of an ideal transformer has 600 turns and the primary has 30 turns. Show that when connected to a 120-V line with a current of 20 A in the primary, the current in the secondary will be 1 A.

25-32. The transformer on a utility pole steps the voltage down from 20,000 V to 120 V for use in a home.
Show that the ratio of primary to secondary turns in the transformer is 167 to 1.

25-33. In the transformer of the previous problem, the home uses 2.0 kW of power.
Show that the primary and secondary currents in the transformer are 0.10 A and 17 A, respectively.

25-34. A velocity selector is designed to pass protons through undeflected only if they have a speed of 5.0×10^5 m/s.
Show that if the magnetic field used is 0.40 T, then the electric field oriented perpendicular to B must have a strength of 2.0×10^5 N/C.

25-35. A circular loop with an area of 0.012 m^2 is in a uniform magnetic field of 0.25 T.
Show that the flux through the loop when the field is oriented perpendicular to the loop is 0.0030 T·m^2.

25-36. A circular loop of radius 0.30 m is positioned in a magnetic field of 0.18 T.
Show that the maximum flux in the loop is 0.051 T·m^2.

25-37. Suppose the loop in the previous problem is in a magnetic field that decreases from 0.18 T to zero in 0.50 s.
Show that the induced voltage in the loop is 0.10 V.

25-38. Suppose that an ammeter is connected to the loop of the previous problem. The total resistance of the loop and the meter is 0.5 ohms.
Show that the current pulse is 0.2 A.

25-39. A coil of 100 turns of square loops of wire with an area of 0.50 m^2 is in a region where a magnetic field increases by 0.30 T in 0.10 s.
Show that the induced voltage in the coil is 150 volts.

25-40. Suppose that an ammeter is connected to the coil of the previous problem. The total resistance of the coil and the meter is 5.5 ohms.
Show that the current is about 27 A.

28 Reflection and Refraction

In this chapter we examine the reflection of light from plane mirrors and the refraction of light when it passes from one medium to another. We also study the behavior of light in simple lenses.

The Law of Reflection

When light is incident at an angle on a reflecting surface, it is reflected at the same angle. This is the law of reflection:

$$\text{Angle of incidence} = \text{angle of reflection}.$$

It is customary to measure the angles from the normal* to the reflecting surface as shown.

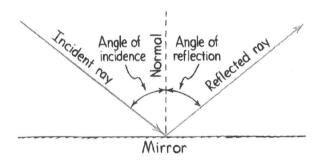

Sample Problem 1

A beam of light is incident on a plane mirror at an angle θ relative to the normal.
(a) What is the angle between the reflected ray and the normal?

Answer: In accord with the law of reflection, the angle of reflection equals the same angle of incidence θ, from the other side of the normal.

(b) What is the angle between the reflected ray and the mirror *surface*?

Answer: The angle with respect to the mirror surface is the complement of θ, which is **$90° - \theta$**. For example, a ray at an angle of 30° to the normal would make an angle of 90°− 30° or 60° with the surface. To avoid confusion, angles are always expressed relative to the normal in optical situations.

---

* *Normal* is another name for "the line perpendicular to the surface." We draw the normal to the surface at the point where the incident ray strikes the surface. At a point on a curved surface, we draw the normal perpendicular to the line tangent to the surface at that point. (See Figure 28.9 on page 489 in your *Conceptual Physics* textbook.)

© Paul G. Hewitt and Phillip R. Wolf

Refraction

When light bends in passing from one medium to another, we call the process *refraction*. Refraction is a consequence of light traveling at different speeds in different media. It is customary to compare the speed of light in a vacuum, c (300,000 km/s or 300,000,000 m/s), with the speed of light in other media. In so doing, we define a ratio called the *index of refraction*, n.

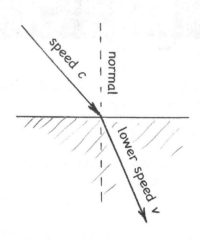

$$n = \frac{c}{v} = \frac{\text{speed of light in a vacuum}}{\text{speed of light in a medium}}.$$

For example, light in a diamond travels at 124,000 km/s, so for

$$\text{diamond } n = \frac{300,000 \text{ km/s}}{124,000 \text{ km/s}} = 2.42.$$

The indices (plural for index) of refraction shown in the table are given for a specific wavelength of light because light of different frequencies has different speeds in a transparent medium. Note that n is always greater than 1 because the speed of light in a vacuum is greater than the speed of light in any medium. Since n is a ratio of speeds, it has no units. The values of n in the table may be useful for the problems that follow.

Indices of Refraction (at $f = 5.0 \times 10^{14}$ Hz)	
Air	1.00029
Water	1.33
Ethyl alcohol	1.36
Fused quartz	1.46
Oil (typical)	1.50
Glass (by type)	1.45-1.70
Diamond	2.42

Sample Problem 2

What is the index of refraction of light in a vacuum?

Answer: $n = \dfrac{c}{v} = \dfrac{c}{c} = 1.$

Snell's Law

In about 1621, just after Pilgrims had landed at Plymouth Rock in America, the Dutch astronomer and mathematician Willebrord Snell discovered a relationship between the indices of refraction of two media and the angles of incidence and refraction as the light passed from one medium into the other. Snell's law for light passing from medium 1 to medium 2 is

$$n_1 \sin \theta_1 = n_2 \sin \theta_2,$$

where n_1 and n_2 are the respective indices of refraction of the media in which the light travels. As always, angles θ are taken with respect to the normal. Snell's law also holds if the light is going in the other direction, from medium 2 to medium 1.

Snell's law can be written to show the relationship between the speed of light in each medium and the angles with the normal that an oblique ray makes as it passes from one medium to another.

$$\frac{\sin \theta_1}{\sin \theta_2} = \frac{v_1}{v_2}.$$

Sample Problem 3

A merman beneath the surface of the sea looks above at a woman standing on a rock. Light from her hair on the way to his eye is incident upon the water at an angle θ_1 to the normal, and is refracted at the surface to his eyes. The index of refraction of water is n_w.

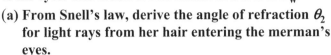

(a) From Snell's law, derive the angle of refraction θ_2 for light rays from her hair entering the merman's eyes.

Focus: Angle of refraction $\theta_2 = ?$

Snell's law provides the necessary physics.

From $n_1 \sin\theta_1 = n_2 \sin\theta_2 \Rightarrow \sin\theta_2 = \dfrac{n_1}{n_2}\sin\theta_1$.

So θ_2 is the angle whose sine is $\dfrac{n_1}{n_2}\sin\theta_1$.

We can write this as $\theta_2 = \sin^{-1}\left(\dfrac{n_1}{n_2}\sin\theta_1\right)$. Here we have $n_1 = 1$ for air, and n_2 is n_w.

So $\theta_2 = \sin^{-1}\left(\dfrac{1}{n_w}\sin\theta_1\right)$.

(b) **Calculate the angle of refraction given that the angle of incidence is 60°. The index of refraction for water is given in the table on the preceding page.**

Solution: $\theta_2 = \sin^{-1}\left(\dfrac{1}{n_w}\sin\theta_1\right) = \sin^{-1}\left(\dfrac{1}{1.33}\sin 60°\right) = \sin^{-1}\left(\dfrac{0.866}{1.33}\right) = \sin^{-1} 0.65$. Depending on your type of calculator, try entering 0.65, and then tap the inverse sine function. You should find the angle to be **41°**. (If not, get help from your instructor. Also be sure that the MODE button on your calculator is set for angles in DEGREES, and not radians.)

(c) **If the merman shone a laser pointer along his line of sight to the woman, would it shine on her hair?**

Answer: **Yes**. The path of light in one direction is the same path in the opposite direction, a rule that applies to all optical systems, however complicated. It is called the *principle of reversibility*.

(d) **If the merman pushed a long straight stick along his line of sight to the woman, would it reach her?**

Answer: **No**. Whereas the path of light changes where the water and air meet, the stick would remain straight and therefore pass above the woman's head. Interestingly, the stick would appear bent to both the merman and the woman.

© Paul G. Hewitt and Phillip R. Wolf

Critical angle

Imagine a lamp shining light upward from the bottom of a calm swimming pool. Light rays hitting the water-air interface at the surface of the pool are refracted to larger angles in the air. According to Snell's law:

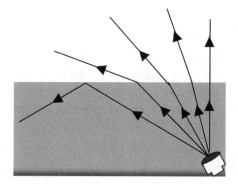

$$n_{water} \sin \theta_{water} = n_{air} \sin \theta_{air} \Rightarrow \sin \theta_{air} = \frac{n_{water}}{n_{air}} \sin \theta_{water}.$$

Since the index of refraction of water is greater than the index for air, the angle between the light ray and normal in air is always greater than the angle that the ray makes in water. As the angle of incidence in water beneath its surface increases, the angle of refraction in air approaches 90°. For some *critical angle* of incidence from water to air, the angle of refraction in air would be 90°. For angles of incidence greater than this *critical angle* there is, in a sense, nowhere left in air for the light to go. Light rays that approach the water-air boundary at angles greater than this *critical angle* are reflected from the boundary as though it were a perfect mirror.

Sample Problem 4

Suppose that the merman in Sample Problem 3 aims his laser beam upward at an angle of 60° with a normal to the surface.
(a) What is the path of the light beam after meeting the surface?

Focus: $\theta_{air} = ?$

From $n_1 \sin \theta_1 = n_2 \sin \theta_2 \Rightarrow \sin \theta_2 = \frac{n_1}{n_2} \sin \theta_1 \Rightarrow \theta_2 = \sin^{-1}\left(\frac{n_1}{n_2} \sin \theta_1\right).$

Here we have $n_1 = n_{water} = 1.33$; $\theta_1 = 60°$; $n_2 = n_{air} = 1.00$ so

$\theta_2 = \sin^{-1}\left(\frac{n_w}{n_{air}} \sin \theta_w\right) = \sin^{-1}\left(\frac{1.33}{1.00} \sin 60°\right) = \sin^{-1}(1.15).$ If you plug this into your calculator, you get an ERROR because no angle can have a sine greater than 1. Physically, this means that the light cannot be refracted into the air. Instead, **the light beam reflects from the surface**, at 60° with the normal in the water (the same angle as the incident beam). So rather than refracting up into air, the beam reflects downward in the water.

(b) What is the critical angle for light passing from water to air?

Focus: $\theta_c = ?$ The critical angle from water to air is the angle in water that would give a 90° angle of refraction in air.

$\theta_c = \sin^{-1}\left(\frac{n_{air}}{n_w} \sin \theta_{air}\right) = \sin^{-1}\left(\frac{1}{n_w} \sin 90°\right) = \sin^{-1}\left(\frac{1}{n_w}\right) = \sin^{-1}\left(\frac{1}{1.33}\right) = 48.8°.$

Any beam of light approaching the water/air boundary at an angle of greater than 48.8° with the normal will be totally internally reflected at the surface.

Focal Point of a Convex Lens

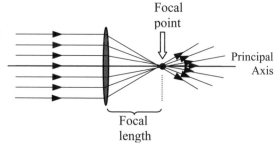

The *principal axis* of a simple lens is a line perpendicular to the lens, passing through its center. Light rays entering a convex lens along a path parallel to the principal axis are all refracted through the *focal point* of the lens. The distance between the lens and the focal point is called the *focal length*. The focal length is a physical property of the lens and depends upon the curvature of the lens and the index of refraction of the lens material.

Ray Diagrams and the Thin-Lens Equation

Image formation by lenses is nicely described using ray diagrams. In the diagram below, all of the light that leaves the candle (the *object* in this case) and passes through the lens converges to form an *image* of the candle on the screen. Any two of three principal rays are sufficient for locating the position of the image. (See pages 125–126 in the *Practicing Physics* book.)

1. Light that enters the lens parallel to its principal axis is refracted through the focal point on the opposite side.
2. Light that enters the lens through the near focal point is refracted through the lens in a direction parallel to the principal axis of the lens.
3. Light entering the center of the lens passes through undeflected.

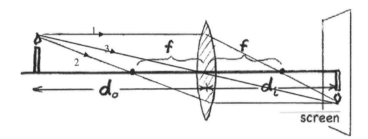

The sketch above shows all three of these principle rays. A quantitative way of relating object distances with image distances is given by the thin-lens equation:

$$\frac{1}{d_o} + \frac{1}{d_i} = \frac{1}{f} \quad \text{or} \quad d_i = \frac{d_o f}{d_o - f}.$$

Here d_o is the distance of the object from the lens, d_i is the distance from the lens to the image, and f is the focal length of the lens. Distances are positive (+) when object and image are on opposite sides of the lens. (See an illustration of this common case in Figure 28.49 of your *Conceptual Physics* textbook.) Distance d_i is negative (−) when the image is on the same side of the lens as the object. (See the convex-lens situation in Figure 28.48 of your textbook.) The *magnification* of a lens (the height of the image h_i compared with the height of the object h_o) is given by

$$M = \frac{h_i}{h_o} = -\frac{d_i}{d_o}.$$

A negative value of M means that the image is inverted. For a negative image distance d_i (image on the same side of the lens as the object) magnification is positive and the image is right side up.

Lenses that converge parallel rays of light are convex and have positive (+) focal lengths. Lenses that diverge parallel rays of light are concave and have negative (−) focal lengths. In either case, the size of the lens, its diameter, or aperture, does NOT affect focal length. Larger size lenses simply catch and refract more light than smaller size lenses. Focal length is determined only by the curvature of each side, the refractive index of the lens material, and the wavelength of light.

Sample Problem 5

A college student amazes her little brother with a demonstration of real images. She uses a strong magnifying glass to project an image of a candle onto a piece of paper several centimeters away. The lens has a focal length of 3.0 cm. The flame of the candle is about 1.0 cm tall and is 5.0 cm in front of the lens.

(a) **Predict the position of the image and its approximate size by drawing a ray diagram showing three rays from the candle flame.**

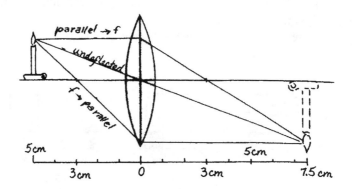

Analysis: The image position is located at the intersection of three rays: one parallel to the axis, that after refracting through the lens goes downward through the opposite-side focal point; another that goes straight through the center of the lens without deflection; and a third ray that passes through the near focal point and exits the lens parallel to the axis.

(b) **Use the thin-lens equation to calculate the distance between the lens and the image.**

Solution: $d_i = \dfrac{d_o f}{d_o - f} = \dfrac{(5.0 \text{ cm})(3.0 \text{ cm})}{(5.0 \text{ cm} - 3.0 \text{ cm})} = 7.5 \text{ cm}$.

(c) **How large will the image of the flame be?**

Focus: $h_i = ?$

From $M = \dfrac{h_i}{h_o} = -\dfrac{d_i}{d_o} \Rightarrow h_i = -\dfrac{d_i}{d_o} h_o = -\dfrac{7.5 \text{ cm}}{5.0 \text{ cm}} \, 1.0 \text{ cm} = -1.5 \text{ cm}.$

The image is 1.5 cm tall. The minus sign tells us that the image is inverted.

The Magnifying Glass

A ray diagram is still useful when an object's location is *less* than one focal length away from a convex lens. As before, light from the object that enters the lens parallel to the principal axis is refracted through the focal point, and light entering the center of the lens passes through without deviation.

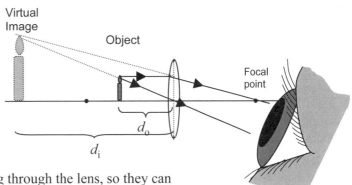

The light rays spread out (diverge) after passing through the lens, so they can never meet to form a real image. But to an observer, who believes that light travels in straight lines, light from the top of the object will *appear* to be coming from the top of a *larger* image farther back from the lens. Since the light really isn't coming from that point, we call this a *virtual image*. Mathematically, the magnification will be a *positive* number (since the image is right side up) and the image distance will be *negative* (since the image is virtual).

Sample Problem 6

In reading fine print in a telephone book, Caryn holds a magnifying glass with focal length f a distance d_o from the fine print ($d_o < f$).

(a) Derive an equation for how much the print is magnified.

Focus: $M = ?$ From $M = \dfrac{h_i}{h_o} = -\dfrac{d_i}{d_o}$, we need to find the image distance d_i. From the thin lens equation

$$\frac{1}{f} = \frac{1}{d_i} + \frac{1}{d_o} \Rightarrow \frac{1}{d_i} = \frac{1}{f} - \frac{1}{d_o} \Rightarrow d_i = \frac{d_o f}{d_o - f} \text{ so}$$

$$M = -\frac{d_i}{d_o} = -\frac{\left(\dfrac{d_o f}{d_o - f}\right)}{d_o} = \frac{f}{f - d_o}.$$

(b) Calculate the magnification if Caryn holds a 5.0-cm focal length lens 3.0 cm away from the fine print on the page.

Solution: $M = \dfrac{f}{f - d_o} = \dfrac{5.0 \text{ cm}}{5.0 \text{ cm} - 3.0 \text{ cm}} = 2.5$. The magnified print is 2.5 times larger than the original print. Since the magnification is positive, the image is right side up (consistent with our experience using a magnifying glass).

(c) Where does the image of the print appear to be?

Focus: The image distance d_i is the distance between image and the lens. We can find d_i from

$M = \dfrac{h_i}{h_o} = -\dfrac{d_i}{d_o} \Rightarrow d_i = -M d_o = -(2.5)(3.0 \text{ cm}) = -7.5$ cm. The negative sign tells us that this is a **virtual image appearing 7.5 cm behind the lens (opposite the side our eye is on) or 4.5 cm farther back from the lens than the print itself**.

The Diverging (Concave) Lens

When parallel rays enter a concave lens, the rays diverge (spread out). To an observer on the far side of the lens, the light appears to be spreading out from a single point, called the *virtual focal point* of the lens ("virtual" because the light isn't really spreading out from that point). Mathematically, we say that the lens has a negative focal length.

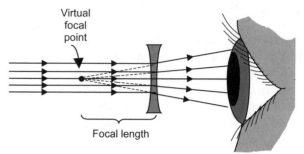

The rules for drawing ray diagrams for a thin concave lens are somewhat different than those for a convex lens:

1. Light from the object that enters the lens parallel to the principal axis is refracted *as though* it originated from the same-side focal point.
2. Light aimed at the *far-side* focal point exits the lens parallel to the principal axis.
3. Light entering the center of the lens passes through unrefracted.

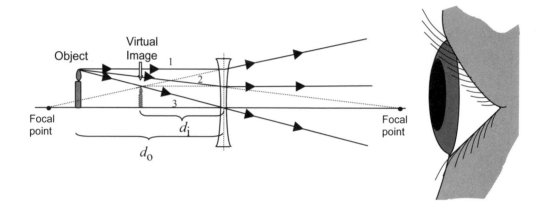

Since the light rays diverge after passing through the lens, they can never meet to form a real image. But to an observer, the light rays from the top of the object all *appear* to be coming from the top of a *smaller* image located closer the lens. Again, because the light isn't really coming from that point, we call this a *virtual image*. Mathematically, the magnification will be a *positive* number (since the image is right side up) and the image distance will be *negative* (since the image is virtual).

Sample Problem 7

Caryn looks through a concave lens with focal length f at a raisin located a distance d_o beyond the lens.

(a) Will the image appear smaller or larger than the actual raisin itself? How much larger or smaller?

Focus: $M = ?$ The ray diagram on the previous page (and the photograph in Figure 28.50 on page 503 of your *Conceptual Physics* textbook) suggests that the image will appear smaller and closer to us than the actual object itself. The math is the same as in previous Sample Problem 5—the thin lens equation applies to all thin lenses, convex or concave. We get $M = \dfrac{f}{f - d_o}$ again, but this time the focal length is negative. The object distance is a positive number, which makes the denominator $f - d_o$ even more negative and gives us $M = \dfrac{\text{a negative number}}{\text{a larger negative number}} = $ a positive number less than 1, which is what we expect from a concave lens—**a smaller, upright image**.

(b) Calculate the height and location of the image of a 1.0-cm tall raisin located 20.0 cm behind a concave lens of focal length –10.0 cm.

Focus: $h_i = ?$ $d_i = ?$ From $M = \dfrac{h_i}{h_o} \Rightarrow h_i = Mh_o$. From above, $M = \dfrac{f}{f - d_o} = \dfrac{-10.0 \text{ cm}}{-10.0 \text{ cm} - 20.0 \text{ cm}} = \dfrac{1}{3}$.

The image height is going to be 1/3 of the object height $= \frac{1}{3} \times 1.0 \text{ cm} = \mathbf{0.33 \text{ cm}}$. From $M = \dfrac{h_i}{h_o} = -\dfrac{d_i}{d_o} \Rightarrow d_i = -Md_o = -\frac{1}{3}(20.0 \text{ cm}) = \mathbf{-6.7 \text{ cm}}$. This is a virtual image of the raisin, smaller and closer to the lens than our original raisin, just as our ray diagram suggests.

Here is a table that should be useful in keeping track of the signs when solving thin lens problems:

LENS	Sign of f	Object is located ...	Image will be ...	M is	Sign of d_i
Convex	+	$> f$ from the lens	real	–	+
		$< f$ from the lens	virtual	>1 and +	–
Concave	–	anywhere	virtual	<1 and +	–

Problems for Reflection and Refraction

28-1. A beam of light is incident on a plane mirror at an angle θ relative to the normal.
  (a) What is the angle between the reflected ray and the normal?
  (b) What is the angle between the incident and reflected rays?

28-2. A light beam reflects to and fro between ordinary parallel mirrors as shown.
  (a) Sketch the path of the beam as it reflects between the mirrors, and estimate the number of reflections that occur before the beam is outside the mirror system.
  (b) How will the strength of the beam exiting the mirror system compare with its strength before entering?

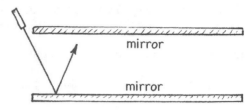

28-3. • A beam of light is incident upon a plane mirror at angle θ. The mirror is then rotated a bit by angle α as indicated in the sketch.
  (a) By how much is the angle of reflection changed?
  (b) A beam of light is incident on a mirror at an angle of 30° with the normal, and reflects onto a vertical screen 10.0 m away. If you rotate the mirror by 2° as shown, how far will the spot of light move on the screen?

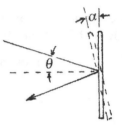

28-4. You walk toward a mirror at speed v.
  (a) How fast do you and your image approach each other?
  (b) In walking away from the mirror at speed v, how fast do you and your image recede from each other?

28-5. Jeanette makes a pinhole camera with a screen located a distance x from the pinhole. She points the camera at a tree of height H a distance d away from the pinhole.
  (a) What is the height h of the image formed on the screen?
  (b) Calculate the height of the image for a screen 15 cm away from the pinhole of a camera pointed at a 6.0-m tall tree 20 m away.

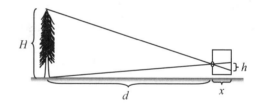

28-6. A pool table has dimensions $W \times L$. A pool ball strikes the cushion at an angle θ to the normal as shown.
  (a) Assuming the ball obeys the law of reflection, at what angle does it bounce from the cushion?
  (b) How far does it travel before it strikes the opposite cushion?

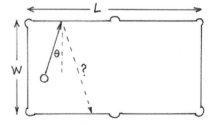

28-7. A beam of light is incident upon a pane of window glass at 60° to the normal.
  (a) What is the angle that the refracted beam makes with the normal inside the glass if the index of refraction of the glass is 1.5?
  (b) At what angle will the beam emerge from the glass? (Assume that the glass surfaces are parallel to each other.)

28-8. Light travels at different speeds in glass and in a diamond.
 (a) Find the ratio of the speed of light in glass ($n = 1.50$) to the speed of light in diamond ($n = 2.42$).
 (b) Once light is inside a diamond, it may reflect from an inner surface. How does the angle of reflection inside the diamond compare with its angle of incidence inside the diamond?

28-9. Flint glass has an index of refraction $n = 1.638$ for red light and $n = 1.675$ for blue light.
 (a) How much faster (in m/s) does red light travel in this glass compared with blue light?
 (b) In air, red light takes 1.00×10^{-10} s to cover 3.00 cm. How long does it take to travel 3.00 cm through flint glass?

28-10. Consider the same flint glass of the previous problem. A beam of white light shines through air onto it at an angle θ with the normal.
 (a) Calculate the angle of refraction for the blue light when the incident beam is 36.00° to the normal.
 (b) Calculate the angle of refraction for the red light in the same beam.

28-11. Rainbows are formed when white light from the Sun meets raindrops in the atmosphere. Water has an index of refraction of $n = 1.3311$ for red light and $n = 1.3330$ for yellow light.
 (a) If the sunbeam shines on a raindrop at an angle of 41.00° with respect to the normal at a particular point of the drop, what is the angle of refraction in the drop for red light refracted at that point?
 (b) For yellow light?
 (c) Light of both colors travels at very nearly the same speed in the air. Which one is slowed down more when it enters the drop? Defend your answer.

28-12. Looking straight downward, Marjorie notices that a coin at the bottom of a glass of water appears closer to her—closer by the index of refraction of air divided by the index of refraction of water.
 (a) Calculate how far below the water's surface the coin appears in water 10.0 cm deep.
 (b) Explain why the coin appears larger, as though it has been magnified.

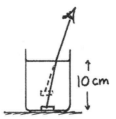

28-13. • A plastic block with index of refraction n and thickness t is placed in the path of a laser beam such that the beam makes an angle θ with the normal to the block. The beam emerges parallel to the original beam and deviated from its original path by an amount x, as shown in the diagram.
 (a) Explain why the shifted beam is parallel to the original beam.
 (b) Derive an expression for d.
 (c) Calculate the deviation distance for a beam that makes a 35° angle with the normal of a 2.5-cm thick plastic block with $n = 1.38$.

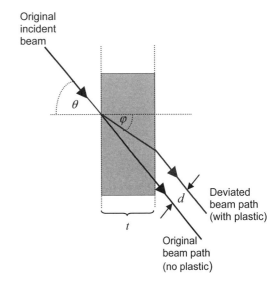

© Paul G. Hewitt and Phillip R. Wolf

28-14. An optical fiber is made from glass with index of refraction n.
(a) What is the maximum value for the angle of incidence (i.e., the critical angle) that will guarantee total internal reflection within the optical fiber?
(b) Calculate this angle for fused silica with $n = 1.47$.

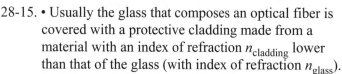

28-15. • Usually the glass that composes an optical fiber is covered with a protective cladding made from a material with an index of refraction $n_{cladding}$ lower than that of the glass (with index of refraction n_{glass}).
(a) What is the maximum value for the angle of incidence that will guarantee total internal reflection within the clad optical fiber?
(b) Calculate this angle for glass with $n_{glass} = 1.47$ and $n_{cladding} = 1.46$.
(c) Is the critical angle larger than or smaller than the critical angle for the unclad fiber?

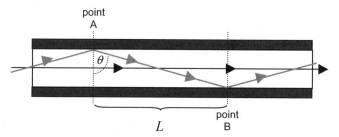

28-16. When a pulse of light is sent through an optical fiber with index of refraction n, not all of the light traveling through the fiber follows the same path. In the diagram, two rays are shown, one (in black), which travels straight along the fiber and another (in gray) that makes an angle θ with the normal at the side of the fiber.
(a) What is the speed of light in the optical fiber?
(b) Consider a length L of the fiber, as shown in the diagram. How long will it take for light traveling straight along the length of the fiber to travel a distance L?
(c) Consider light that makes an angle θ with the normal, as shown in the diagram. How far does it travel in going from point A to point B? How long will this take?
(d) Calculate the ratio $\dfrac{t_{\text{angled path}}}{t_{\text{straight path}}}$. *
(e) Consider a pulse of light traveling along a 10.0-km-long optical fiber with $n = 1.47$, where the angled beam makes an 85° angle with the normal. By what factor is the travel time of the angled path longer than the travel time for the straight path?

---

* This ratio tells us how much the pulse "spreads out." For example, a light pulse that travels during 1 nanosecond (10^{-9} s) will be "wider" at the other end of the optical fiber because light that starts at one end of the fiber can take different amounts of time to reach the other end of the optical fiber. Because we don't want initially separate pulses to overlap at the other end, this provides a practical limit to how many pulses (= how much information) we can send along a single optical fiber. Physicists and engineers have figured out clever ways to address this problem by using different colors of light at the same time (different colors for different sets of data), or using optical fibers made with several layers so that as we get farther from the center of the fiber, the index of refraction decreases. That way, even though light following the angled path has a longer distance to go, the lower index of refraction means that the light on the angled path is traveling faster than the light going along the straight path. This results in a closer travel time along the angled path and along the straight path, and reduces the spreading of the light pulse, which, in turn, means we can send more pulses (= data) on the same fiber.

28-17. In each half of a pair of binoculars, there are two 45°–45°–90° prisms as shown in the diagram. Total internal reflection directs the incident light to the eyepiece.
(a) What is the minimum value of index of refraction that the glass can have?
(b) Suppose that you tried to build this with two equilateral prisms as shown in the diagram. What would be the minimum index of refraction required in this case?

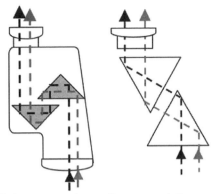

28-18. Experimental Ernie holds a converging lens out in the sunlight over a piece of paper and the paper ignites.
(a) What is the likely focal length of the lens if the paper ignites when the lens is held 15 cm above it?
(b) Why would the paper not ignite at other distances from the lens?

28-19. When an object is located a distance x to the left of a lens, the image is formed at a distance $0.5x$ to the right of the lens.
(a) What is the focal length of the lens?
(b) If the object were instead placed $2x$ to the left of the same lens, where would the image appear?

28-20. Because Justine is "far-sighted" she cannot distinctly see any object closer than 50 cm away. The glasses that her optometrist prescribes will have lenses that will allow her to clearly see an object that is only 25 cm away. (This means that the lens must be able to form a clear image at 50 cm of an object placed 25 cm from the lens.)
(a) Do you think the image is going to be right side up or upside down?
(b) Draw a diagram showing the location of the object, the image, and the lens. Will this lens be converging or diverging?
(c) Will the image be real or virtual?
(d) Calculate the focal length of the lens that is needed.

28-21. Because Claudia is "near-sighted," she can't clearly see an object if it is more than 65 cm away. The glasses that her optometrist prescribes will have lenses that will allow her to see clearly an object that is 6.0 m away. (This means that the lens must be able to form a clear image at 65 cm of an object located 6.5 m from the lens.)
(a) Do you think the image is going to be right side up or upside down?
(b) Draw a diagram showing the location of the object, the image, and the lens. Will this lens be converging or diverging?
(c) Will the image be real or virtual?
(d) Calculate the focal length of the lens that is needed.

28-22. Chelsea places an object of height h_o at a distance x to the left of a converging lens with a focal length f.
(a) What is the image distance?
(b) What is the height of the image?
(c) Calculate values for (a) and (b) for a 5.0-cm tall object located 7.0 cm to the left of a lens of focal length 15 cm.
(d) Calculate the values if instead she uses a diverging lens of focal length –15 cm.

28-23. A diverging lens is used to view an object placed on the opposite side of the lens. The focal length of the lens is –33 cm and the object is located 25 cm from the lens.
  (a) Draw a ray diagram illustrating this situation, showing where the image should appear. Will the image be right side up or upside down?
  (b) Calculate the image distance and the magnification.
  (c) What will you see when you look through the lens at this object?

28-24. A lens for your uncle's 35-mm film camera has a focal length 45 mm.
  (a) How close to the film should the lens be to form a sharp image of a 50-cm tall object that is 10.0 m away?
  (b) What is the magnification of the image on the film? How tall will the image on the film be?

28-25. Continuing with the problem above, your photographer uncle replaces the normal camera lens with a telephoto lens with a 135-mm focal length.
  (a) How close to the film should the lens be to form a sharp image of the same 50-cm tall object that is 10.0 m away? (Do you understand now why these lenses protrude so much from the camera body?)
  (b) What is the magnification of the image on the film? How tall will the image be on the film?

28-26. A detective examines evidence with a standard converging magnifying glass. He holds the lens close enough to the evidence so that it produces an enlarged virtual image.
  (a) How far from the evidence should the detective hold the lens if the focal length of the lens is 15 cm, and the object position is adjusted so that the image distance will be –22 cm (that is, on the same side of the lens as the object)?
  (b) What is the magnification?
  (c) What is the advantage of having the object located within the focal length?

28-27. When a physics instructor focuses on an exam paper, the lens in his or her eye forms a real image of the paper on the retina in the back of the eye. A wondrous series of physiological events occurs that turns the paper into a grade.
  (a) If the paper is 28 cm in front of the instructor's eye, and the lens of the eye is 3.0 cm in front of the retina, what is the focal length of the lens?
  (b) Is the eye lens a converging or a diverging lens?

28-28. The heat lamp in the bathroom has been absentmindedly left on. The light shines across the room, striking a slender perfume bottle with convex sides that focus the light into a small dot that slowly but surely burns a hole in the wall behind the bottle.
  (a) If the lamp is 1.9 m from the bottle, which is 9.8 cm from the wall with the smoldering hole, what is the focal length of the bottle acting as a lens?
  (b) If the diameter of the bulb in the heat lamp is 14 cm, what is the diameter of its damaging image on the wall?

28-29. As part of an art exhibition, you want the Sun to shine through a small stained-glass window. You want to use a giant lens to project the image of the window, four times taller and wider than real size, onto a wall a horizontal distance L from the window.
(a) Will the image on the wall be a real image or a virtual image?
(b) What will be the magnification in this situation? Will the magnification be positive or negative?
(c) How large will the image distance be in comparison to the object distance?
(d) Make a sketch of the situation, showing approximately the lens location relative to the window and the wall. Calculate values for d_o and d_i in terms of L. (What do you know about the sum of $d_o + d_i$ in this situation?)
(e) Use the thin lens equation to determine the appropriate type and focal length of lens that you should use.
(f) Calculate values for d_o, d_i, and f for a window-to-wall distance L of 4.5 meters.
(g) Draw a reasonably accurate ray diagram outlining the situation.

28-30. Curious Carlos is playing with a large convex lens before class and notices that when he holds the lens a distance h above the lab table he sees on the lab table a clear image of the classroom's fluorescent light bulbs, which are distance H above the lens.
(a) What is the focal length of the lens?
(b) Calculate the focal length of the lens if a clear image of lights 1.50 m above the lens is formed on the lab table 30 cm below the lens.

28-31. A jeweler's loupe is a small magnifying glass of short focal length that is held close to the eye to enable a very close look at a gem or watch part.
(a) If the loupe has a focal length of 2.2 cm, and a gem being inspected is just far enough away from the lens so that the image distance is –22 cm from the lens (on the same side of the lens as the gem), how far from the lens is the gem being held?
(b) What will be the magnification?

28-32. • When a magnifying glass is held x cm above a piece of ruled notebook paper, three notebook paper spaces fit into the same height as one magnified space.
(a) Derive an equation for the focal length of the lens in terms of x.
(b) Calculate the focal length of the lens if x is 6.0 cm.
(c) Using this same lens, at what height above the paper would two notebook paper spaces fit into one magnified space?

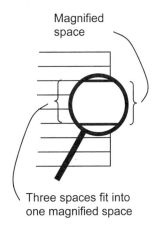

28-33. One way to find the focal length of a convex lens using an object, a lens, and a screen is to arrange them (by trial and error) so that the image is in focus on the screen with the object distance and image distance equal to one another.

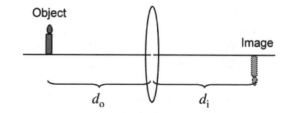

(a) Draw a ray diagram illustrating this situation.
(b) What is the focal length of the lens?
(c) Calculate the focal length of the lens if the object and image distances are 25.0 cm.

28-34. • One way to find the focal length of a *diverging* lens is to cover the lens with a piece of paper that has a 1-cm diameter hole in it. You take the lens outside and point it at the Sun, and adjust the position of a screen until there is a bright spot 2 cm wide on the screen. Call the distance between the screen and the lens d.

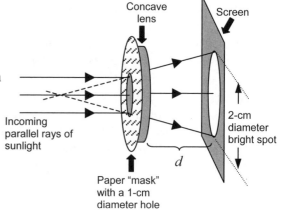

(a) Explain how this method gives you the focal length of the lens.
(b) What is the focal length of the lens?

Show-That Problems for Reflection and Refraction

28-35. A student 1.8 m tall stands in front of a plane mirror.
Show with a ray diagram that the smallest mirror to show his full image is 0.9 m tall. Further show that the distance from the mirror is not a factor.

28-36. A bee stands 0.3 m in front of a small mirror with a wall mirror 0.9 m behind him.
Show that the image of his stinger appears a distance 2.4 m in front of him.

28-37. Light slows when it enters a diamond, $n = 2.42$.
Show that the speed of light in a diamond is about 41% its speed in air.

28-38. Snell's law is introduced in this chapter in two ways.
Show how $\dfrac{\sin \theta_1}{\sin \theta_2} = \dfrac{v_1}{v_2}$ becomes $n_1 \sin \theta_1 = n_1 \sin \theta_2$

28-39. A beam of light in air is incident upon a piece of crown glass ($n = 1.52$) at an angle of 37.0° with the normal.
Show that the angle of refraction in the glass is 23.3°.

28-40. A laser beam in air is incident on a block of glass ($n = 1.50$) at an angle of 45°.
Show that the angle of refraction in the glass is 28°.

28-41. In the lab you shine a beam of white light at the middle of the flat side of a semicircular piece of glass.
(a) Explain why the light beam is refracted when it enters the glass but *not* when it exits the glass.
(b) The refracted light exits the glass and is projected onto a screen 1.000 meters away, as shown in the diagram. Show that the index of refraction of the red light is 1.51, while the index of refraction for violet light is 1.53.

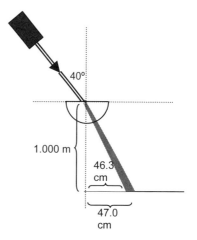

28-42. A laser beam beneath the surface of water is aimed upward toward the surface. Beyond a critical angle (where the light would refract along the water surface at 90° to the normal), light undergoes total internal reflection rather than refracting into the air above.
Show that the critical angle for the water-air interface is 48.8°.

28-43. The index of refraction for a diamond is 2.42.
Show that the critical angle for a diamond in air, beyond which internal reflection occurs, is 24.4°.

28-44. An object is placed 40.0 cm to the left of a lens and produces a real image at a distance of 80.0 cm to the right of the lens.
Show that the focal length of the lens is approximately 27 cm.

28-45. An object is placed 30 cm to the left of a converging lens and produces a real image at a distance of 15 cm to the right of the lens.
Show that the focal length of the lens is 10 cm.

28-46. A converging glass lens has an index of refraction of 1.60. Its focal length is 30.0 cm.
Show that an object 2.5 m in front of the lens will appear in sharp focus when the screen is 34 cm in back of the lens.

28-47. An object is placed 16 cm in front of a converging lens with a focal length 11 cm.
Show that the image distance is 35 cm.

28-48. An object is placed 16 cm in front of a diverging lens with a focal length –11 cm.
Show that the image distance is – 6.5 cm.

28-49. A candle is placed 50.0 cm in front of a converging lens of focal length 10.0 cm.
Show that the magnification of the candle image is – 0.25.

28-50. A converging lens with a focal length of 20.0 cm produces an image on a screen that is 2.00 m from the lens.
Show that the object distance is 22 cm.

28-51. A converging lens is used to produce a real image of an object. The object distance is twice the image distance. The object is 5.0 cm tall.
Show that the image is 2.5 cm tall and inverted.

28-52. An image that is 1.0 cm tall is formed on a screen behind a converging lens when an object 2.0 m tall is located 8.0 m in front of the lens.
Show that the distance between the lens and the screen is 0.040 m.

28-53. A candle 10 cm tall is placed 30 cm from a thin converging lens. The crown glass lens has a focal length, f, of 20 cm, and an index of refraction, n, of 1.52.
Show that the image of the candle is 60 cm from the lens.

28-54. A box 5.0 cm wide is placed 10.0 cm from a diverging lens. The image of the box, at a negative image distance, is one-fifth as wide as the box.
Show that the focal length of the lens is –2.5 cm.

28-55. In the biology lab you want to examine a piece of tissue at a magnification of +5.00.
Show that if the tissue is 5.00 cm from the lens, then the focal length of the lens is 6.25 cm.

33 The Atomic Nucleus and Radioactivity

So far we've considered interactions between molecules (in solids, liquids, and gases, for example) and rearrangement of atoms within molecules such as occurs in chemical reactions. All of these interactions are fundamentally *electrical* in nature—they involve attractions and repulsions between the nuclei and the electrons surrounding them. Whatever changes occur, the atoms themselves are left largely unchanged.

Nuclear reactions, on the other hand, involve additional forces and usually lead to the formation of an entirely new nucleus. The electrical repulsion between the protons in the nucleus tends to push the nucleus apart, while the strong nuclear force acting between pairs of adjacent nucleons holds the nucleus together. Hence a nucleus needs a certain balance of neutrons and protons for stability—nearly equal numbers for light nuclei, and a preponderance of neutrons for heavy nuclei (where the mutual repulsion of the protons plays an important role). If a nucleus is unstable because of "too many" protons, it will undergo a spontaneous change (a decay) that *decreases* the ratio of protons to neutrons. If it is unstable because of "too many" neutrons, it will decay in such a way as to *increase* the ratio of protons to neutrons.

There are five basic ways that a nucleus can spontaneously change its identity. These changes constitute *radioactive decay*. The effects of different types of radioactive decay are summarized in the table below:

Type of decay	What happens	Effect on the nucleus	Example
Alpha decay (α)♦	Nucleus ejects a helium nucleus (2 p, 2 n)	Nucleus has two fewer protons and four fewer nucleons overall	$^{235}_{92}U \rightarrow {}^{231}_{90}Th + {}^{4}_{2}\alpha$
Beta decay (β)*	Electron emitted, a neutron is changed into a proton†	Nucleus has one more proton, same number of nucleons as before	$^{239}_{92}U \rightarrow {}^{239}_{93}Np + {}^{0}_{-1}e$
Gamma decay (γ)	Nucleus "falls" from a higher to a lower state and emits a high-energy photon	Nucleus has the same number of protons and neutrons as before	$^{99}_{43}Tc^* \rightarrow {}^{99}_{43}Tc + {}^{0}_{0}\gamma$ (the * indicates that the nucleus is in an excited state).
Electron Capture (EC)	Orbital electron captured, a proton is changed into a neutron*	Nucleus has one fewer proton, same number of nucleons as before	$^{40}_{19}K + {}^{0}_{-1}e \rightarrow {}^{40}_{18}Ar$
Positron emission (β⁺)	Positron emitted, a proton is changed into a neutron*	Nucleus has one less proton, same number of nucleons as before	$^{23}_{12}Mg \rightarrow {}^{23}_{11}Na + {}^{0}_{1}e$

♦ If we are looking at the *atom* level, then we write $^{4}_{2}He$ rather than $^{4}_{2}\alpha$, since $^{235}_{92}U$ has 92 electrons, $^{231}_{90}Th$ has 90 electrons, and the alpha particle must pick up two electrons to become $^{4}_{2}He$ in order to balance out the charge.

* Sometimes called beta-minus decay (β⁻) to distinguish it from the similar decay in which a positron is released (β⁺)

† Another particle, called an *antineutrino*, is also formed. It has no charge, is very light (much less massive than an electron), and carries some of the momentum and energy from a nuclear decay. Neutrinos interact so weakly with matter that, for our purposes, we can ignore them.

* A neutrino is also formed.

© Paul G. Hewitt and Phillip R. Wolf

Radioactivity and half-life

Various degrees of "instability" exist among radioactive nuclei. Some radioactive samples will decay in less than a second; other samples will hardly decay at all in a person's lifetime. One way to characterize the degree of instability of a radioactive material is to report its *half-life*—the time for half of a radioactive sample to decay into something else. For instance, if you had 10 grams of a particular radioactive material, after one half-life only 5 grams of the original radioactive material would remain (the other 5 grams would have been converted to a different material). After two half-lives, 2.5 grams of the original radioactive material would remain. A material with a shorter half-life is less stable than one with a longer half-life—each nucleus within the less-stable sample has a higher probability of decaying in a unit time.

Suppose that you have a thousand dice and you roll them. Approximately one-sixth of them will come up as a one. If you remove all of the dice that come up as ones and roll the remaining dice again, the odds of any particular die coming up a one on the next roll hasn't changed—the odds are still one in six. But over time there will be fewer and fewer ones coming up because after each dice roll (and the removal of all of the dice that come up as one) there are fewer dice remaining to be rolled.

In this respect, radioactive nuclei are like dice. If a certain nucleus doesn't decay in one time interval, it has the same probability of undergoing decay in the next time interval. Over time, the number of decays from a given radioactive sample will decline, because the number of nondecayed nuclei remaining in the sample has declined.

When a material decays, most of the mass changes to another substance. Some mass is changed to energy in accord with the principle of mass-energy equivalence—next chapter.

Sample Problem 1

A bismuth-199 nucleus $\left(^{199}_{83}\text{Bi}\right)$ undergoes an alpha decay. What nucleus results?

Solution: In alpha decay the nucleus ejects an alpha particle—a helium nucleus composed of two protons and two neutrons. The overall positive charge in the nucleus is reduced by 2, and the overall number of nucleons is reduced by 4. Let X represent the unknown resulting nucleus.
Then $^{199}_{83}\text{Bi} \rightarrow\, ^{(199-4)}_{(83-2)}\text{X} +\, ^{4}_{2}\alpha$, where $^{195}_{81}\text{X}$ is the unknown. A numerical check confirms that both the number of nucleons and the overall charge are conserved. From the periodic table in your *Conceptual Physics* textbook (Figure 11.9 on page 203), you can see that the element with 81 protons is thallium, which has the symbol Tl. The new nucleus is $^{195}_{81}\text{Tl}$, thallium-195.

Sample Problem 2

A nucleus of barium-142, $^{142}_{56}\text{Ba}$, undergoes a beta-minus decay. What nucleus results?

Solution: In beta-minus decay, a neutron within the nucleus turns into a proton and ejects an electron (the β^- particle) from the nucleus. This increases the positive charge in the nucleus by 1, while the overall number of particles in the nucleus remains the same.
So $^{142}_{56}\text{Ba} \rightarrow\, ^{142}_{56+1}\text{X} +\, ^{0}_{-1}\text{e}$. The new nucleus is $^{142}_{57}\text{La}$, lanthanum-142. Again, a check shows that both the number of nucleons and the overall charge are conserved.

Sample Problem 3

When a sample of boron-10 ($^{10}_{5}B$) is bombarded with neutrons, the nucleus can be transmuted to a new nucleus by first absorbing a neutron, and then decaying into a new element by alpha decay. What nucleus is ultimately formed?

Solution: The reaction is $^{10}_{5}B + ^{1}_{0}n \xrightarrow{\text{neutron capture}} ^{?}_{?}X \xrightarrow{\alpha \text{ decay}} ^{?}_{?}Y + ^{4}_{2}\alpha$.

The first reaction is $^{10}_{5}B + ^{1}_{0}n \rightarrow ^{11}_{5}B$ since the nucleus adds a nucleon without changing its charge. Then the alpha decay takes away two protons and a total of four nucleons from the boron-11 nucleus, so $^{11}_{5}B \rightarrow ^{7}_{3}Li + ^{4}_{2}\alpha$. The final nucleus formed is $^{7}_{3}Li$.

Sample Problem 4

Watch dials are often painted with a compound containing tritium, $^{3}_{1}H$, which has a half-life of 12.3 years. When the tritium atom undergoes beta decay, the emitted electron strikes a fluorescent pigment, which then glows.

(a) If your tritium-dial watch is giving off a certain amount of light, how many half-lives must pass until the watch gives off only one-eighth as much light?

Focus: $t = ?$

When the watch dial is producing one-eighth as much light as initially, there must be one-eighth as many beta particles being emitted and striking the fluorescent pigments on the dial. Since the odds that any given tritium atom will decay haven't changed, there must be only one-eighth of the original number of tritium atoms still present.

During each half-life the number of tritium nuclei in the sample (and the number of particles being emitted by the sample) is reduced by half:

Number of half-lives that have occurred	Fraction of the original sample remaining	Fraction of original number of particles being emitted
1	½	½
2	¼	¼
3	⅛	⅛
4	1/16	1/16

It will take **3 half-lives**, or 3 × 12.3 yr ≈ 37 yrs for the watch to dim to one-eighth of its original brightness.

(b) What nucleus is formed when the tritium undergoes β decay?

Answer: In beta decay a neutron within the nucleus becomes a proton as an electron is ejected. The number of protons in the nucleus increases by 1 while the total number of nucleons remains the same.

So $^{3}_{1}H \rightarrow ^{?}_{?}X + ^{0}_{-1}e \Rightarrow ^{?}_{?}X = ^{3}_{2}He$, an isotope of helium.

(c) Why do the beta rays emitted by the face of the watch pose no health hazard?

Answer: The emitted electrons don't have enough energy to penetrate the watchcase or crystal. (See Figure 33.4 in the textbook.) Hence they pose no health hazard.

Sample Problem 5

Uranium-235 dating is useful in estimating the age of ancient rocks. Uranium-235, through a series of decays, transforms to lead-207. The half-life for this decay process is 704 million years. By measuring the ratio of lead-207 to uranium-235 in the rock, one can determine how long ago the rock was formed, assuming that the rock initially contained no lead-207.

(a) Complete the table below.

Number of half-lives that have passed since the rock formed	Ratio of lead-207 atoms to uranium-235 atoms is
1 half-life	1:1
2 half-lives	?
3 half-lives	?

Solution: Let's let N_0 represent the original number of uranium-235 atoms in the sample. At the end of one half-life, half of the uranium-235 atoms have decayed to produce lead-207. At that time, the ratio lead-207 atoms: uranium-235 atoms is

$$\tfrac{1}{2}N_0 : \tfrac{1}{2}N_0 \Rightarrow \textbf{a 1:1 ratio}.$$

After two half-lives, half of the remaining uranium-235 atoms have decayed into lead-207 and only one-quarter of the original uranium-235 atoms remain. The ratio lead-207 atoms : uranium-235 atoms is

$$\tfrac{3}{4}N_0 : \tfrac{1}{4}N_0 \Rightarrow \textbf{a 3:1 ratio}.$$

After three half-lives, half of the remaining uranium-235 atoms have decayed, so only one-eighth of the original uranium-235 atoms remain. Seven-eighths have been transformed to lead-207. The ratio lead-207 atoms : uranium-235 atoms is then

$$\tfrac{7}{8}N_0 : \tfrac{1}{8}N_0 \Rightarrow \textbf{a 7:1 ratio}.$$

The completed table looks as follows:

Number of half-lives that have passed since the rock formed	Ratio of lead-207 atoms to uranium-235 atoms is
1 half-life	1:1
2 half-lives	3:1
3 half-lives	7:1

(b) A geologist finds an ancient volcanic rock where the ratio of lead-207 atoms to uranium-235 atoms is 2.9:1. Approximately how old is the rock?

Answer: We determined that if the ratio were 3:1, the rock would be 2 half-lives old, which is 2 × 704 million yr = 1.408 billion years. Since the ratio of lead-207 to uranium-235 atoms is almost, but not quite 3:1, we'd estimate the rock to be just a bit younger, approximately **1.4 billion years old**.

Radioactivity Problems

33-1. What nucleus results when $^{81}_{36}\text{Kr}$ undergoes electron capture?

33-2. What nucleus results when $^{27}_{14}\text{Si}$ undergoes positron emission?

33-3. Complete the following table:

	Initial nucleus	Decay mode	Equation describing the decay	Resulting nucleus
(a)	$^{232}_{90}\text{Th}$	α		
(b)	$^{140}_{56}\text{Ba}$	β^-		
(c)	$^{51}_{25}\text{Mn}$	β^+		
(d)	$^{109}_{49}\text{In}$	EC		
(e)	$^{60}_{27}\text{Co}^*$	γ		
(f)	$^{74}_{33}\text{As}$	β^+		
(g)	$^{3}_{1}\text{H}$	β^-		
(h)	$^{239}_{94}\text{Pu}$	α		
(i)	$^{240}_{96}\text{Cm}$	α		
(j)	$^{95}_{40}\text{Zr}$	β^-		
(k)	$^{210}_{85}\text{At}$	EC		

33-4. A uranium-238 nucleus $\left(^{238}_{92}\text{U}\right)$ in a nuclear reactor can absorb a neutron and then experience two beta decays. What nucleus is formed?

33-5. A nitrogen-14 nucleus $\left(^{14}_{7}\text{N}\right)$ can absorb a high-energy neutron resulting from interactions between cosmic rays and the atmosphere, and then eject a proton from its nucleus to form the nucleus of a new element. What element is formed?

33-6. When oxygen-16 $\left(^{16}_{8}\text{O}\right)$ absorbs a neutron and then undergoes an alpha decay, it forms a nucleus that is useful for studying chemical reactions in living organisms. What nucleus is formed?

33-7. Nuclear reactors are sometimes started up with neutrons originating from a beryllium-9 $\left(^{9}_{4}\text{Be}\right)$ target that has been bombarded by alpha particles. When a beryllium-9 nucleus absorbs an alpha particle, this produces a new nucleus and an ejected neutron. What is the identity of the new nucleus?

33-8. Americium-241 $\left(^{241}_{95}\text{Am}\right)$ is used in smoke detectors. It undergoes an alpha decay with a half-life of 432 years. The alpha particles ionize some of the air near the americium-241 source, and these charged ions are collected on a metal plate that has a small voltage applied to it. The collected particles cause a small electric current to flow in the detector. If smoke particles get in between the americium-241 source and the detector and reduce the current, the alarm is triggered.

  (a) Americium-241 is formed in a nuclear reactor when plutonium-239 $\left(^{239}_{94}\text{Pu}\right)$ absorbs two neutrons and then undergoes beta decay. Write the nuclear reaction for the formation of americium-241.

  (b) A typical smoke detector experiences 370,000 alpha decays each second. How much time is needed for the number of decays to reduce to 93,000 decays per second?

33-9. Plutonium-238 ($^{238}_{94}$Pu) is useful as a power source for heart pacemakers and for spacecraft. Heat from its nuclear decay is used by specialized semiconductor chips to produce electric current in a device called a Radioisotope Thermoelectric Generator (RTG). Plutonium-238 is manufactured in nuclear reactors by bombarding neptunium-237 ($^{237}_{93}$Np) with neutrons. The neptunium-237 nucleus absorbs a neutron and then undergoes beta decay to form plutonium-238.
(a) Write the nuclear reaction for the formation of plutonium-238.
(b) Plutonium-238 has a half-life of about 88 years. How much time would have to elapse for the electric current in your RTG to fall to $1/16$ of its original value?
(c) Would you have to worry about replacing the power source of a plutonium-238 powered pacemaker? Why or why not?

33-10. Strontium-90 ($^{90}_{38}$Sr) undergoes beta-minus decay with a half-life of about 29 years. The Russian navy used it to power remote navigational beacons and lighthouses. A strontium-90 based Radioisotope Thermoelectric Generator (RTG) can produce up to 128 watts of electrical power from the heat of 9.6×10^{15} radioactive decays per second.
(a) If a strontium-90 based RTG was initially producing 128 watts, how much time would have to elapse for it to produce only 2 watts of electric power (assuming that all of the other electronic bits were still functioning)?
(b) How many decays would be occurring each second at this point?
(c) What element results from strontium-90 decay?

33-11. Fluorine-18 ($^{18}_{9}$F) has a half-life of about 110 minutes and is formed when a target nucleus absorbs a proton, and then emits a neutron. The fluorine-18 decays via positron emission. When the fluorine-18 is incorporated into a sugar-like molecule, it is absorbed in parts of the body that are metabolizing sugar rapidly, such as cancerous tumors or active parts of the brain. The emitted positron collides with an electron and both are annihilated, producing two gamma rays. By seeing where the gamma rays originate (a process called positron emission tomography, or PET), doctors can localize a tumor, or they can study what parts of the brain are active in different kinds of thought processes.
(a) What is the target nucleus for the formation of fluorine-18?
(b) After a patient is injected with fluorine-18, how long will it take for the amount of fluorine-18 to decline to $1/64$ of its initial value?

33-12. Potassium-40 makes up only 0.012% of all potassium atoms found in nature. It can decay by two principal modes, either beta decay or electron capture followed by a gamma decay.
(a) What is the decay product from beta decay?
(b) What is the decay product from electron capture?

33-13. • Continuing the previous problem, potassium-40 has a half-life of 1.26 billion years. That means that the probability that any particular potassium-40 nucleus will decay is 1.74×10^{-17} in any given second. Only 11% of these decays will follow the electron capture path and result in a gamma ray being emitted.
(a) If a large banana has 600 mg of potassium (equivalent to about 9×10^{21} potassium atoms overall), how many gamma rays will be coming out of the banana every second?
(b) If bananas had been around 2.5 billion years ago, how many gamma rays would a large one have emitted every second back then?

33-14. Iodine-131 has a half-life of 8.02 days. In the body it tends to concentrate in the thyroid. For patients who have had a cancerous thyroid gland removed, iodine-131 is often administered to kill any thyroid cells that the surgery may have missed because the beta particles that iodine-131 emits are sufficiently energetic to kill the surrounding cells.
   (a) Suppose that you anticipate needing enough radioactive iodine to produce 1.0×10^{12} decays per second 16 days from now. How many decays should your sample be producing today?
   (b) What element results from iodine-131 decay?

33-15. Amanda does some baking and has an interesting idea on how to tell when her cookies have passed their "sell-by" date and should be removed from shelf. She plans to bake her cookies using iodized salt, where some of the added iodine is radioactive iodine-133 with a half-life of 21 hours.
   (a) If Amanda's cookies originally contain enough iodine-133 to each give off 200 decays per minute, about how many decays per minute will she measure from a cookie that remains on the shelf $2\frac{1}{2}$ days later?
   (b) What will be the decay count after twice this time?

33-16. Strontium-90 (half-life, 29 years) was produced as part of the nuclear fallout from atomic bomb tests conducted in the 1950s.
   (a) Approximately what fraction of the strontium-90 created in 1955 is still present in the environment?
   (b) How much of the strontium-90 created in 1955 will remain in 2042?

33-17. The amount of carbon-14 in a once-living sample decreases with time.
   (a) What fraction of the carbon-14 originally present in a sample remains after 9 half-lives have passed?
   (b) What fraction remains after 12 half-lives have passed?

33-18. Carbon-14 has a half-life of 5760 years. A 1-gram sample of an ancient pine spear contains $\frac{1}{8}$ as much carbon-14 as is present in a modern pine branch.
   (a) Approximately how old is the spear?
   (b) If a spear were 100,000 years old, why would carbon dating be inadequate for dating it?

33-19. The half-life of carbon-14 is 5760 years. A piece of flesh from a frozen mammoth discovered in Siberia was recently sampled for carbon-14. It contained one-fourth as much carbon-14 as a same-mass piece of flesh taken from a recently deceased elephant.
   (a) Approximately how long ago did the mammoth live?
   (b) If instead the mammoth's flesh had only $\frac{1}{8}$ as much carbon-14 as does the flesh of a living elephant, how long ago did the mammoth die?

33-20. Suppose that a carbon-14 sensor can only measure down to 0.2% of the original amount of carbon-14 present.
   (a) What is the oldest sample that can be dated by radiocarbon dating?
   (b) Would radiocarbon dating be useful for finding the age of meteorites? Explain.

33-21. A rock contains element A, which decays to element B with a half-life of 30 million years. Assume that the rock had no element B to begin with, and now it has 15 times as many B atoms as A atoms.
   (a) How long ago was the rock formed?
   (b) If the rock initially did contain some element B, would its age be younger than or older than your answer for part (a)? Defend your answer.

33-22. Uranium-238, which has a half-life of 4.5 billion years, decays through a series of steps to form lead-206. Suppose that a piece of an asteroid has a lead-206 atoms : uranium-238 atoms ratio of 1:1.
  (a) Assuming that all of the lead-206 in the asteroid came from the uranium-238, how old is the asteroid?
  (b) If some lead-206 was already present in the asteroid when it was formed, is the asteroid younger than or older than your answer for part (a)? Defend your answer.

Show-That Problems for Radioactivity

33-23. Cobalt-60 undergoes beta decay.
  Show that the beta decay of cobalt-60 $\left(^{60}_{27}\text{Co}\right)$ results in nickel-60 $\left(^{60}_{28}\text{Ni}\right)$.

33-24. When a target of aluminum-27 $\left(^{27}_{13}\text{Al}\right)$ is bombarded with neutrons, a target nucleus can absorb a neutron and eject an alpha particle.
  Show that the target material then contains sodium-24 $\left(^{24}_{11}\text{Na}\right)$.

33-25. When a target of oxygen-16 $\left(^{16}_{8}\text{O}\right)$ is bombarded with protons, a target nucleus can absorb a proton and then eject an alpha particle.
  Show that the target material then contains nitrogen-13 $\left(^{13}_{7}\text{N}\right)$.

33-26. Radon-212 $\left(^{212}_{86}\text{Rn}\right)$ is a radioactive gas with a half-life of 24 minutes.
  Show that when radon-212 undergoes alpha decay, polonium-208 is formed $\left(^{208}_{84}\text{Po}\right)$.

33-27. • Referring to the previous problem, show that if you start with 96 milligrams of radon-212, you will have 91 mg of polonium-208 2 hours later (and not 93 milligrams!).

33-28. Uranium-238 absorbs a neutron and then emits a beta particle.
  Show that the result is neptunium-239.

33-29. Plutonium-239 has a half-life of approximately 24,000 years.
  Show that it will take about 190,000 years for the amount of plutonium-239 in a sample to decrease to $1/256$ of its present amount.

33-30. A meteorite contains the radioactive isotope X, which undergoes beta decay into a stable isotope Y. Suppose that a recently discovered meteorite contains 45 mg of X and 315 mg of Y.
  First show that if it initially contained no Y, the meteorite originally contained 360 mg of X. Then show that the age of the meteorite is three times the half-life of X.

33-31. One form of cancer treatment is called *brachytherapy*. A pencil-lead-thin titanium-encased "seed" of either iodine-125 $\left(^{125}_{53}\text{I}\right)$ with a half-life of 60 days, or palladium-103 $\left(^{103}_{46}\text{Pd}\right)$, with a half-life of 17 days, is inserted into cancerous tissue. The radioactive nuclei undergo electron capture decays, followed by the emission of an x-ray, which occurs when an electron farther from the nucleus falls into an inner electron energy level to replace the electron that was captured. Show that the end products of the decays are tellurium-125 and rhodium-103.

33-32. Uranium-238 with a half-life of 4.5 billion years decays in a multistep process.
  Show that the following series of decays will result in the formation of lead-206.

$$^{238}_{92}\text{U} \xrightarrow{\alpha} \xrightarrow{\beta} \xrightarrow{\beta} \xrightarrow{\alpha} \xrightarrow{\alpha} \xrightarrow{\alpha} \xrightarrow{\alpha} \xrightarrow{\beta} \xrightarrow{\alpha} \xrightarrow{\beta} \xrightarrow{\alpha} \xrightarrow{\beta} \xrightarrow{\alpha} \xrightarrow{\beta} {}^{206}_{82}\text{Pb}$$

34 Nuclear Fission and Fusion

In nuclear fission, considerable energy is released when a large nucleus splits into smaller parts. In nuclear fusion, energy is released when smaller nuclei combine to form a larger nucleus. In both cases, the protons and neutrons (the nucleons) after the reaction are more tightly bound to one another with less overall mass than before.

Because a nucleus is a tightly bound arrangement of protons and neutrons, its mass is less than the sum of the masses of the individual nucleons that make it up.[*] The difference between the mass of a nucleus and the sum total of the masses of its constituent nucleons is called the *mass defect*. The more tightly the nucleons are bound in the nucleus, the greater the mass defect. If in a reaction the total mass defect becomes larger, as it does in fission and fusion, energy is released. According to Einstein's celebrated equation, $E = \Delta mc^2$, the energy released is the *change* in the mass defect, Δm, multiplied by c^2.

Recall from Chapters 11 and 33 that atomic masses are given in atomic mass units (amu), where 1 amu is defined to be exactly one-twelfth the mass of a carbon-12 atom. In kilogram units, 1 amu = 1.66×10^{-27} kg. One amu converted to energy units is equivalent to 1.494×10^{-10} J = 931.5 MeV (mega electron-volts, a conveniently sized unit when dealing with nuclear calculations).

The table below will be useful in solving problems in this chapter. All but the first three entries (and the entry for the alpha particle) give the masses of the *atoms* (including electrons) rather than of just the nuclei, but *changes* in atomic masses are the same as changes in nuclear masses.

Particle	Mass (amu)	Particle	Mass (amu)
electron $\left(_{-1}^{0}e\right)$	0.000549	carbon-12 $\left(_{6}^{12}C\right)$	12.00000
proton	1.007276	magnesium-24 $\left(_{12}^{24}Mg\right)$	23.985045
neutron $\left(_{0}^{1}n\right)$	1.008665	calcium-40 $\left(_{20}^{40}Ca\right)$	39.962591
hydrogen-1 $\left(_{1}^{1}H\right)$	1.007825	krypton-91 $\left(_{36}^{91}Kr\right)$	90.92344
deuterium $\left(_{1}^{2}H\right)$	2.014102	strontium-94 $\left(_{38}^{94}Sr\right)$	93.915367
tritium $\left(_{1}^{3}H\right)$	3.016049	xenon-140 $\left(_{54}^{140}Xe\right)$	139.992144
helium-3 $\left(_{2}^{3}He\right)$	3.016029	barium-142 $\left(_{56}^{142}Ba\right)$	141.916448
α-particle $\left(_{2}^{4}\alpha\right)$	4.000407	thallium-208 $\left(_{81}^{208}Tl\right)$	207.9820047
helium-4	4.002603	bismuth-83 $\left(_{83}^{212}Bi\right)$	211.991272
lithium-6 $\left(_{3}^{6}Li\right)$	6.015122	uranium-235 $\left(_{92}^{235}U\right)$	235.043923

---

[*] This relationship is illustrated graphically in Figure 34.16 on page 608 of *Conceptual Physics*.

Sample Problem 1

In a typical fission reaction, a slow neutron strikes a U-235 nucleus to produce two daughter nuclei and three neutrons. One such fission reaction is shown below:

$$_0^1 n + {}_{92}^{235}U \rightarrow {}_{36}^{91}Kr + {}_{56}^{142}Ba + 3\left({}_0^1 n\right).$$

(a) How many joules of energy are released in this fission reaction?

Solution: We use $E = \Delta m c^2$, where Δm is the change in mass in the reaction: the initial mass of uranium and the incident neutron minus the masses of the product nuclei and the three ejected neutrons. This is equivalent to the difference between the mass of the uranium atom and that of the daughter nuclei and *two* ejected neutrons.

$$E = [m_{U\text{-}235} - m_{Kr\text{-}91} - m_{Ba\text{-}142} - 2m_n]c^2$$

$$= [235.043923 - 90.92344 - 141.916448 - 2(1.008665)] \text{amu} \times 1.66 \times 10^{-27} \frac{\text{kg}}{\text{amu}} \left(3.00 \times 10^8 \frac{\text{m}}{\text{s}}\right)^2$$

$$= 2.79 \times 10^{-11} \text{ J}.$$

(b) There are 2.56×10^{21} U-235 nuclei in 1 gram of U-235. How many grams of U-235 would have to be fissioned to produce enough energy to heat a cup of water (250 g) from 20°C to 100°C to make tea?

Solution: $m = ?$ The energy needed to raise the temperature of the water $= c_w m_w \Delta T_w$, which is $4.18 \frac{\text{J}}{\text{g} \cdot \text{C}°}(250\text{g})(80°\text{C}) = 83600$ J. From part (a), the energy released when one uranium-235 nucleus fissions is 2.79×10^{-11} J. The energy released when 1 *gram* of U-235 fissions is

$$\left(2.56 \times 10^{21} \frac{\text{U--235 nuclei}}{\text{gram}}\right)\left(2.79 \times 10^{-11} \frac{\text{Joules}}{\text{fissioning U--235 nucleus}}\right) = 7.14 \times 10^{10} \frac{\text{J}}{\text{g}}.$$

We want to fission enough grams of U-235 so that

$$m \text{ (grams)} \times 7.14 \times 10^{10} \frac{\text{J}}{\text{g}} = 83600 \text{ J}.$$

$$\Rightarrow m = \frac{83600 \text{ J}}{7.14 \times 10^{10} \frac{\text{Joules}}{\text{g U} - 235}} = 1.2 \times 10^{-6} \text{g} = 1.2 \ \mu\text{g}.$$

This would be a cube of uranium 40 μm on a side. (40 μm is about half the diameter of a human hair).

(c) Natural gas has an energy content of about 50 kJ per gram. How many grams of natural gas would have to be burned to heat the same 250 grams of water?

Solution: $m = ?$ Using the same reasoning as above, $m = \dfrac{83600 \text{ J}}{50 \times 10^3 \frac{\text{Joules}}{\text{g}}} = 1.7 \text{ g}.$

Note that this is 1.4 million times as much as the amount of uranium needed for the same amount of energy. The energy content of nuclear reactions relative to chemical reactions is truly extraordinary.

Sample Problem 2

A neon-22 ($^{22}_{10}$Ne) atom has a mass of 21.991384 amu.

(a) Calculate the mass defect of a neon-22 atom.

Solution: Neon-22 has 10 protons, 12 neutrons, and 10 electrons. Its mass defect is equal to the difference in mass between the protons and neutrons and electrons that make it up, and the mass of the final atom. That is

$$\text{Mass defect} = \left[(10m_{\text{proton}} + 12m_{\text{neutron}} + 10m_{\text{electron}}) - m_{\text{Ne-22 atom}}\right]$$
$$= [10(1.007276 \text{ amu}) + 12(1.008665 \text{ amu}) + 10(0.000549 \text{ amu}) - 21.991384 \text{ amu}]$$
$$= 0.190846 \text{ amu}.$$

(b) How much energy would be released if you could assemble a neon-22 atom from 10 protons, 12 neutrons, and 10 electrons?

Solution: In effect, we fuse the individual nucleons to form the neon nucleus. The amount of energy is equal to the corresponding mass defect multiplied by c^2.

$$E = \Delta mc^2 = 0.190846 \text{ amu} \times \frac{931.5 \text{ MeV}}{\text{amu}} = 177.8 \text{ MeV}$$

or $\quad E = \Delta mc^2 = 0.190846 \text{ amu} \times \dfrac{1.494 \times 10^{-10} \text{ J}}{\text{amu}} = 2.851 \times 10^{-11} \text{ J}.$

This may not seem like much energy, but the energy released to produce just 1 kilogram of neon-22 in this way is about 7.7×10^{14} J, enough to melt a cube of ice over 130 m on each side!

Sample Problem 3

Bismuth-212 ($^{212}_{83}$Bi) spontaneously undergoes alpha decay to form a thallium-208 nucleus ($^{208}_{81}$Tl).

(a) Write the reaction for alpha decay of a bismuth-212.

Answer: $^{212}_{83}\text{Bi} \rightarrow {}^{4}_{2}\alpha + {}^{208}_{81}\text{Tl}.$

The bismuth-212 nucleus loses 2 protons and 2 neutrons. The final nucleus has $83 - 2 = 81$ protons, and $212 - 4 = 208$ nucleons overall. This nucleus is thallium-208.

(b) Show that the products of the decay have a smaller mass than the original bismuth atom.

Solution: The mass difference between the original nucleus and the decay products is the same as the mass difference between a bismuth-212 atom (bismuth nucleus + 83 electrons) and a helium-4 and thallium-208 atom (our decay products plus 2 and 81 electrons, respectively).[*] Using data from the table on the page 283,

$$\Delta m = m_{\text{Bi-212}} - (m_{\text{Tl-208}} + m_{\text{He-4}}) = 211.991272 \text{ amu} - (207.9820047 + 4.002603) \text{ amu}$$
$$= 0.0066643 \text{ amu}.$$

---

[*] In part (a) we are looking at the *nuclear* reaction, so we write it in terms of the nuclei. To do part (b), we use the *atomic* masses, which include the mass of the electrons. Bi-212 has 83 electrons and Tl-208 has 81 electrons. We account for the mass of the other two electrons by using the *atomic* mass of He-4 (which includes two electrons) rather than the mass of the helium nucleus (that is, the alpha particle), which does not.

Problems for Nuclear Fission and Fusion

34-1. Amanda sees that the element lithium has a lower atomic mass than iron, and hence is to the left of the bottom of the curve of Figure 34.16 in the textbook.
   (a) Would lithium be a candidate for releasing energy by fission, or by fusion?
   (b) Calculate the mass defect for a lithium-6 $\left(^{6}_{3}\text{Li}\right)$ atom.

34-2. One of the many ways uranium-235 can fission when hit with a neutron is to form xenon-140 $\left(^{140}_{54}\text{Xe}\right)$ and strontium-94 $\left(^{94}_{38}\text{Sr}\right)$ as daughter nuclei.
   (a) Write this nuclear fission reaction. (Remember to include the neutron at the beginning, and the appropriate number of neutrons at the end.)
   (b) Calculate the energy released in one fission reaction.

34-3. • Nuclear power plants are typically about 33% efficient in converting thermal energy to electricity.
   (a) The fission of a single U-235 nucleus typically releases an average of 3.2×10^{-11} J. How much energy is released from the fission of 1.00 kg of U-235?
   (b) How many kg of U-235 are "burned" every day in a 1000 MW electric power plant?
   (c) If the power plants were more efficient, would more or less uranium fuel be needed for the same power output? Defend your answer.

34-4. One reaction scheme for running a fusion reactor involves firing a neutron at a lithium-6 nucleus to produce tritium (hydrogen-3) and then fusing the tritium to a deuterium (hydrogen-2) nucleus to form helium and another neutron:
   (1) $^{1}_{0}\text{n} + {}^{6}_{3}\text{Li} \rightarrow {}^{3}_{1}\text{H} + {}^{4}_{2}\text{He}$
   (2) $^{3}_{1}\text{H} + {}^{2}_{1}\text{H} \rightarrow {}^{4}_{2}\text{He} + {}^{1}_{0}\text{n}$
   (a) How much energy is produced in each step? (Use the atomic masses in the table at the outset of this chapter.)
   (b) How much energy is produced in the overall reaction?
   (c) Calculate the amount of energy produced by the fission of 3.00 kg of lithium-6 and the subsequent fusion of 1.5 kg of tritium with 1.00 kg of deuterium.
   (This is 3.0×10^{26} atoms of each material.)

34-5. Assume that energy produced in part (c) of the fusion reaction of the previous problem was spread out over a period of 1 day.
   (a) What would be the average thermal power output (in MW) of this reactor?
   (b) If 40% of this heat energy could be converted to electrical energy, what would be the electrical power output of this fusion reactor?

34-6. Continuing the above problem, an efficient coal-burning power plant consumes about 10 tons of coal to produce a megawatt of electrical power for 1 day.
   (a) How many tons of coal would have to be burned to produce the same amount of electrical energy as would be obtained from the fusion of 3 kg of lithium-6 and 1 kg of deuterium in a 40% efficient power plant? [Refer to your results from Problem 34-5 part (b).]
   (b) A railroad car can transport about 110 tons of coal. How many railroad cars of coal are required per day to produce the same power obtained from the fusion of 4 kg of fusion fuel per day?

34-7. The biggest fusion bomb ever detonated was the one-of-a-kind *Tsar Bomba*, which was exploded on July 10, 1961, in a test by the former Soviet Union. The bomb had an explosive yield of about 50 megatons of TNT (each megaton being 4.2×10^{15} joules). The bomb itself was a cylinder about 8 meters long and 2 meters in diameter.
   (a) How much mass was converted to energy in this bomb?
   (b) If an asteroid moving at 18 km/s were to have this much kinetic energy in striking Earth, what would be the approximate mass of the asteroid?

34-8. A "kiloton" is the amount of energy released when 1000 tons of TNT explodes, about 4.2×10^{12} J. Suppose that a single uranium fission reaction releases about 3.2×10^{-11} J.
   (a) Approximately how many uranium nuclei had to fission in the 14-kiloton bomb that was dropped on Hiroshima?
   (b) There are about 2.6×10^{21} uranium nuclei in a gram of uranium. How many grams of uranium fissioned in the Hiroshima bomb?
   (c) Uranium has a density of about 19 g/cm^3. If the fissioned uranium were to take the shape of a cube, what would be the length of its sides?

34-9. Californium-252 has a half-life of 2.65 years. It is used industrially and medically as a fast neutron source because it will spontaneously fission in 3% of its decays.
   (a) One of the fission reactions for californium-252 is
   $$^{252}_{98}\text{Cf} \rightarrow {}^{94}_{38}\text{Sr} + {}^{154}_{60}\text{Nd} + \text{some neutrons.}$$
   How many neutrons are produced in this reaction?
   (b) The fission of californium-252 produces, on average, 3.8 neutrons. A 1-μg sample of californium-252 produces 170 million neutrons per minute. How many fissions is this each minute?

34-10. • Continuing with the previous problem, note that fission represents only 3% of all decays for californium-252.
   (a) How many decays are there *overall* every minute?
   (b) How many minutes does it take for half of all of the californium-252 atoms in a 1 μg sample to decay? (*Hint*: What does the half-life of a substance represent?)
   (c) Use your answers to (a) and (b) to *estimate* roughly how many atoms there are in a 1-μg sample of californium-252. (*Hint*: Knowing how many decays per minute there are initially, you can find the number of decays per minute one half-life later. Then figure how many minutes are involved.)

34-11. Protons and electrons within the Sun are fused in a multistep reaction to produce a helium nucleus, a pair of neutrinos (v), and gamma radiation (γ). The overall reaction is:
   $4{}^{1}_{1}p + 2{}^{0}_{-1}e \rightarrow {}^{4}_{2}\alpha + 2{}^{0}_{0}v + 6{}^{0}_{0}\gamma.$ The neutrinos escape and the gamma rays interact with matter within the Sun, their energy eventually radiating as lower frequency light.
   (a) Calculate the energy released by the overall fusion reaction. (Ignore the neutrinos and the gamma rays in the calculation because they have no appreciable mass.)
   (b) The Sun's luminosity is about 3.9×10^{26} watts. How much mass is being converted to energy every second?

34-12. Continuing from the previous problem, the mass of the Sun is 2.0×10^{30} kg and it is about 5 billion years old.
   (a) Assuming that the Sun has had the same luminosity for its entire lifetime, about what percentage of the Sun's mass has already been converted to energy?
   (b) How would you expect this loss of mass to affect Earth's orbit around the Sun?

Show-That Problems for Nuclear Fission and Fusion

34-13. The difference between the mass of a nucleus and the sum total of the masses of its constituent nucleons is called the *mass defect*.
   Show that the mass defect for magnesium-24 $\left(^{24}_{12}\text{Mg}\right)$ is 0.212835 amu.

34-14. Within the Sun, three helium atoms can fuse to form carbon-12:
   $^{4}_{2}\text{He} + ^{4}_{2}\text{He} + ^{4}_{2}\text{He} \rightarrow ^{12}_{6}\text{C}$.
   Show that the fusion reaction releases 1.17×10^{-12} J for each carbon-12 atom formed.

34-15. One possible fusion reaction is the combination of two deuterium (hydrogen-2) nuclei to form helium-3 and a neutron: $^{2}_{1}\text{H} + ^{2}_{1}\text{H} \rightarrow ^{3}_{2}\text{He} + ^{1}_{0}\text{n}$.
   Show that this reaction produces 3.27 MeV per fusion.

34-16. Burning 1 metric ton (1000 kg) of dry wood releases about 20 GJ (gigajoules = 10^9 J) of energy. Show that this is equivalent to converting 0.22 milligrams of mass to energy.

34-17. Fermium-256 $\left(^{256}_{100}\text{Fm}\right)$ spontaneously fissions to produce xenon-140 $\left(^{140}_{54}\text{Xe}\right)$ and palladium-112 $\left(^{112}_{46}\text{Pd}\right)$. Show that 4 neutrons are released in this fission.

34-18. The total U.S. consumption of electricity in 2008 was approximately 4×10^{12} kilowatt-hours, or approximately 1.4×10^{19} J.
   Show that this much energy is equivalent to converting approximately 160 kg of mass to energy (roughly speaking, the mass of a small motorcycle).

34-19. A uranium-235 nucleus will fission if it gains 4.6 MeV of energy. When a uranium-235 nucleus adds a neutron, it becomes uranium-236: $^{235}_{92}\text{U} + ^{1}_{0}\text{n} \rightarrow ^{236}_{92}\text{U}$. A uranium-235 atom has a mass of 235.043923. A uranium-236 atom has a mass of 236.045562.
   Show that the mass "lost" in the process of adding a neutron to uranium-235 provides more than enough energy for the nucleus to fission.

34-20. A uranium-238 nucleus will fission if it gains 5.5 MeV. When a uranium-238 nucleus adds a neutron, it becomes uranium-239: $^{238}_{92}\text{U} + ^{1}_{0}\text{n} \rightarrow ^{239}_{92}\text{U}$.
   A uranium-238 atom has a mass of 238.050783 amu. A uranium-239 atom has a mass of 239.054288 amu.
   Show that the mass "lost" in the process of adding a neutron to uranium-238 does *not* provide enough energy for the nucleus to fission. (Uranium-238 can be made to fission, but only if the neutrons are very energetic.)

35 Special Theory of Relativity

If you look into a rocket ship whizzing past you at very high speed, you'll see that its clocks run slower than yours do. You'll also see that the rocket ship and objects inside it appear shortened in the direction of motion. These changes are *relativistic effects*—the consequences of high-speed motion. Passengers on the rocket ship, interestingly, see things differently. They consider *themselves* to be at rest and *you* to be moving. To them, you are the one with the slow clocks and shrunken metersticks. In the language of relativity, you don't sense relativistic effects within your own non-accelerating frame of reference (which you consider to be "at rest"). Relativistic effects are always attributed to the frame of reference of "the other guy." Still, you and the other guy *will* agree on your relative speed. You will also agree on the speed of light, c (although you will disagree about the light's frequency, or color).

Solving special relativity problems requires identifying the frames of reference—designating one as the "stationary" frame and the other as the "moving" frame—and then transforming measurements made in one frame of reference to equivalent measurements in the other frame of reference.

Imagine that Bonnie is wearing a watch and holding onto a foot-long submarine sandwich, all inside of a cruising spaceship. Bonnie, the ship, her watch, and the sandwich all inhabit the same unaccelerated (or *inertial*) frame of reference—they are all at rest with respect to one another. Her measurement of the length of the sandwich (one foot in this case) is called its *proper length*, L_0. Her watch's measurement of the time it takes for her to eat the sandwich is called the *proper time*, t_0. (The term *proper*, from the Latin *proprius*, meaning "one's own," here means "measured in a frame of reference that is, at rest, relative to the thing being measured.")

Suppose that Tom is also wearing a watch but is standing on Earth observing Bonnie and her ship go by. Tom, Earth, and his watch are in a reference frame *different* than Bonnie's. Tom and Bonnie will agree that they are moving with velocity v relative to one another. But Tom will say that he is in a stationary frame of reference and that Bonnie is in a moving frame of reference. Using his own yardstick and clock, Tom will measure that Bonnie's moving sandwich is less than one foot long and that her moving watch seems to be ticking slowly—Tom's watch measures a longer time for Bonnie to finish her sandwich than Bonnie's watch does.

Although their measurements of time intervals and length will disagree, both the stationary observer (Tom) and the moving observer (Bonnie) *will* agree that their measurements are linked by the factor gamma: $\gamma \equiv \dfrac{1}{\sqrt{1-\left(\dfrac{v}{c}\right)^2}}$. The relationship between their time measurements is:

$$t = \gamma t_0 = \dfrac{t_0}{\sqrt{1-\left(\dfrac{v}{c}\right)^2}}.$$ (Note from the definition of γ that it is always greater than 1.)

Time intervals as measured by Tom's "stationary" watch (t) will be *longer* than time intervals as measured on Bonnie's "moving" watch (t_0) by the factor γ. The moving clock runs slow.

The relationship between their measurements of length or distance is:

$$L = \dfrac{L_0}{\gamma} = L_0\sqrt{1-\left(\dfrac{v}{c}\right)^2}.$$

The length that the stationary observer Tom measures for the sandwich (L) is *shorter* than what the moving observer Bonnie measures (L_0) by a factor γ. Moving objects appear contracted.

Of course, Bonnie will say that *she* is the stationary observer, quietly eating her sandwich in her stationary ship while Tom and Earth cruise by in the other direction, and that *Tom's* clocks run slow and *his* yardsticks are shortened. Special relativity gives us a way of reconciling these two different views of the same situation.♠

Sample Problem 1

While you stay on Earth, your friend Albert hops into a spaceship and zips away at 0.87c to visit Sirius, approximately 8.7 light-years away.[*][†]

(a) How long do you say that his trip takes?

Focus: $t_{\text{you measure}}$ = ? We will call Earth the stationary frame of reference and call Albert's ship the moving frame of reference. Accordingly, we will use t and L to represent measurements you make on Earth and t_0 and L_0 to represent measurements that Albert makes. Albert has to travel L = 8.7 light-years at 0.87 times the speed of light, going from here to there like any other object.

From $v = \dfrac{d}{t}$ $\Rightarrow t = \dfrac{L}{v} = \dfrac{8.7 \text{ ly}}{0.87c} = \dfrac{8.7 \, c \cdot \text{year}}{0.87 \, c} = $ **10.0 years**.

(b) How long a time will Albert's watch measure for the trip?

Focus: t_0 = ? We are still considering you to be the stationary observer and Albert to be moving, so we use t to represent time measured on your clocks and t_0 to represent time as measured by Albert's clocks. Albert's clocks will run slower than your clocks by the factor γ.

From $t = \gamma t_0 = \dfrac{t_0}{\sqrt{1-\left(\frac{v}{c}\right)^2}}$ $\Rightarrow t_0 = t\sqrt{1-\left(\dfrac{v}{c}\right)^2} = 10.0 \text{ y}\sqrt{1-\left(\dfrac{0.87c}{c}\right)^2} = $ **4.9 y**.

(c) What explanation can Albert provide to account for the trip taking only 4.9 years?

Answer: t = ?

Now we consider *Albert* to be the stationary observer, and the Earth-Sirius system to be the moving frame of reference. So now we will designate measurements in Albert's frame of reference as L and t, while measurements made in the Earth-Sirius frame of reference will be designated L_0 and t_0.

♠ To measure the length of something, you have to note the location of its two ends *at the same time*. Since the observers in the two frames (in this case, Bonnie and Tom) won't agree on what "at the same time" means, each will claim that the other is getting a different length because he or she is measuring the location of one end of the object, letting it move, and *then* measuring the other end of it.

[*] If Albert hopped in a ship and it suddenly accelerated to 0.87c, he would be flattened against the bulkhead and would be in no position to enjoy the trip. A "real" starship would have to accelerate slowly enough for Albert to withstand the forces on him. However, "thought experiments" like the one in this problem, used by Einstein himself, provide a good way to understand the meaning of relativity.

[†] A *light-year* (abbreviated ly) is a unit of distance, equal to how far light travels in one year at its speed c (3×10^8 m/s). If we're willing to accept answers that give time in units of years and speeds as some fraction of the speed of light (or light-years per year), we don't need to convert to the more standard meters and seconds.

You (and Earth and Sirius) are at rest relative to one another. The distance that you measure between Earth and Sirius is the proper length, $L_0 = 8.7$ ly. Albert sees you and the Earth-Sirius system as the moving frame of reference, moving by at $0.87c$. So Albert measures the *moving* Earth-Sirius distance to be contracted by a factor $\sqrt{1-\left(\frac{v}{c}\right)^2}$ compared with the distance (L_0) that you measure. That is,

$$L = L_0\sqrt{1-\left(\frac{v}{c}\right)^2} = 8.7\text{ ly}\sqrt{1-\left(\frac{0.87c}{c}\right)^2} = 4.3\text{ ly}.$$

For Albert, the time for the voyage is the time between Earth passing his ship and Sirius passing his ship.

From $v = \dfrac{\text{distance}}{\text{time}} \Rightarrow t = \dfrac{L}{v} = \dfrac{4.3\text{ ly}}{0.87c} = \dfrac{4.3\ c\cdot\text{year}}{0.87c} = 4.9\text{ y}!$

Notice that you and Albert will *agree* that he'll measure a trip time of 4.9 years even though you'll *disagree* as to why. You'll say he measures a shorter time because his clocks are running slow. He'll say that his clocks are fine, but that the time is shorter because the distance is shorter. The beauty of special relativity is that it gives both of you a way to understand what the other will measure.

Sample Problem 2

You measure your space speeder to be 36 m long, but observers on the space station you fly past measure it to be 27 m long. How fast are you moving relative to the space station?

Focus: $v = ?$

Observers in the space station will be in the stationary frame of reference (L, t). You and your ship are in the moving frame of reference (L_0, t_0).

You are at rest relative to your ship, so the ship length that you measure is the ship's proper length $L_0 = 36$ m. From the space station's frame of reference, your *moving* ship has length $L = 27$ m.

We start out with $L = L_0\sqrt{1-\left(\frac{v}{c}\right)^2}$, from which we want to get v. To isolate v, we'll need to get both L's on one side of equation, square both sides to get rid of the square root sign, get $\left(\frac{v}{c}\right)^2$ by itself, and then take the square root. Here we go:

$$\frac{L}{L_0} = \sqrt{1-\left(\frac{v}{c}\right)^2} \Rightarrow \left(\frac{L}{L_0}\right)^2 = 1-\left(\frac{v}{c}\right)^2 \Rightarrow \left(\frac{v}{c}\right)^2 = 1-\left(\frac{L}{L_0}\right)^2$$

$$\Rightarrow \frac{v}{c} = \sqrt{1-\left(\frac{L}{L_0}\right)^2} \Rightarrow v = c\left(\sqrt{1-\left(\frac{L}{L_0}\right)^2}\right) = \sqrt{1-\left(\frac{27\text{ m}}{36\text{ m}}\right)^2}\,c = 0.66\ c.$$

Problems for Special Relativity

35-1. Thomas, a rhino, is 2.5 meters long when at rest.
   (a) How long will you measure him to be when he runs past you at 0.80c?
   (b) How much time will you say it takes for him to pass you?

35-2. A ship zips by at 0.60c. Someone aboard is making a "3-minute egg" for breakfast.
   (a) What cooking time will you measure for the egg?
   (b) Why should you not be surprised when the egg turns out to be perfectly cooked, rather than overcooked? (*Hint*: How much does the chef age during the time the egg is cooking?)

35-3. You measure the length of a moving meterstick to be 0.40 m.
   (a) How fast is it going? (For this and the rest of the problem, assume that lengths are measured parallel to the direction of motion.)
   (b) Using $L = L_0/\gamma$, show that when both the observer and the meterstick being observed are in the same reference frame, $v = 0$ and $L = L_0$.

35-4. Before leaving the planet Hislaurels for a starship voyage, you pack a meterstick in your luggage. After the ship has settled down to a steady speed of 0.50c, you take the meterstick out of your bag and point it along the direction the ship is moving.
   (a) How long will you measure the meterstick to be?
   (b) How long does the moving meterstick appear to an observer at rest on Hislaurels?

35-5. You are standing facing forward on the floor of your starship, which is moving at 0.80c relative to Earth. Before you left Earth, you measured your feet to be 25 cm long.
   (a) How long will people on Earth measure your feet to be now?
   (b) Do you need to be concerned now that the shoes that you packed for the trip will be too big? Defend your answer.

35-6. Pinocchio is concerned that Geppetto will see his long nose and realize that Pinocchio has been lying. So Pinocchio decides to run past Geppetto so fast that his 10-inch long nose will appear to Geppetto to be only 2 inches long.
   (a) How fast must Pinocchio run?
   (b) Will Geppetto appear contracted to Pinocchio?

35-7. Lizzie is scooting down the Interstate at 18.0 percent of the speed of light and measures the distance between mileposts to be less than 5280 feet.
   (a) What is the distance that Lizzie measures between the mileposts?
   (b) What would be the distance she measures at twice her speed?

35-8. A proton from the Sun heads toward Earth, a distance of 500 light-seconds, at 0.995c.
   (a) How much time does an Earth observer measure for the proton to travel from the Sun to Earth?
   (b) If the proton carried a clock, how much elapsed time would it record for the trip?
   (c) What Sun-Earth distance will the proton measure (in light-seconds)?

35-9. • Recent Hubble Space Telescope measurements put the distance to the open star cluster called the Pleiades (or the "Seven Sisters") to be 440 light-years.
   (a) If your ship could travel there at $0.95c$, would you live long enough to survive the trip?
   (b) If you wanted to arrive at the Pleiades 44 years older than when you left Earth, how fast would your ship have to go?

35-10. A financial products salesman has devised a "Get-Rich-Fast" scheme. You invest $20,000 at 4% compound interest for 20 years on Earth while you are taken for a ride on a starship at $0.995c$.
   (a) When you return, 20 years will have passed on Earth. How much older will you be?
   (b) How much money will you have in the bank when you return?*

35-11. You take your starship on a 5-year mission to boldly collect interstellar gas samples. You leave Earth at $0.95c$. After $2\frac{1}{2}$ years have passed (according to the ship's clock), you stop and quickly turn around. You take another $2\frac{1}{2}$ years (also on your clock) to return to Earth.
   (a) When $2\frac{1}{2}$ years have passed according to clocks on your ship, how much time has passed according to clocks on Earth?
   (b) How far away from Earth will you be when you turn your ship around, according to observers on Earth?
   (c) How much time will have passed on Earth when you return 5 of your years later?
   (d) The star Proxima Centauri is 4.22 light-years away from Earth as measured in Earth's frame of reference. Would you be able to pass by it during your voyage?

35-12. A space probe with some wooden parts leaves Earth for a very long round-trip at $0.80c$. When the probe returns to Earth, scientists on Earth determine that the proportion of carbon-14 in the wooden parts of the probe is half as much as the proportion of carbon-14 in a tree on Earth growing outside of their lab. The half-life of carbon-14 is 5730 years.
   (a) How much time has passed for the space probe?
   (b) How much time has passed on Earth?
   (c) According to Earth observers, how far did the space probe travel before it turned around to come back to Earth?

---

* After 1 year you'd have $20,000 \times (1.04)^1$. After 2 years you'd have $20,000 \times (1.04)^2$. After 3 years you'd have $20,000 \times (1.04)^3$. After 20 years…

35-13. • One verification of special relativity theory came from measurements of muons, subatomic particles produced near the top of the atmosphere by cosmic-ray bombardment. A muon's half-life, in its own frame of reference, is 2.2×10^{-6} s, or 2.2 μs. Suppose that we set up two labs, one at 3300 m above sea level and the other at sea level at the bottom of the same mountain, and that we measure 640 muons per hour at the higher-altitude lab.

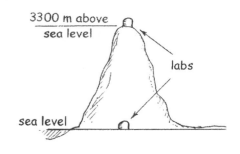

(a) Assuming that the muons are moving at a speed very close to the speed of light c (i.e., at nearly 3.00×10^8 m/s), how long will we say that it takes for them to make the trip from the higher altitude lab to the lab at sea level?
(b) How many muon half-lives is this?
(c) Ignoring special relativity effects, how many muons per hour should remain by the time they arrive at sea level? (Remember from Chapter 33 that a half-life is the time for half of the sample of the muons to decay.)
(d) Experimentally, researchers find that 320 muons per hour arrive at the lower lab. How many half-lives have passed in the muon's frame of reference?
(e) What is the value of the ratio $\dfrac{t_{\text{muon's frame}}}{t_{\text{lab frame}}} = \dfrac{t_0}{t}$?
(f) What is the speed of the muons expressed as a fraction of the speed of light?
(g) In the muon's frame of reference, they are at rest and Earth is rushing toward them. How far will they measure the distance between the upper lab and the lower lab to be?
(h) Traveling at nearly the speed of light, how much time will they measure between a point where the first lab passes them and a point where the second lab passes them?
(i) In the muon's frame of reference, if there are 640 particles when the upper lab passes by, how many should remain when they encounter the lower lab?

35-14. • You decide to visit *Wolf 359*, a star 7.8 light-years distant from Earth.
(a) If you travel there from Earth at $0.90c$, how long will the folks on Earth say that it takes for your ship to arrive at *Wolf 359*?
(b) How much older will you be when you arrive than when you depart?
(c) According to you, what is the distance to *Wolf 359*?
(d) Immediately upon arrival, you transmit a radio message to Earth.* How long after leaving Earth will the folks on Earth hear about your arrival?
(e) You spend a year studying *Wolf 359*, send another radio message telling Earth that you are leaving, and then immediately head back home, again at $0.90c$. From Earth's point of view, how long after they receive your "I'm coming home" message will you arrive home?

---

* A radio wave is an electromagnetic wave. Hence its speed is the same as the speed of light.

35-15. • In 1849 Armand Fizeau measured the speed of light using a rapidly spinning toothed wheel. In the experiment, a light beam passes through a gap in the wheel, reflects from a far-away mirror, and then heads back toward the wheel. If the rotational speed of the wheel and the distance between the wheel and the mirror are just right, the reflected light will pass through the next gap in the wheel. The time it takes for the wheel to turn by one tooth would be the round-trip time for the light. If the speed of the wheel is not just right, a tooth of the wheel will block the reflected light beam.*

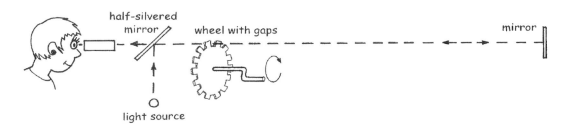

(a) Suppose that the wheel turns 150 times per second. How much time elapses during each complete rotation of the wheel?

(b) If there were 200 equally spaced teeth and gaps around the perimeter of the wheel, how long would it take for the wheel to rotate by one tooth?

(c) If the distant mirror is located 5.0 km away from the toothed wheel, what is the round-trip distance for the light from the wheel to the mirror and back?

(d) If the reflected light nicely passes through the adjacent gap in the wheel, calculate the speed of light from the data.

---

* Fizeau used a system of lenses and mirrors to focus the light. The diagram above, not surprisingly, is a simplified representation of his apparatus.

Show-That Problems for Special Relativity

35-16. A 1.00-km long ship moves past you at $0.20c$.
Show that you will measure the ship to be 980 m long.

35-17. Earth has a diameter of approximately 12,700 km.
Show that a ship passing by Earth at $0.40c$ will measure Earth to be about 11,600 km across.

35-18. Albert Sneedley had 15 minutes of fame on October 17, 1953.
Show that a ship passing by Earth at $0.70c$ would have measured his fame to last for 21 minutes.

35-19. Sirius is 8.7 light-years from Earth.
Show that an astronaut making the trip at $0.50c$ will measure the one-way trip to take about 15 years.

35-20. Castor stays on Earth, while his twin Pollux goes for a round-trip journey at $0.80c$. Castor is 30 years older when Pollux returns to Earth.
Show that Pollux will be 12 years younger than Castor.

35-21. An electron in a particle accelerator is moving at $0.9999c$.
Show that in the frame of reference of the accelerator, the electron will take 1.00×10^{-5} s to travel 3.00 km.

35-22. Refer to the 3.00-km trip taken by the electron in the previous problem.
Show that the electron experiences the 3.00-km trip as being 42.4 m long.

35-23. *Barnard's Star* is a dim red dwarf approximately 6.0 light-years from Earth.
Show that traveling there at a speed of 0.95c, you'd judge its distance from Earth to be approximately 1.9 light-years.

35-24. In your trip to Barnard's Star in the previous problem, your time of travel differs from the time that Earth observers would measure.
Show that you measure the travel time to be 2.0 years.

35-25. Marshall takes the super-train of the future from Los Angeles to Chicago, a trip of 3,600 km, at a speed of 1.8×10^8 m/s (60 percent of light speed). After a visit in hometown Chicago he takes the same super-train back to L.A. (Assume he can somehow make this round trip without being flattened by the accelerations.)
Show that Marshall experiences the trip as taking a mere 16 milliseconds to arrive in Chicago and another 16 milliseconds to return.

35-26. Referring to Marshall's round-trip voyage between Los Angeles and Chicago in the previous problem, his age at the end of the trip would be different from his age if he had instead stayed at home.
Show that by taking the high-speed trip, Marshall ages 8 milliseconds less than if he doesn't take the trip.